The Biological Resources of Model Organisms

The Biological Resources of Model Organisms

Edited by
Robert L. Jarret
Kevin McCluskey

CRC Press is an imprint of the
Taylor & Francis Group, an **informa** business

CRC Press
Taylor & Francis Group
6000 Broken Sound Parkway NW, Suite 300
Boca Raton, FL 33487-2742

First issued in paperback 2021

ISBN-13: 978-1-03-209095-5 (pbk)
ISBN-13: 978-1-138-29461-5 (hbk)

Publisher's Note
The publisher has gone to great lengths to ensure the quality of this reprint but points out that some imperfections in the original copies may be apparent.

Library of Congress Cataloging-in-Publication Data

Names: Jarret, R. L., editor. | McCluskey, Kevin, 1961- editor.
Title: The Biological Resources of Model Organisms / [edited by] Robert L. Jarret, Kevin McCluskey.
Description: Boca Raton, Florida : CRC Press, [2019] | Includes bibliographical references and index.
Identifiers: LCCN 2019009290 | ISBN 9781138294615 (hardback : alk. paper) | ISBN 9781315100999 (ebook)
Subjects: | MESH: Biological Specimen Banks | Databases, Genetic | Collections as Topic
Classification: LCC QH447 | NLM QU 24.51 | DDC 572.8/629--dc23
LC record available at https://lccn.loc.gov/2019009290

Visit the Taylor & Francis Web site at
http://www.taylorandfrancis.com

and the CRC Press Web site at
http://www.crcpress.com

Contents

Foreword

Biology is built on collections. All over the world, biological resource centers keep alive thousands of collections of yeast, bacteria, viruses, plasmids, cell cultures, animals, and plants used in scientific research. These institutions curate such collections in ways that crucially shape the practices and progress of biological science. But their expertise and organization are largely invisible, even to their scientist users. For several years in the early 2000s I worked as a fruit fly geneticist, regularly ordering mutant *Drosophila* through the post, which would arrive in our lab within a matter of days. But the ease with which I could order them online and the rapidity of their arrival by post meant that I rarely questioned how this was possible. In my new career as a historian of science I have had cause to reflect further on just how little we know about how stock centers were established, who has worked in them, and what differences they have made to biology. Work carried out at such institutions is rarely acknowledged in publications and has been inconsistently recorded in archives. This remarkable volume is a testimony to the complex nature of the operations that take place at such institutions, and to the dedication of the people who maintain them. The chapters that follow are written by these scientists, managers, and curators, who describe in detail the history of these centers, their practices of curation and care, and the rich array of expertise and biological knowledge that these sites have accumulated.

The collections described here are astonishingly diverse, but in different ways. Some are composed of a rich diversity of species; others contain hundreds of strains or mutants of a single species. Testifying to the wide array of organisms used as models, discussed here are collections of *Ambystoma* (axolotl) salamanders, the model plant *Arabidopsis*, aquatic rotifers, the nematode *C. elegans*, the green algae *Chlamydomonas*, the zebrafish (*Danio* sp.), the fruit fly *Drosophila*, a wide array of fungi and bacteria, the deer mouse *Peromyscus*, the eukaryotic ciliate *Tetrahymena*, *Xenopus* frogs, *Xipophora* fish, as well as cell cultures, plasmids, plant tissue cultures, protozoa, yeast, viruses, and molecular genomic tools. Some of these collections were first established over 100 years ago; some were founded in the twenty-first century. All originated in research labs, where a scientist (or group of scientists) with expertise on a particular organism chose to establish a resource collection for their respective research community. This raises fascinating questions as to why and how a scientist might choose and manage such a task, which several of the following chapters reflect on. Today, the centers described here are scattered throughout the United States—from Kentucky to Kansas, from New York to Texas—and send stocks to laboratories across the world.

The following chapters underline the profound (if obvious) fact that much scientific research depends on highly specialized practices of care that occur at biological resource centers. Care is always tailored to the unique characteristics of organisms. Some are relatively straightforward to look after. *Arabidopsis* seeds can be kept long term in ultra-cold storage. Rotifer eggs are dried, powdered, and stored for decades. Most sporulating fungi can be stored on silica gel or freeze dried. *Bacillus* cultures and *C. elegans* worms are often deep-frozen. All of these depend on technologies of preservation, but demand limited interim maintenance. Other collections are far more labor intensive. *Xipophorus* stock keepers must feed their fish 2–3 times a day on different foods, check their health daily, and rigorously sterilize all equipment. *Xenopus* workers need to pay careful attention to the water quality, health, and diet of frogs. Surprising to many is the fact that the *Drosophila* life cycle cannot be suspended by freezing embryos; instead, thousands of fruit fly cultures must be regularly fed and checked for viability by stock keepers generation after generation. The kinds of workers who carry out these tasks vary, from teams of undergraduates at the Chlamydomonas Resource Center to the 50-member strong team of professional *Drosophila* stock keepers at the Drosophila Stock Center in Bloomington, Indiana. Scholars in the history of biology are increasingly interested in the fact that although 'care' is often overlooked, research could not happen without it.

Curators and collection managers are vigilant in maintaining the composition of collections and often negotiate with users and advisory boards to decide which stocks and cultures to introduce into collections and which to jettison. Continued support for such collections often depends on their continued relevance to the communities that they serve—even when the needs of a community change over time. Keeping biological resources alive takes time and money, and stock centers must routinely assess which stocks they can afford to keep. Nevertheless, there can be important advantages to retaining cultures and stocks that do not appear to have an obvious current or future role in research. Serendipitous (and important) discoveries have been made using seemingly unremarkable strains kept by stock centers. For example, *Taq* polymerase was isolated from an apparently commonplace bacterial culture accessioned into the American Type Culture Collection in 1967. In another instance at the same collection, a strain of the Zika virus first deposited in 1953 has suddenly become medically relevant and urgently important.

Documentation matters. A crucial aspect of curatorial work for any biological collection is the rigorous labeling of stocks and the management of clear relationships to their entries in databases. For more than 80 years, the maintenance of living research organisms has been intimately connected to technologies for documenting and circulating information. In the 1940s, the US National Research Council (NRC) convened a meeting—the Conference on the Maintenance of Pure Genetic Strains—to recommend guiding principles for the provision of standardized organisms. One of its conclusions was that the most successful research communities used regular newsletters to distribute up-to-date information about mutants, new findings, and the location of stocks. At the time, the best known of these were the *Maize Genetics Cooperation News Letter* and the *Drosophila Information Service*. Delegates agreed that timely, accurate, and standardized information was crucial for the distribution of reliable, standardized research organisms.

This is still true today. A crucial aspect of curating genetic resources is the careful documentation and management of identifiers and names associated with each accession, tasks that often require negotiation with community database managers and researchers beyond those centers. Many of the essays here describe the processes used to label stocks and to link those labels with databases that are cross-referenced with other kinds of genetic information (examples mentioned include FlyBase, WormBase, and The Arabidopsis Information Resource). When a center accessions a stock, managers assess how it will be entered into its catalog, sometimes changing its name so that it conforms to nomenclatural standards. For some organisms—such as the *Peromyscus* deer mice—collection managers also need to keep a careful record of birth date, gender, weaning date, and litter size, so that pedigrees can be traced back to foundation stocks. The following chapters on the National Xenopus Resource, the Tetrahymena Stock Center, and the Ambystoma Genetic Stock Center discuss their participation with the Research Resource Identification (RRID) Initiative, a project that aims to allow researchers to unambiguously identifying strains, reagents, and tools used in research.

A striking feature of the centers described here is that the care of living biological collections can position an institution as a site not only of rich resources but also of expertise and authority. Many of the institutions in this volume have invested time, physical facilities, knowledge, and other resources in developing educational tools for high-school students, undergraduates, and teachers. Of particular note in this regard are the Arabidopsis Resource Center, the Chlamydomonas Resource Center, and the Tetrahymena Stock Center. Others have fashioned additional roles for themselves as centers of expertise for their respective communities. The Zebrafish International Resource Center offers diagnostic testing and consultation services, the Fungal Genetics Stock Center provides cloning vectors and cloned genes, the Tetrahymena Stock Center maintains the Tetrahymena Genome Database and offers fee-based genetic transformation services, the Xenopus Genetic Stock Center hosts genome assemblies, while the National Xenopus Resource partly serves as a site for researchers to convene and collaborate.

Finally, many of the resource centers described here carry out research alongside their other responsibilities. The Caenorhabditis Genetics Center is developing new strains and molecular tools; the Fungal Genetics Stock Center is studying the characteristics of its cultures; the Zebrafish International Resource Center is developing new methods for detecting pathogens; the Bloomington Drosophila Stock Center has generated new kinds of genetically useful stocks; while the American Type Culture Collection has a research programme to find more effective ways of identifying, characterizing, and authenticating the biological materials it distributes.

Collections can tell us what we value, and what we have valued, in biology. They offer a trace of past endeavors, struggles for scientific authority, colossal investments of money and time, and the accumulation of knowledge about organisms that are at once unique and generative of fundamental biological knowledge. Moreover, the collections described in this book are made not just the living organisms themselves, but also of the apparatus, people, and practices that keep them alive. These are the material representations of the collaborative and negotiated character of the biological sciences. When historians look back at early twenty-first century biology, thousands of carefully managed and documented living collections will testify to the curators, collections managers, stock keepers, and database curators that make science possible.

Jenny Bangham
Department of History and Philosophy of Science
University of Cambridge

Editors

Robert L. Jarret was born and raised in Franklin, Massachusetts and graduated from the local high school. After a tour with the US Navy, he attended Bridgewater State College majoring in Biology. Graduate studies were completed at Purdue University (M.S. and Ph.D.) and Colorado State University (MBA). He worked for a period of time at the Centro Agronomico Tropical de Investigacion y Ensenanza (CATIE) in Costa Rica and subsequently at the University of Florida (Homestead), eventually moving to his present position. His research activities combine field and laboratory activities and are often multi-disciplinary and typically focused on the conservation and characterization of genetic resources/diversity. In addition to conducting research, he curates collections of various crop and crop-related taxa. He currently resides in Griffin, Georgia.

Kevin McCluskey obtained his Bachelor's and Master's degrees at Stanford University. After working in an MIT lab developing applications for a prototype Positron Emission Tomography system, he obtained his Doctorate in Botany and Plant Pathology at Oregon State where he pioneered the application of pulsed field gel electrophoresis to study the genomes of plant pathogenic fungi. Following a post-doctoral fellowship studying Fusarium at the University of Arizona, he accepted the position of Curator of the Fungal Genetics Stock Center. As curator, he developed the FGSC website and the databases that allow clients to identify and request materials. He became a leader in the national and international culture collection communities and was elected twice to the Executive Board of the World Federation for Culture Collections. After two terms on the Executive Board, he was elected Vice President of the WFCC. Dr. McCluskey was the Principal Investigator of the NSF Research Coordination Network grant that established the US Culture Collection Network. He has also served on the Convention on Biological Diversity Global Taxonomy Initiative and also the CBD ad hoc Technical Expert Group on Digital Sequence Information. He is a scientific member of the US National Genetic Resources Advisory Council and also served two terms on the American Phytopathological Society Public Policy Board. Dr. McCluskey has published over 50 articles and chapters on fungal genetics and genomics. He retired from the FGSC in 2018 and is currently working in biotechnology.

Contributors

Erik C. Andersen
Department of Molecular Biosciences
Northwestern University
Evanston, Illinois

Jelena Brkljacic
Arabidopsis Biological Resource Center
Center for Applied Plant Sciences
The Ohio State University
Columbus, Ohio

Christopher S. Calhoun
Arabidopsis Biological Resource Center
Center for Applied Plant Sciences
The Ohio State University
Columbus, Ohio

Donna Cassidy-Hanley
Tetrahymena Stock Center
Microbiology and Immunology
C5 152 Veterinary Medical Center
Cornell University
Ithaca, New York

D. Mariola Castrejon
Arabidopsis Biological Resource Center
Center for Applied Plant Sciences
The Ohio State University
Columbus, Ohio

Theodore Clark
Tetrahymena Stock Center
Microbiology and Immunology
C5 152 Veterinary Medical Center
Cornell University
Ithaca, New York

Deborah K. Crist
Arabidopsis Biological Resource Center
Center for Applied Plant Sciences
The Ohio State University
Columbus, Ohio

Raymond H. Cypess
American Type Culture Collection
(ATCC)
10801 University Boulevard
Manassas, Virginia

Aric L. Daul
Department of Genetics
University of Minnesota
Minneapolis, Minnesota

Paul Doerder
Tetrahymena Stock Center
Microbiology and Immunology
C5 152 Veterinary Medical Center
Cornell University
Ithaca, New York

Samantha L. Fenn
American Type Culture Collection
(ATCC)
10801 University Boulevard
Manassas, Virginia

April Freeman
Zebrafish International Resource Center
Institute of Neuroscience
University of Oregon
Eugene, Oregon

Erich Grotewold
Arabidopsis Biological Resource Center
Center for Applied Plant Sciences
and
Department of Molecular Genetics
The Ohio State University
Columbus, Ohio

and

Department of Biochemistry and Molecular Biology
Michigan State University
East Lansing, Michigan

Amanda Havighorst
Department of Drug Discovery and Biomedical Sciences
College of Pharmacy
University of South Carolina
Columbia, South Carolina

Manzour Hernando Hazbón
American Type Culture Collection (ATCC)
10801 University Boulevard
Manassas, Virginia

Ron Holland
Zebrafish International Resource Center
Institute of Neuroscience
University of Oregon
Eugene, Oregon

Marko E. Horb
The National *Xenopus* Resource
Marine Biological Laboratory
Eugene Bell Center for Regenerative Biology and Tissue Engineering
Woods Hole, Massachusetts

Jen-Jen Hwang-Shum
Zebrafish International Resource Center
Institute of Neuroscience
University of Oregon
Eugene, Oregon

Vimala Kaza
Peromyscus Genetic Stock Center
University of South Carolina
Columbia, South Carolina

Hippokratis Kiaris
Department of Drug Discovery and Biomedical Sciences
College of Pharmacy
and
Peromyscus Genetic Stock Center
University of South Carolina
Columbia, South Carolina

Emma M. Knee
Arabidopsis Biological Resource Center
Center for Applied Plant Sciences
The Ohio State University
Columbus, Ohio

David Lains
Zebrafish International Resource Center
Institute of Neuroscience
University of Oregon
Eugene, Oregon

Matthew Laudon
Department of Plant and Microbial Biology
University of Minnesota
St. Paul, Minnesota

Paul A. Lefebvre
Department of Plant and Microbial Biology
University of Minnesota
St. Paul, Minnesota

Benson E. Lindsey
Arabidopsis Biological Resource Center
Center for Applied Plant Sciences
The Ohio State University
Columbus, Ohio

Yuan Lu
Department of Chemistry & Biochemistry
Xiphophorus Genetic Stock Center
Texas State University
San Marcos, Texas

James W. Mann
Arabidopsis Biological Resource Center
Center for Applied Plant Sciences
The Ohio State University
Columbus, Ohio

Jennifer Matthews
Zebrafish International Resource Center
Institute of Neuroscience
University of Oregon
Eugene, Oregon

Kevin McCluskey
Fungal Genetics Stock Center
Department of Plant Pathology
Kansas State University
Manhattan, Kansas

Sean McNamara
The National *Xenopus* Resource
Marine Biological Laboratory
Eugene Bell Center for Regenerative Biology and Tissue Engineering
Woods Hole, Massachusetts

Julie A. Miller
Arabidopsis Biological Resource Center
Center for Applied Plant Sciences
The Ohio State University
Columbus, Ohio

Katrina Murray
Zebrafish International Resource Center
Institute of Neuroscience
University of Oregon
Eugene, Oregon

Chris Muzinic
Department of Neuroscience
Spinal Cord and Brain Injury Research Center
and
Ambystoma Genetic Stock Center
University of Kentucky
Lexington, Kentucky

Laura Muzinic
Department of Neuroscience
Spinal Cord and Brain Injury Research Center
and
Ambystoma Genetic Stock Center
University of Kentucky
Lexington, Kentucky

Eva Nagy
Arabidopsis Biological Resource Center
Center for Applied Plant Sciences
The Ohio State University
Columbus, Ohio

Andrzej Nasiadka
Zebrafish International Resource Center
Institute of Neuroscience
University of Oregon
Eugene, Oregon

Eduardo Orias
Tetrahymena Stock Center
Microbiology and Immunology
C5 152 Veterinary Medical Center
Cornell University
Ithaca, New York

Courtney G. Price
Arabidopsis Biological Resource Center
Center for Applied Plant Sciences
The Ohio State University
Columbus, Ohio

Erin Quinn
Zebrafish International Resource Center
Institute of Neuroscience
University of Oregon
Eugene, Oregon

Marco A. Riojas
American Type Culture Collection (ATCC)
10801 University Boulevard
Manassas, Virginia

Ann E. Rougvie
Department of Genetics
University of Minnesota
Minneapolis, Minnesota

Markita Savage
Department of Chemistry & Biochemistry
Xiphophorus Genetic Stock Center
Texas State University
San Marcos, Texas

Carolyn Silflow
Department of Plant and Microbial Biology
University of Minnesota
St. Paul, Minnesota

Frank P. Simione
American Type Culture Collection (ATCC)
10801 University Boulevard
Manassas, Virginia

Terry W. Snell
School of Biological Sciences
Georgia Institute of Technology
Atlanta, Georgia

R. Keith Slotkin
Arabidopsis Biological Resource Center
Center for Applied Plant Sciences
and
Department of Molecular Genetics
The Ohio State University
Columbus, Ohio

and

Donald Danforth Plant Science Center
St. Louis, Missouri

and

Division of Biological Sciences
University of Missouri
Columbia, Missouri

Zoltan M. Varga
Zebrafish International Resource Center
Institute of Neuroscience
University of Oregon
Eugene, Oregon

S. Randal Voss
Department of Neuroscience
Spinal Cord and Brain Injury Research Center
and
Ambystoma Genetic Stock Center
University of Kentucky
Lexington, Kentucky

Ronald B. Walter
Department of Chemistry & Biochemistry
Xiphophorus Genetic Stock Center
Texas State University
San Marcos, Texas

Monte Westerfield
Zebrafish International Resource Center
Institute of Neuroscience
University of Oregon
Eugene, Oregon

Cale Whitworth
Bloomington Drosophila Stock Center
Department of Biology
Indiana University Bloomington
Bloomington, Indiana

Marcin Wlizla
The National *Xenopus* Resource
Eugene Bell Center for Regenerative Biology and Tissue Engineering
Woods Hole, Massachusetts

Daniel R. Zeigler
Department of Microbiology
The Ohio State University
Columbus, Ohio

1 Introduction to the Laboratory Axolotl and the *Ambystoma* Genetic Stock Center

S. Randal Voss, Laura Muzinic and Chris Muzinic

CONTENTS

Abstract: Laboratory populations of the Mexican axolotl have been sustained for over 150 years in support of biological research. With recent advances in genetic and genome resource development, the axolotl is attracting considerable attention from new researchers, especially in the area of tissue regeneration. To reduce the learning curve and generally facilitate broader use of the axolotl, we introduce the *Ambystoma* Genetic Stock Center (AGSC), a research resource center that sustains and makes axolotl stocks available to researchers nationally and internationally. In order to assist researchers who are unfamiliar with axolotl biology, we describe our methods of husbandry, highlighting water quality and extrinsic environmental variables that the stock center controls and monitors to ensure axolotl health and well-being.

INTRODUCTION

In this chapter we introduce the primary salamander model in laboratory research, the Mexican axolotl (*Ambystoma mexicanum*). We also introduce the *Ambystoma* Genetic Stock Center (AGSC), which is funded by the National Institutes of Health

(NIH) to provide axolotls in support of biomedical research. Our objective is to provide useful information to those that use, or are planning to use, axolotls in research and educational activities.

THE AXOLOTL MODEL ORGANISM

The axolotl has a deep and rich history as a laboratory model organism. Present day laboratory populations trace their ancestry back to an original collection of 34 axolotls that were shipped to Paris in 1863 from aquatic habitats near present day Mexico City (Reiß et al. 2015). Over the next few decades, axolotls were propagated in laboratories across Europe and used to study questions related to development and evolution, and these studies helped to originate the field of experimental zoology. During the twentieth century, axolotls factored prominently in studies of embryonic and post-embryonic development, sex determination, cloning, and tissue regeneration (Smith and Smith 1971, Voss et al. 2009). Today, axolotls are attracting increased attention from biomedical researchers due to their unrivaled ability to regenerate entire organs. They share the body plan of tetrapod vertebrates (Figure 1.1) and are unique in their ability to regenerate a broad spectrum of damaged organs throughout life, including limbs, spinal cord, brain, lens, skin, ovary, and heart (Amamoto et al. 2016, Cano-Martínez et al. 2010, Erler et al. 2017, Haas and Whited 2017, Ponomareva et al. 2015, Suetsugu-Maki et al. 2012, Tazaki et al. 2017, Yokoyama et al. 2018). Understanding the cellular and genetic mechanisms

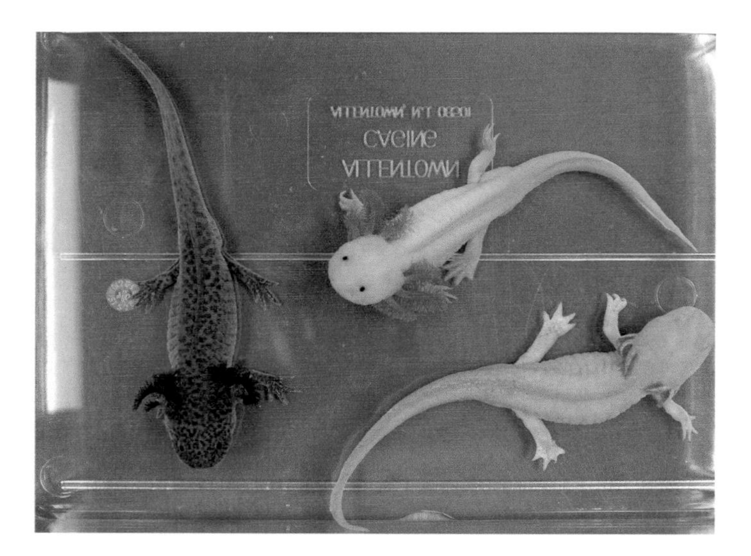

FIGURE 1.1 (See color insert.) Three different Mexican axolotl stocks that differ in their pigment pattern. The wildtype axolotl has three different pigment cells (xanthophores, melanophores, and iridophores) that combine to yield a dark greenish coloration (left). The white axolotl (center) is homozygous for a mutated *endothelin 3* gene that results in the loss of dark melanocytes soon after hatching. The golden albino (right) has an abundance of yellow xanthophores but lacks a functional *tyrosinase* gene for melanin production. (Courtesy of Lee Thomas.)

by which axolotls regenerate tissues could have clinical significance for treating human trauma, disease, and aging.

In several ways, the axolotl was pre-adapted to become a laboratory model. Axolotls are members of the Tiger salamander complex, a group of ambystomatids that exhibit considerable variation in life history and modes of development (Shaffer and Voss 1996, Voss et al. 2015). Some species undergo a metamorphosis after an aquatic larval phase and gain traits for terrestrial life. Other species like the axolotl are paedomorphic and remain in the aquatic habitat throughout their life cycle. While it is possible to rear metamorphic forms in the lab, the axolotl's totally aquatic life history greatly simplifies laboratory culture. Moreover, axolotls are capable of breeding more than one time a year and produce considerably more offspring per spawn than metamorphic forms, which are seasonal breeders. Axolotls are ideal laboratory models because they can be propagated as captive managed populations that are self-sustaining and capable of providing living stocks (embryos, larvae, juveniles, and adults) to meet the needs of a research community. In contrast, other salamanders that are used in biomedical research are annual breeders that are harvested from natural populations. In particular, newts (*Notophthalmus*) and Tiger salamanders (*A. tigrinum*) continue to be harvested from natural populations in the US for regeneration and retina electrophysiology preparations, respectively. Because amphibians in general are declining around the world, it is difficult to justify the collection of salamanders from natural populations for laboratory studies when axolotl stocks are available from a sustainable, captive-bred population (Baddar et al. 2015).

Axolotls are interesting research models because they maintain juvenile morphology throughout their life. Unlike their ancestors, axolotls typically do not undergo metamorphosis and transition to land during their life cycles. This mode of development is called paedomorphosis (Gould 1977) and the evolution of paedomorphosis is associated with changes in neuroendocrine axes that regulate metamorphosis (Johnson and Voss 2013). Paedomorphosis evolved so recently in axolotls that the ancestral metamorphic mode of development can be induced by adding thyroid hormone to an axolotl's rearing water (Page and Voss 2009).

Because of the recent development of a chromosome scale genome assembly (Smith et al. 2019) and tools to manipulate gene function (Fei et al. 2017, Flowers et al. 2014, Khattak and Tanaka 2015, Khattak et al. 2014, Woodcock et al. 2017), it is now possible to investigate the molecular basis of thyroid hormone signaling and paedomorphosis. Indeed, the impact of having an axolotl genome assembly on future research efforts cannot be overstated. The assembly of the enormous ~30 Gb axolotl genome enables sequence-based methods of inquiry that are typical of genetic model organisms. For example, researchers can access gene models from which to identify mutants and design molecular tools for PCR, reporter constructs, and genome editing. Additionally, information from coding and noncoding regions enriches our understanding of amphibian genome structure and function, and provides context for annotating and interpreting transcriptomic, genomic, and epigenetic datasets. Further improvements in the quality of the assembly and databases from which to access information will facilitate broader use of the axolotl in biomedical research.

THE *AMBYSTOMA* GENETIC STOCK CENTER

Almost all of the domesticated axolotls in the world trace their ancestry to the AGSC at the University of Kentucky, which is funded by the NIH (P40-OD019794) to provide axolotls in support of research and educational efforts. This collection is irreplaceable because decades of inbreeding have yielded homogeneous genetic stocks that thrive in the lab (Voss and Kump 2016). The collection provides standard living stocks (embryos, larvae, and adults) and supplies to culture axolotls in investigator labs. By supplying living material from a single facility, the AGSC minimizes the need for smaller satellite collections. This effectively reduces the cost of animal use for investigators, and yearly expenses for the entire research community are greatly reduced because axolotl material is obtainable on demand. Many AGSC users are located at institutions that do not have the equipment, expertise, or dedicated space to rear axolotls. For investigators that maintain relatively large numbers of axolotls, the AGSC serves as a back-up facility to replenish stocks and as a source of genetic variation to maintain the vigor of strains. For investigators that need early stage embryos, the AGSC provides breeders with advice about performing crosses and maintaining axolotls. Thus, the AGSC serves the needs of investigators in a variety of ways, and it is the historical with contemporary source of axolotl stocks in the world.

The AGSC has approximately 3000 sq ft for maintaining approximately 1000 juvenile and adult axolotls, and thousands of embryos and larvae. The AGSC primarily functions as a custom order business as it is too costly to maintain all axolotl stocks in sufficient numbers to meet immediate user needs. Thus, it is important that users plan their experiments well ahead of time as it can take several months to produce some stocks and a year to produce adults (Figure 1.2).

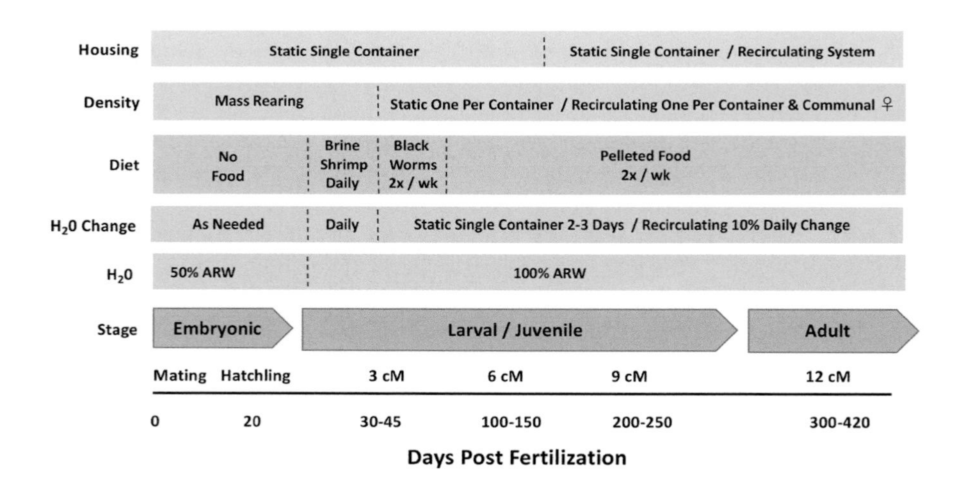

FIGURE 1.2 Diagram illustrating the integration of the axolotl life history in the laboratory with husbandry variables.

WEBSITE

Users initiate the purchasing process by first completing an online registration form that is available from the AGSC webpage (http://www.ambystoma.org/genetic-stock-center). After registering, users are put into contact with AGSC staff to discuss their order and the timeframe for receiving purchased stocks.

In addition to supplying axolotl stocks and associated husbandry supplies, the AGSC serves as an informatics hub where investigators obtain information about the collection, technical procedures, potential collaborators, and research findings. Several times each week, AGSC staff respond to user questions and every effort is made to provide answers to queries within 24 hours. An independent website was developed for the AGSC to ensure flexibility and control of web and database design, networking, and file storage. Information on the AGSC website provides researchers with a list of available stocks, pricing, and instructions on how to order axolotls and related supplies. Also, the AGSC website provides information concerning axolotl husbandry, development, and genetics. The AGSC website is accessible through Sal-Site (Baddar et al. 2015) which was developed under NIH-funding (R24-OD010435) to distribute genome-based informatics to the research community. A newsletter (*Axolotl*) is distributed annually to keep the community informed about axolotl services, new advances in the field, NIH policies and opportunities, and techniques (Keinath et al. 2018, Voss and Kump 2016).

AXOLOTL STOCKS

The AGSC maintains a variety of axolotl stocks of all life stages, as well as mutants and transgenics. Each stock is associated with a unique Research Resource Identifier (RRID) number (Table 1.1) as recommended by the Resource Identification Initiative at SciCrunch (SciCrunch.org). The RRID is a database-driven initiative that seeks to 'barcode' all of the critical reagents and tools that are used in the course of scientific research with RRIDs. In addition to wild types, there are three pigment mutants (*melanoid, white, albino*), two lethal mutants (*cardiac* and *short toes*), one sterile mutant (*eyeless*), and two transgenics (green fluorescent protein and red fluorescent protein) that were donated by Elly Tanaka (IMP, Vienna). The color mutants and ubiquitously expressing transgenic axolotls facilitate tissue grafting and cell lineage experiments. Three of the mutants were recently cloned (Smith et al. 2019, Woodcock et al. 2017) and others are in various stages of gene identification. For example, the albino mutant traces to an interspecific hybridization in 1962 that introduced a mutated, *A. tigrinum tyrosinase* allele into the AGSC population, as well as additional *A. tigrinum* genomic DNA. The number of stocks in the AGSC is expected to increase over the next few years as a result of community efforts in making transgenics and mutants.

TABLE 1.1
Research Resource Identifier (RRID) Numbers for the Various Axolotl Stocks Available or Not Available (NA) from the *Ambystoma* Genetic Stock Center

			Life Stage			
Stock	**Embyro**	**Hatchling**	**Larva**	**Juvenile**	**Sub-Adult**	**Adult**
Wildtype	C_100E	C_100H	C_100L	C_100J	C_100S	C_100A
White	C_101E	C_101H	C_101L	C_101J	C_101S	C_101A
Albino	C_102E	C_102H	C_102L	C_102J	C_102S	C_102A
Melanoid	C_103E	C_103H	C_103L	C_103J	C_103S	C_103A
GFP-white	C_110E	C_110H	C_110L	C_110J	C_110S	C_110A
RFP-white	C_112E	C_112H	C_112L	C_112J	C_112S	C_112A
Cardiac[a]	C_104E	NA	NA	NA	NA	NA
Short toes[a]	C_106E	NA	NA	NA	NA	NA
Eyeless[a]	C_108E	NA	NA	NA	NA	NA
Cardiac-C[b]	NA	NA	NA	C_104J	C_104S	
C_104A Short toes-C[b]		NA	NA	NA	C_106J	
C_106S		C_106A Eyeless-C[b]	NA	NA	NA	
C_108J		C_106S		C_108A		

[a] Available as a complete spawn with 25% frequency of mutants.
[b] Carrier.

EXTRINSIC ENVIRONMENTAL VARIABLES—ENSURING REPRODUCIBLE AXOLOTL RESEARCH

There is growing appreciation for the need to monitor and report extrinsic environmental variables that can affect the reproducibility of scientific experiments (Federation of American Societies for Experimental Biology 2016). This includes all of the variables that are standardized in the rearing of animals in stock centers like the AGSC. It seems likely that new standards for reporting extrinsic environmental variables will be codified by granting agencies and journals in the near future. Thus, to aid axolotl users going forward, we address the most important extrinsic environmental variables for ensuring reproducibility of scientific research using the axolotl. This information will be useful to those that obtain axolotl stocks from the AGSC or want to establish satellite axolotl facilities with parallel standard operating procedures. These variables include water quality, temperature, housing, diet, and light.

Water Quality

Axolotl health and well-being is critically tied to water quality. Axolotls are freshwater organisms that thrive in a dilute saline solution. The AGSC uses reverse osmosis (RO) water to make Axolotl Rearing Water (ARW) that contains 1.75 g NaCl, 100 mg $MgSO_4$, 50 mg $CaCl_2$, and 25 mg KCl per liter. The water is buffered by manual or automatic-dosing of $NaHCO_3$ to achieve a pH in the range of 7.1–7.6. Larvae, juveniles, and adults are reared in 100% ARW while 50% ARW is used to rear embryos.

Recognizing that many investigators might not have access to RO water, we note that it is possible to use other water sources and filtration systems to prepare ARW. In the past, we have used distilled water and charcoal filtered municipal water to make ARW, and some labs have reported success in rearing axolotls in minimally conditioned well-water and municipal water. Water chemistry, including pH, can vary dramatically when using municipal water and it is important to mitigate municipal water additives (e.g. chorine, chloramine, and ammonia) that are toxic to axolotls. Chlorine and chloramine can be removed by charcoal filtering or by aging water with an air bubbler. Ammonia can be removed by adding conditioners like *Amquel Plus* (Kordon). When using municipal water, pH, ammonia, and chlorine/chloramine should be carefully and frequently monitored to ensure axolotl health. Ammonia is less of a concern when rearing axolotls statically in bowls because the frequency of water changes can be scheduled to mitigate ammonia buildup. However, it is important to carefully and frequently measure ammonia when rearing axolotls in recirculating systems or large tanks with filters. When measured alongside nitrites and nitrates, ammonia provides an indicator of water quality. When ammonia levels exceed 2.0 ppm within a recirculating system, the system is not in equilibrium in regards to the nitrogen cycle. This can be a serious problem if the pH is also high because high pH increases ammonia toxicity. To mitigate high ammonia and pH (>8.0) in recirculating systems, 10% of the water is replaced daily with fresh, 100% ARW. Additional water changes may be needed to bring non-equilibrium systems below pH 8.00, at which time an ammonia detoxifier (e.g. Amquel Plus) can be added to alleviate ammonia buildup. However, a second reason for high ammonia levels in a recirculating system is inefficient biological filtration. To mitigate this problem, nitrifying bacteria can be added to a system to increase ammonia processing through the nitrogen cycle. The AGSC uses ProLine Nitrifying Bacteria (Pentair Aquatic Eco-Systems) following manufacturer dosing instruction for both initial set up and maintenance of recirculating systems.

Housing and Diet

Several methods are used to house axolotls in research and educational settings. The three primary methods are static housing (still water that is changed frequently), filtered aquaria, and recirculating systems. The AGSC uses static housing and recirculating systems to house all stocks (Figure 1.3). An advantage of static housing is that a

FIGURE 1.3 *Ambystoma* Genetic Stock Center axolotl adults housed using both automated recirculating systems (left) and statically maintained containers on rolling carts (right).

larger number of animals can be maintained within a small foot print. For example, 66 adult animals can be maintained on a single 36″ × 21″ × 60″ mobile rack. However, static housing requires frequent water changes and bowl cleaning and thus significantly more labor. Recirculating systems require less hands-on cleaning because water changes are automatic and continuous, and tanks only need to be cleaned every two weeks. However, recirculating systems require a larger footprint. Each double-sided recirculating system in the AGSC supports 60 animals but occupies the same space as three mobile racks (198 animals). Also, recirculating systems require a significant upfront financial investment and greater technical expertise to maintain. These trade-offs should be considered well in advance of establishing a facility to rear axolotls.

As a general rule, the size of a housing container should scale with animal number and body size. In the AGSC, husbandry begins with embryos that are housed in circular 1.5-L glass bowls with no more than 200 individuals per bowl. The water is manually stirred once daily to ensure oxygenation and dead embryos are removed. When necessary to slow down or speed up development, embryos are reared for short periods of time at 6°C and 24°C respectively. After larvae hatch, they are transferred to clean 1.5-L glass bowls with 100% ARW and reared in mass. When larvae make air bubbles at the surface of the water to indicate the onset of feeding behavior, they are fed newly hatched brine shrimp. Setting up a brine shrimp hatchery is fairly simple with a kit available from the AGSC (Table 1.2) and instructions provided on the AGSC website.

Regardless of what method is used to rear brine shrimp, unhatched eggs and egg shells should be removed before feeding because brine shrimp shells and unhatched eggs can block the intestines of larvae. Larvae are maintained at moderate-to-low densities to prevent bite injuries to limbs and tails; however, such injuries are inevitable unless individuals are housed separately. The vast majority of larvae in bowls incur bite injuries that are repaired by regeneration (Thompson et al. 2014). The AGSC offers a pay-for-fee service to rear individuals separately for investigators that require non-bite injured axolotls.

When larvae reach 3 cm in length, individuals are moved into 2-L plastic bowls with approximately 1 L of water. At this time, individuals are fed California blackworms to increase their growth rate and better facilitate the transition from live food to pelleted food, although axolotls can be reared throughout the adult phase on

TABLE 1.2
Supplies Available from the *Ambystoma* Genetic Stock Center for Axolotl Husbandry

Item	Catalog No.
Axolotl pellets—4 mm–1/2 lb	AGSC_PL_L
Axolotl pellets—5 mm–1/2 lb	AGSC_PL_S
Brine shrimp eggs–1 oz	AGSC_BS
Brine shrimp hatchery kit	AGSC_BSH
100% Axolotl rearing water salts–40 g	AGSC_ARW
AmQuel Plus–2 oz	AGSC_AMQ
NovAqua Plus–2 oz	AGSC_NOV

blackworms. The pelleted food is available in two sizes (4 mm for 4–8 cm animals and 5 mm for larger animals) and is aged for 6 months prior to use. Aging the food is necessary because the pellets contain an ingredient (most likely thyroid hormone from bovine blood) that can induce spontaneous metamorphosis. Individuals are fed 2–3 pellets twice a week. When individuals reach 8–9 cm in length they are moved into larger 4-L bowls containing 2.5 L of water. The AGSC sells pelleted food and salts to make ARW to better ensure reproducibility of axolotl husbandry (Table 1.2).

The method of husbandry (static vs. recirculating system) determines the frequency of cleaning axolotl housing containers. During the embryonic period, water is changed as needed. Embryos do not produce a lot of waste but dead embryos and residual feces from the mating parents can foul water. During the larval period, all the individuals in a bowl are strained into a net and the net is placed into a temporary holding container while the bowl is cleaned with a dilute bleach solution (5%) or baking soda, and then thoroughly rinsed. After the addition of fresh 100% ARW, larvae are returned to their original bowl. This procedure ensures that larvae from different spawns are not mixed during the cleaning process. When larvae/juveniles/adults are reared individually in single bowls, water is changed every 3–4 days using the same cleaning method described above. In general, the frequency of cleaning needs to be optimized for animal and container size, and as was mentioned earlier, water quality. Animals that foul their bowls before their scheduled cleaning should be attended to immediately.

Aquarius (Aquatic Enterprises) recirculating systems are used within the AGSC to house the majority of the adult breeding population. Each system is constructed of powder coated steel or aluminum racks with molded polycarbonate boxes. The life support system includes an 80-gallon sump per rack (8 per system) with screen filters, a self-cleaning rotating drum filter, high efficiency pumps, a 100-W UV sterilizer, carbon filter, with an automatic water change system, drawing water from a 100-gallon reservoir tank. Each system is automated using a NEMA protected ProFilux touchscreen monitoring system that includes sensors for pH, conductivity, temperature, and water level. Dosing tanks are maintained at pH 7.5 and conductivity at 4.2 µS (based on 100% ARW). Males are housed individually in 8-L containers, while females are housed three per 23-L container. As was mentioned previously, approximately 10% of the water in each system is changed daily and containers are cleaned every 2 weeks.

Although the AGSC utilizes static and recirculating housing methods, it is possible to house axolotls in filtered aquaria. Ideally, no more than two adults should be housed together in a 10-gallon tank. The use of substrate is discouraged as axolotls will ingest small rocks and gravel during feeding. Low current power filters with biological filtration can efficiently mitigate ammonia and nitrates; however, it is still necessary to perform weekly 10% water changes and monitor water chemistry.

Temperature and Light

To maintain a relatively constant temperature of 15°C–17°C the AGSC relies upon both the building HVAC system and auxillary air-cooling units. A 12 hr light/12 hr dark photoperiod is maintained throughout the year and the facility is lit by cool-white fluorescent lighting that is typical of research and educational buildings.

CONCLUSIONS

In this chapter we introduced the Mexican axolotl and the primary stock center that provides axolotls to researchers and educators. The axolotl has the deepest laboratory pedigree of all animal models and with the recent development of a genome assembly its use will likely expand in coming years, especially in the area of tissue regeneration. This will present new opportunities and challenges for the AGSC. To meet the needs of an expanding research community it will be important to prioritize production of the most useful stocks and develop methods to cryopreserve an increasing number of transgenic and mutant lines that are being produced within the community. It will also be important to sustain the central role that the AGSC plays in facilitating axolotl research and education by providing homogeneous stocks, supplies, services, and useful information about axolotl husbandry.

REFERENCES

Amamoto, R., V.G. Huerta, E. Takahashi et al. 2016. Adult axolotls can regenerate original neuronal diversity in response to brain injury. *eLife* 5:e13998. doi:10.7554/eLife.13998.

Baddar, N.W., M.R. Woodcock, S. Khatri, D.K. Kump and S.R. Voss. 2015. Sal-site: Research resources for the Mexican axolotl. *Methods in Molecular Biology* 1290:321–336.

Cano-Martínez, A., A. Vargas-González, V. Guarner-Lans, E. Prado-Zayago, M. León-Oleda and B. Nieto-Lima. 2010. Functional and structural regeneration in the axolotl heart (*Ambystoma mexicanum*) after partial ventricular amputation. *Archives de Cardiologia de Mexico* 80:79–86.

Erler, P., A. Sweeney and J.R. Monaghan. 2017. Regulation of injury-induced ovarian regeneration by activation of oogonial stem cells. *Stem Cells* 35:236–247.

Federation of American Societies for Experimental Biology. 2016. Available at http://www.faseb.org/Portals/2/PDFs/opa/2016/FASEB Response to NOT-OD-17-011_FINAL_Letterhead.pdf.

Fei, J.F., M. Schuez, D. Knapp, Y. Taniguchi, D.N. Drechsel and E.M. Tanaka. 2017. Efficient gene knockin in axolotl and its use to test the role of satellite cells in limb regeneration. *Proceedings of the National Academy of Sciences* (USA) 114:12501–12506.

Flowers, G.P., A.T. Timberlake, K.C. McLean, J.R. Monaghan and C.M. Crews. 2014. Highly efficient targeted mutagenesis in axolotl using Cas9 RNA-guided nuclease. *Development* 141:2165–2171.

Gould, S.J. 1977. *Ontogeny and Phylogeny.* Cambridge, MA: Belknap Press.

Haas, B.J. and J.L. Whited. 2017. Advances in decoding axolotl limb regeneration. *Trends in Genetics* 33:553–565.

Johnson, C.K. and S.R. Voss. 2013. Salamander paedomorphosis: Linking thyroid hormone to salamander life history and life cycle evolution. *Current Topics in Developmental Biology* 103:229–258.

Keinath, M.C., N.Y. Timoshevskaya, D.L. Hardy, L. Muzinic, S.R. Voss and J.J. Smith. 2018. A PCR based assay to efficiently determine the sex of axolotls. *Axolotl* 2:5–7.

Khattak, S., M. Schuez, T. Richter et al. 2014. Germline transgenic methods for tracking cells and testing gene function during regeneration in the axolotl. *Stem Cell Reports* 2:243. doi:10.1016/j.stemcr.2013.03.002.

Khattak, S. and E.M. Tanaka. 2015. Transgenesis in axolotl (*Ambystoma mexicanum*). *Methods in Molecular Biology* 1290:269–277.

Page, R.B. and S.R. Voss. 2009. Induction of metamorphosis in axolotls (*Ambystoma mexicanum*). *Cold Spring Harbor Protocols.* doi:10.1101/pdb.prot5268.

Ponomareva, L.V., A.T. Athippozhy, J.S. Thorson and S.R. Voss. 2015. Using *Ambystoma mexicanum* (Mexican axolotl) embryos, chemical genetics, and microarray analysis to identify signaling pathways associated with tissue regeneration. *Comparative Biochemistry and Physiology, Part C* 178:128–135.

Reiß, C., L. Olsson and U. Hoßfeld. 2015. The history of the oldest self-sustaining laboratory animal: 150 years of axolotl research. *Journal of Experimental Zoology, Part B* 324:393–404.

SciCrunch. https://scicrunch.org.

Shaffer, H.B. and S.R. Voss. 1996. Phylogenetic and mechanistic analysis of a developmentally integrated character complex: Alternate life history modes in ambystomatid salamanders. *American Zoologist* 36:24–35.

Smith, H.M. and R.B. Smith. 1971. *Synopsis of the Herpetofauna of Mexico. Vol. 1: Analysis of Literature on the Mexican Axolotl.* Augusta, WV: Eric Lundberg.

Smith, J.J., N. Timoshevskaya, V.A. Timoshevskiy, M.C. Keinath, D. Hardy and S.R. Voss. 2019. A chromosome-scale assembly of the enormous (32 Gb) axolotl genome. *Genome Research* (In press).

Suetsugu-Maki, R., N. Maki, K. Nakamura et al. 2012. Lens regeneration in axolotl: New evidence of developmental plasticity. *BMC Biology* 10:103. doi:10.1186/1741-7007-10-103.

Tazaki, A., E.M. Tanaka and J.F. Fei. 2017. Salamander spinal cord regeneration: The ultimate positive control in vertebrate spinal cord regeneration. *Developmental Biology* 432:63–71.

Thompson, S., L. Muzinic, C. Muzinic, M.L. Niemiller and S.R. Voss. 2014. Probability of regenerating a normal limb after bite injury in the Mexican axolotl (*Ambystoma mexicanum*). *Regeneration* 1:27–32.

Voss, S.R., H.H. Epperlein and E.M. Tanaka. 2009. *Ambystoma mexicanum*, the axolotl: A versatile amphibian model for regeneration, development, and evolution studies. *Cold Spring Harbor Protocols*. doi:10.1101/pdb.emo128.

Voss, S.R., M.R. Woodcock and L. Zambrano. 2015. Tale of two axolotls. *Bioscience* 65:1134–1140.

Voss, S.R. and K. Kump. 2016. A genotyping assay to identify carriers of *albino.Axolotl* 1:5–7.

Woodcock, M.R., J. Vaughn-Wolf, A. Elias et al. 2017. Identification of mutant genes and introgressed tiger salamander DNA in the laboratory axolotl, *Ambystoma mexicanum*. *Scientific Reports* 7:6. doi:10.1038/s41598-017-00059-1.

Yokoyama, H., N. Kudo, M. Todate, Y. Shimada, M. Suzuki and K. Tamura. 2018. Skin regeneration of amphibians: A novel model for skin regeneration as adults. *Development, Growth and Differentiation* 60:316–325.

2 The Genetic Resources of *Arabidopsis thaliana*

The Arabidopsis Biological Resource Center

Christopher S. Calhoun, Deborah K. Crist, Emma M. Knee, Courtney G. Price, Benson E. Lindsey, D. Mariola Castrejon, Eva Nagy, James W. Mann, Julie A. Miller, Erich Grotewold, R. Keith Slotkin and Jelena Brkljacic

CONTENTS

Abstract: Over the course of the past 40 years, *Arabidopsis thaliana* has risen through the ranks to become one of the best studied and most popular research models. In addition to features like small size, self-pollination, and short life cycle which facilitate its maintenance and propagation, *Arabidopsis* has been an unprecedented model for both classical and molecular genetics. The stock centers, including the Arabidopsis Biological Resource Center, have been essential in supporting an increased range and number of tools and resources developed by the research community. Here we present a short research history of this model, followed by its taxonomy and the significance of its relationship with other closely related species. We describe best practices for the acquisition, maintenance, and distribution of *Arabidopsis* and other species from the Brassicaceae family. We provide a few examples of successful translation of *Arabidopsis* research and describe the power of this model for genomic and functional genomics studies. Finally, we present our vision for collection improvements tailored to continue supporting cutting-edge *Arabidopsis* research that enables the next-generation of plant biotechnology.

INTRODUCTION TO *ARABIDOPSIS*

The establishment of *Arabidopsis* as a model plant can be traced back to the 1940s when the potential of this species for genetic studies was recognized by Friedrich Laibach (Laibach 1943, Meyerowitz 2001, Rédei 1975). Laibach's seminal findings were expanded during the 1950s by the discovery that *Arabidopsis* could be utilized to investigate biochemical genetics (Langridge 1955). This notion gained more traction in the 1970s as a consequence of its use in tissue culture applications (Negrutiu et al. 1978). It was not, however, until the 1980s that *Arabidopsis* became solidly established as a plant model system due to the recognition of its small genome size and relatively low content of repetitive DNA (Meyerowitz 2001, Pruitt and Meyerowitz 1986). Its short generation time, low frequency of outcrossing, and its production of a large number of seeds per plant made *Arabidopsis* ideal for advanced genetic screens (Koornneef and Meinke 2010; Somerville and Ogren 1980, 1981, 1982, 1983; Somerville et al. 1982). Further support for the adoption of *Arabidopsis* as a model by the scientific community was gained with the development of an *Agrobacterium*-mediated plant transformation protocol (Feldmann and Marks 1987). This breakthrough technique helped establish the first of many transfer DNA (T-DNA) insertion mutant populations that were ultimately deposited in the stock center collection (Feldmann et al. 1989; Meinke and Scholl 2003).

The *Arabidopsis* community recently celebrated 50 years since the first conference on *Arabidopsis* research which took place in 1965 in Göttingen, Germany. In the decades that have passed since this inaugural conference, the research community has witnessed the publication of over 54,000 research papers involving *Arabidopsis* (Provart et al. 2016). This productivity would not have been possible without the programs committed to the storage, propagation, and distribution of *Arabidopsis* genetic resources that have served the needs of an estimated 30,000 researchers worldwide.

SYSTEMATICS OF *ARABIDOPSIS*

The most widely recognized value of *A. thaliana* lies in its use as a model organism. The genomes of over 1,000 natural accessions have been sequenced to reveal the history of the species and the wealth of its genetic diversity which contributes to the natural variation reflected at both the morphological and molecular level (Genomes Consortium 2016).

The Genus

The genus *Arabidopsis* is taxonomically classified as a member of the Brassicaceae family, order Brassicales, which belongs to the eudicot clade of flowering plants. *Arabidopsis* species are classified botanically as herbs and exist as annuals, perennials, and biennials. They are native to Europe and Asia and can be found in a wide range of environments. The genus *Arabidopsis* contains primarily self-incompatible species, as is the case for most members of the Brassicaceae family (Yamamoto and Nishio 2014). In 2002, the list of species comprising the genus *Arabidopsis* was updated based on molecular phylogeny (Al-Shehbaz and O'Kane 2002). This widely accepted classification recognized nine species and eight subspecies in the genus *Arabidopsis*. Using genome-wide sequencing, the genus was organized into four common species, three species with limited geographic distributions and two allotetraploid species (Novikova et al. 2016).

The Principal Species

Unlike most of the other species in the genus, *Arabidopsis thaliana* (thale cress or mouseear cress) is a self-compatible winter annual. While *A. thaliana* is native to Eurasia (Beck et al. 2008, Francois et al. 2008, Genomes Consortium 2016, Pico et al. 2008, Sharbel et al. 2000) and Africa (Brennan et al. 2014, Durvasula et al. 2017), it has since been naturalized around the globe. This weedy plant, which has no direct economic value or agricultural use, is found in a variety of environments such as disturbed land, fields, meadows, river banks, roadsides, rocky slopes and sandy soil.

Related Taxa of Current or Potential Significance

The Brassicaceae family, also known as the mustard or the crucifer family, is of economic importance as it contains cruciferous vegetables (i.e., cabbage, cauliflower, broccoli, horseradish, turnip, and radish), various ornamental plants (i.e., *Aurinia saxatilis, Iberis umbellata, Lunaria annua,* and *Malcolmia maritima*), as well as species that are used for vegetable oil production (i.e., *Brassica carinata, Brassica juncea, Brassica napus,* and *Brassica rapa*). *Brassica napus*, *Brassica rapa*, *Camelina sativa* (Camelina), *Thlaspi arvense* (field pennycress), and other Brassicas are being evaluated as a source of oil for biofuel production (Gesch et al. 2015, Lu et al. 2011, Moser et al. 2009). *Thellungiella salsuginea* has been suggested as a model system for extremophile plants (Amtmann 2009, Wu et al. 2012, Zhang and Malhi 2010). The value of cruciferous vegetables for cancer prevention is also under investigation (Bosetti et al. 2012).

THE ABRC COLLECTION—HISTORICAL PERSPECTIVE

In response to the growing requirements of the international community for sharing resources, two stock centers were established in the early 1990s. The *Arabidopsis* Biological Resource Center (ABRC) was established in 1992 as a North American seed and clone repository for *Arabidopsis thaliana*. The operation of the Nottingham *Arabidopsis* Stock Centre (NASC) started a few months earlier as a European-based seed repository for this species. The seed collections at ABRC and NASC mirror each other and serve as a secure backup mechanism. ABRC primarily distributes seeds to North, Central, and South America, while NASC distributes seeds to Europe. Researchers in Africa, Asia, and Australia can order stocks from either center. Several other resource centers distribute *Arabidopsis* resources. These include the Riken BioResource Research Center, which focuses on distribution of research materials produced in Japan, and the Institute National de Recherche Agronomique (INRA) Biological Resource Centre of Versailles *Arabidopsis thaliana* Resource Centre, which focuses on distribution of resources developed at INRA.

The small genome of *Arabidopsis* makes it especially suitable for sophisticated genetic screens. These have led to the development of diverse DNA libraries and insertion mutant populations that have been made available to the international community through the stock centers. Additional resources such as a full-length cDNA collection (Seki et al. 2002), characterized mutants, and resources representing genetic diversity (Genomes Consortium 2016) have been added to the collection as they were developed and provided by researchers. In line with the changing interests of the research community, both stock centers continue to add new specialized resources. ABRC's current holdings include cell cultures, plasmids, antibodies, protein chips, and educational materials in addition to seeds. Non-seed resources are distributed worldwide and represent a unique collection not duplicated at NASC. *Arabidopsis thaliana* resources continue to encompass the majority of the collection. However, other species are also represented, including members of the Brassicaceae family as well as other plant species, bacterial and yeast host strains, and both monocot and dicot cloning vectors.

Seeds of *Arabidopsis thaliana* are the largest and most frequently requested items in the collection. The current inventory includes almost 530,000 individual seed lines of natural variants, reference (lab) strains, recombinant inbred lines, mutant and transgenic lines, and chromosomal variants. These seed and other resources are distributed to researchers in more than 80 countries around the world. The distribution of *Arabidopsis* seed accounts for approximately 90% of the more than 160,000 samples shipped annually. Plasmids account for the bulk of the remaining 10%, with plasmids containing *Arabidopsis* DNA comprising more than half of this subset. The ten most popular seed resources, as evidenced by the number of distributions, are shown in Table 2.1.

TABLE 2.1
The Top Ten Seed/Education Stocks Ordered from the ABRC

Stock No.	Donor[a]	Donor Stock Name/No.	AGI[b]	No. Times Ordered
CS60000/CS70000	J. Ecker	Col-0[c]	NA	1,197
CS20	M. Koornneef	Ler-0[d]	NA	871
CS3071	J. Ecker	*ein2-1*	AT5G03280[e]	519
CS22	M. Koornneef	*abi1-1*	AT4G26080[f]	506
CS237	M. Koornneef/A. Bleecker	*etr-1*	AT1G66340[g]	470
CS25	M. Koornneef	*ag-1*	AT4G18960[h]	423
CS8104	Ruth Finklestein	*abi4-1*	AT2G40220[i]	405
CS23	M. Koornneef	*abi2-1*	AT5G57050[j]	390
CS19987	ABRC	"Think Green" education module[k]	NA	380
CS3726	X. Dong	*npr1-1*	AT1G64280[l]	374

[a] Joseph Ecker, Salk Institute for Biological Studies, La Jolla, CA; Martin Koornneef. Department of Plant Sciences, Wageningen University, The Netherlands; Anthony Bleecker (deceased); Ruth Finkelstein, University of California, Santa Barbara, Santa Barbara, CA; Xinnian Dong, Trinity College of Arts and Sciences, Duke University, Durham, NC.

[b] Arabidopsis Gene Identifier.

[c,d] Reference strain.

[e] Involved in ethylene signal transduction by mediating direct regulation of histone acetylation.

[f] Protein phosphatase that acts like a negative regulator of the abscisic acid (ABA) responses.

[g] Ethylene receptor.

[h] Floral homeotic gene encoding a MADS domain transcription factor; Specifies floral meristem, carpel and stamen identity.

[i] Encodes a member of the ERF/AP2 transcription factor family. Involved in ABA signal transduction.

[j] Protein phosphatase that acts like a negative regulator of the ABA responses.

[k] Students use to investigate how plant genotype and natural variation influence plant responses to environmental conditions and how this affects survival.

[l] Salicylic (SA) acid receptor that acts like a transcriptional co-activator to promote SA-mediated defense responses.

SPECIAL COLLECTIONS IN THE ABRC

Reverse genetics is the undisputed core strength of *Arabidopsis* biology. This entails the morphological and/or molecular phenotyping of mutant individuals to dissect the function of a mutated gene or group of genes. Mutant collections for reverse genetics have been generated using chemical or insertional mutagenesis. Targeting Induced Local Lesions in Genomes (TILLING) populations, obtained by EMS mutagenesis and usually containing single nucleotide changes in multiple genes, have been in wide use since 2003 when they were first generated (Till et al. 2003). Due to the ease of transformation, and the randomness of transgene insertion site selection, several large-scale projects have generated publically available collections of T-DNA lines with potentially disruptive insertions within genes (Alonso et al. 2003). SALK, SAIL, GABI-Kat, WiscDsLox, FLAG, SK, and other T-DNA and transposon insertion collections are available from the ABRC, NASC, or other T-DNA collection repositories (O'Malley and Ecker 2010, O'Malley et al. 2015).

Of particular note is the SALK_C "*Confirmed*" collection, for which T-DNA insertions are homozygous in the corresponding seed lines (O'Malley and Ecker 2010). The SALK Institute donated 51,530 confirmed T-DNA lines to the ABRC, representing 24,858 or 90% of the loci in the *Arabidopsis thaliana* Col-0 genome. The genotype of the homozygous lines must be confirmed before a phenotype is analyzed. Insertion lines in a gene of interest can be identified by searching TAIR or SIGnAL (http://signal.salk.edu) genome browsers and ordered through the appropriate repository. Therefore, T-DNA mutants, especially the confirmed lines, can be quickly obtained and investigated if the user has one or a small set of genes that need to be functionally analyzed. Alternatively, entire collections of T-DNA insertions can also be screened for phenotypes/traits of interest through a forward genetics approach, using a T-DNA insertion as the mutagen (Alonso and Ecker 2006). This approach has the distinct advantage of easily identifying the T-DNA flanking DNA to potentially identify the mutated gene without otherwise required mapping. To make this comprehensive screening resource readily available, the ABRC distributes sets of SALK confirmed lines.

The ABRC maintains a large collection of T-DNA insertion mutants. The earliest groups of T-DNA lines received by the ABRC were essentially uncharacterized (Feldmann et al. 1989, Meinke and Scholl 2003). These have been largely used for forward genetic screens in which a large pool of mutant lines is grown to identify specific lines displaying a phenotype of interest. When such lines have been identified, techniques such as Thermal Asymmetric Interlaced PCR (TAIL-PCR) are used (Liu et al. 1995) to identify the location of the T-DNA insertion and thus the gene associated with the mutation.

The earliest pools of T-DNA lines represented an important resource for the *Arabidopsis* research community. However, their use required significant time and resources in order to yield sought-after discoveries. Advances in technology have since facilitated the molecular characterization of T-DNA lines. Many additional T-DNA insertion collections, totaling almost 300,000 sequenced T-DNA lines (O'Malley and Ecker 2010), have been donated to the ABRC. These include the SALK, SAIL, WiscDsLox, and GABI-Kat collections. In contrast to earlier collections, the entire SALK collection was screened using adapter-ligated PCR as an alternative approach to TAIL-PCR, followed by Sanger sequencing to map the location of T-DNA insertions in each line. These lines now provide a valuable resource for both reverse and forward genetic studies (O'Malley et al. 2007). As an illustration of their utility, the T-DNA lines represent the largest portion of the ABRC distributions, ranging from 60% to 70% of the total stocks ordered for the past 15 years. The SALK collection, and several other T-DNA collections, have been further enhanced through the use of whole genome deep sequencing to screen for additional T-DNA insertions in the existing mutant populations. The deep sequencing data has been used to identify many additional T-DNA insertions within the existing seed lines, thereby revealing additional research potential.

ABRC houses the "1001 Genomes Project" collection (Genomes Consortium 2016) (http://tools.1001genomes.org/polymorph/), a group of over 1,000 *Arabidopsis* accessions collected from around the world as a sampling of the genetic diversity of the species. As part of this project, over 1,100 lines were sequenced using genome-wide deep sequencing technology to molecularly characterize genetic diversity and link it to both molecular and phenotypic data (Genomes Consortium 2016).

RNA-mediated silencing approaches, including RNAi and artificial microRNAs (amiRNAs), have been designed as an additional reverse genetic tool to address the issue of functional redundancy (Hauser et al. 2013). Many of these resources have been donated to the ABRC. Thanks to the most recent advances in genome engineering, including Transcription Activator-Like Effector Nucleases (TALEN) and Clustered Regularly Interspaced Short Palindromic Repeats (CRISPR/Cas9), multiple genes can now be targeted for mutagenesis simultaneously in different backgrounds (Cermak et al. 2017). CRISPR/Cas9 vectors and other resources to generate mutations in any gene of interest are now also available for many plant species from the ABRC (https://abrc.osu.edu).

ABRC COLLECTION MANAGEMENT PRACTICES

Acquisition: The resources provided by the ABRC continue to grow in order to meet the needs of the research community. The ABRC receives thousands of stocks each year that are generated in laboratories worldwide. Donations to the ABRC from these programs range from small, one-time deposits to collections from "super-donors" who contribute thousands of stocks from large-scale genomics projects. The ease of long-term storage for both seed and DNA stocks allows the ABRC to have liberal acceptance criteria for new stocks. Thus, almost all donated stocks are accepted, creating the most comprehensive collection of stocks with unique attributes possible.

The ABRC also actively pursues donations to ensure that new stocks are regularly incorporated into the collection. Potential donations are solicited through e-mail communications, presentations, announcements, and other forms of advertisement at research and education conferences, as well as through social media. ABRC initiated a call for journals to require authors to deposit their *Arabidopsis* resources into the collection before publication. This mechanism has already been implemented by several publishing groups. As a result, stocks become available immediately after publication, contributing significantly to the availability of novel lines and the overall quality of the ABRC collection.

To facilitate and encourage donations, the ABRC provides free express shipping for both domestic and international donors whenever possible, as well as an "EZ Seed Donation Form." As a further incentive, the ABRC offers a "Stocks for Stocks" rewards program in which donors receive points that can be redeemed for free stocks. Donors' names are permanently associated with the stocks they contribute. Donors benefit from ABRC's expert maintenance practices and the security of a backup maintained at NASC.

In addition to *Arabidopsis* resources, the ABRC accepts donations of seed of other Brassicaceae species (Figure 2.1), and of DNA resources from any plant species. In support of its education and outreach initiatives, the ABRC also collects and curates *Arabidopsis*-based educational resources to promote plant science instruction at both the K-12 and college levels.

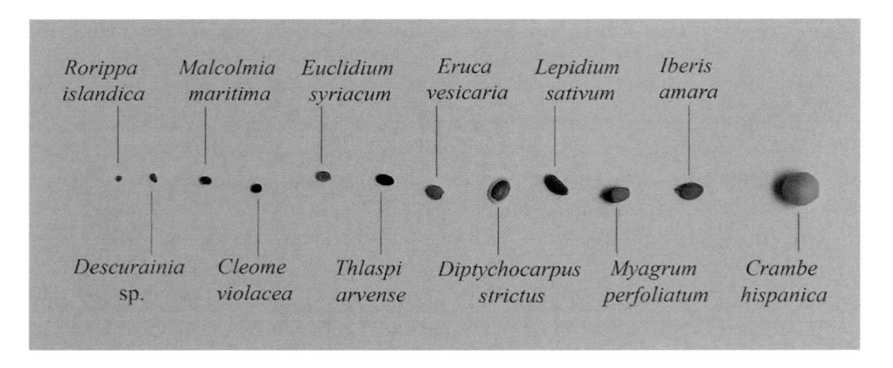

FIGURE 2.1 Seeds from twelve Brassicaceae species illustrating the diversity of size and shape.

Maintenance

In many cases, donated seed stocks need to be amplified to generate distributable quantities of seed. Lines are grown under strict (growth chambers) or moderately strict (greenhouse) environmental conditions. Conditions such as temperature and humidity are monitored for quality control (QC) purposes. Special care is taken to prevent or minimize fungal contamination and to control common insect pests such as Western thrips and fungus gnats. This is accomplished using specific watering protocols and the use of beneficial insects. Potted plants are isolated using plastic floral sleeves after bolting to prevent seed cross-contamination. After harvesting, seeds are dried, inspected for the presence of mold or other contaminants, and tested for germination rate. Seeds are stored at low temperature and low humidity as both distribution and backup copies. An aliquot of each line is sent to NASC. Stocks identified during the curation process as requiring special storage conditions are stored in a –20°C freezer. All cold rooms and freezers are supported by backup emergency power generators and are monitored by an alarm system. Protocols outlining best practices and conditions for growing *Arabidopsis* plants to amplify seeds are available on the ABRC website (https://abrc.osu.edu/resources/growing-stocks) and in more detail in Rivero et al. (2014).

The ABRC maintains both *Arabidopsis thaliana* and maize cell culture lines. *Arabidopsis* cell lines T87 and PSB-L/-D originated from seedling and stem explants, respectively, of two reference strains. These cell lines are maintained either by subcultures in liquid or on solid media, or as cryostocks stored in liquid nitrogen. The Black Mexican Sweet (BMS) maize cell line is derived from the cambial tissue in the hypocotyl of that genotype. The cell suspension culture is subcultured every seven days from three-week old cultures. Backup (callus) cell cultures are maintained on solid medium and subcultured at four- to six-week intervals. BMS cultures are maintained in constant darkness at 24°C. A detailed protocol for the maintenance of all cell cultures at the ABRC is available at https://abrc.osu.edu/resources/growing-stocks.

Distribution Policies

The ABRC ships stocks of *Arabidopsis* and other species in its collection worldwide. Distribution requirements for *Arabidopsis* seeds, plasmids, cell lines, and other resources vary depending on the item being shipped and the requirements of the recipient country. No permit is required for interstate transportation of *Arabidopsis thaliana* seeds, plasmids, cell lines, or other derivative materials, as long as the materials are shipped in accordance with the packaging requirements listed in 7CFR part 340.8(b). Import permits, phytosanitary certificates, customs documents, and declaration letters may be required when materials are shipped internationally. Import permits are issued in accordance with the regulations of the recipient country. Phytosanitary certificates are issued by the United States Department of Agriculture's Animal and Plant Health Inspection Service.

CHARACTERIZATION OF *ARABIDOPSIS* RESOURCES

Morphological

The *Arabidopsis* community has generated a large variety of morphological mutants through the use of various mutagens. While generally used to investigate the function of a mutated gene, mutants acquired by the ABRC have also been used in science education. The collection contains many examples of mutations resulting in modified plant color, leaf morphology, plant size, flower morphology, or in a change in seed size, shape, or color (Meyerowitz 1987).

The ABC model of flower development introduced several *Arabidopsis* mutants with a greatly modified flower structure, including the ones affecting ABC proteins; APETALA 1, APETALA 3, and AGAMOUS (Figure 2.2) (Bowman et al. 2012). This model explained how floral organs were specified by a combination of synergistic and antagonistic effects of the ABC proteins, functioning within precise organ boundaries. Additional mutants were identified and the model was later expanded to include ABCE or ABCDE genes. These studies contributed greatly to the understanding of the evolution of flowers (Chanderbali et al. 2016).

Mutants affecting hormone synthesis/signaling have helped researchers utilize *Arabidopsis* to dissect the regulation of these processes. There are currently nine plant hormones known to regulate plant development and responses to abiotic/biotic stimuli. Mutants affected in hormone synthesis, transport, or signaling display a myriad of changes in plant morphology. For example, gibberellic acid (GA) mutants are reduced in size and will not germinate without exogenous GA (Koornneef and van der Veen 1980), while mutants that lack abscisic acid have reduced seed dormancy and increased transpiration resulting in a wilting phenotype (Koornneef et al. 1982). Even though auxin and brassinosteroid have different roles in regulating plant development, their mutants are dwarf plants with reduced fertility (Clouse et al. 1996, Lincoln et al. 1990). Hormone mutants lend themselves to studies linking a change in plant morphology to gene function. Hormone mutants have historically been among the most popular stocks at the ABRC (Table 2.1).

FIGURE 2.2 **(See color insert.)** Floral organ specification mutants (*ap1-1*, *ap3-1*, and *ag-1*) and Ler-0 reference strain.

Molecular

In addition to the morphological data on various *Arabidopsis* mutants available at the ABRC, a wealth of information is also available on molecular characteristics. Available molecular data range from that acquired in early genetic mapping studies to present-day deep-sequencing and high-throughput metabolomics studies. In some instances, the early genetic mapping studies of a single mutant have been extended to high-throughput techniques, such as Genome Wide Association Studies (GWAS) and whole genome sequencing.

A small genome with few paralogs and the ability to easily generate and screen for mutants have made *Arabidopsis* a favorite for the identification of genes underlying interesting traits or phenotypes. The relative lack of paralogous genes (compared to other larger plant genomes) results in less gene redundancy, and thus simplifies mutational analysis of individual gene function. Many *Arabidopsis* genes are conserved in other plants, and hence serve to help understand fundamental plant processes in crops and non-model organisms. As far back as 1975, George Rédei proposed to use *Arabidopsis* to screen for auxotrophic mutations, based on a study describing a conditional metabolic mutation (Langridge 1955, Rédei 1975). Since those early studies, the use of classical and molecular genetics has allowed for the identification of mutants and the cloning of the underlying genes involved in key aspects of plant development, physiology, and metabolism, as well as the discovery of important components of the pathways that permit plants to interact with the biotic and abiotic environment. Examples of molecular pathways that were comprehensively dissected through the identification of mutants with phenotypes, followed by gene cloning, include the initiation of flowering, the memory of winter (vernalization) (Song et al. 2013, Wellmer and Riechmann 2010), flower development (Chanderbali et al. 2016, Ng and Yanofsky 2000), light and hormone signaling (Lau and Deng 2010, Wang 2005), circadian regulation (Nohales and Kay 2016), and multiple metabolic pathways including those responsible for the formation of seed coat proanthocyanidins (Lepiniec et al. 2006) and glucosinolates (Burow and Halkier 2017). Similarly, *Arabidopsis* has been central to elucidating the molecular components that participate in plant-pathogen interactions such as the Resistance (R) genes that provide important tools for plant biotechnology (Pandolfi et al. 2017).

Opportunities for the translation of research utilizing *Arabidopsis* to important crops are numerous. For example, as a member of the Brassicaceae, *Arabidopsis* shares many attributes with emerging biofuel crops such as the oilseed crop plants *Camelina sativa* and field pennycress. Knowledge gained in *Arabidopsis* about seed development and fatty acid biosynthesis is being swiftly transferred to *C. sativa* and field pennycress (*Thlaspi arvense*) to improve the quality and quantity of seed lipids (Chhikara et al. 2018, Tao et al. 2017).

Screening Studies

The *Arabidopsis* model lends itself to various types of screening studies that can be broadly categorized as forward, quantitative, and reverse genetic screens. Forward genetic screens are performed via mutagenesis and phenotype/trait screening, followed by mapping and identification of the mutated locus (reviewed in greater detail in Ostergaard and Yanofsky 2004). Because of the ease and ubiquity of the first two steps (mutagenesis and screening), most laboratories perform this analysis in-house. Many research laboratories create their own mutagenized population to best suit their screen design. Lehle Seeds (Round Rock, TX) sells ready-to-plant gamma-ray, fast neutron and ethyl methanesulfonate (EMS)-mutagenized screening populations (http://www.arabidopsis.com/main/cat/!ct.html). Once mutants have been identified, thousands of available PCR markers can be used to quickly and efficiently map the mutation or quantitative trait loci (Lukowitz et al. 2000).

PCR markers to distinguish between polymorphic chromosomes of two accessions are widely available and searchable through TAIR and the *Arabidopsis* Mapping Platform (Hou et al. 2010). As a result of the "1001 Genomes Project" (Genomes Consortium 2016), marker density is near the theoretical maximum given the natural polymorphisms present in the species range. This high marker density can be leveraged by either PCR screening, which can typically narrow a mutation to a specific chromosome, or by whole-genome sequencing to assay all mappable polymorphisms simultaneously (Austin et al. 2011, James et al. 2013). In addition, due to the low level of heterochromatin near genes in the *Arabidopsis* genome, mutations are readily mapped to an interval with only one or a few genes present. The *Arabidopsis* model requires a relatively short time to map and molecularly identify mutated genes.

DOCUMENTATION OF *ARABIDOPSIS* RESOURCES

The ABRC stock data includes information such as phenotype, genotype, digital images, and vector maps, as well as protocols for educational resources. These data are provided by the donor of the stock and acquired by ABRC staff from the literature and online resources. Data related to reproduction, such as planting date, location (room and bench), environmental conditions and harvesting date, and quality control testing dates, methods, and results, are also recorded. All stock data is stored in an in-house database. Detailed stock information is uploaded to the ABRC online database (https://abrc.osu.edu) and provided to NASC (arabidopsis.info).

Information is also available via other resources such as the *Arabidopsis* Information Resource (TAIR, arabidopsis.org), the *Arabidopsis* Information Portal (Araport, araport.org), and the Salk Institute Genomic Analysis Laboratory (SIGnAL, signal.salk.edu) upon request. Order history and stock donor and user information are accessible through the ABRC website. Quality control data provided by researchers utilizing ABRC stocks can be entered in the ABRC database by the researcher or by ABRC staff following communication with the researcher.

CURRENT AND FUTURE RESEARCH EFFORTS IN SUPPORT OF COLLECTION HOLDINGS

Since the donation of the first T-DNA collection in 1992, efforts to produce high-quality research resources have been closely correlated with the number of stocks contributed to collection holdings. The generation of large-scale reverse genetic tools (Alonso et al. 2003) or the construction and annotation of a full-length cDNA collection (Seki et al. 2002) had a large impact on the size and diversity of the collection, facilitated by the NSF's "2010 Project" (Chory et al. 2000). Resources resulting from recent research endeavors have been shared through stock centers, and their utilization is facilitating the development of advanced, next-generation resources. A few examples include:

- The acquisition of stocks from the "1001 Genomes Project" whose genome-wide sequencing (deep sequencing) revealed their genomic and epigenomic diversity (Genomes Consortium 2016, Kawakatsu et al. 2016, Pisupati et al. 2017); The ABRC has also actively participated in one of the first largescale studies characterizing the global population of *Arabidopsis* natural accessions (Anastasio et al. 2011, Platt et al. 2010).
- The development of a T-DNA homozygous mutant collection as a genome-wide functional tool for reverse genetic analyses that is distributed as individual lines and sets representing 25,000 *Arabidopsis* genes or over 90% of the genome (O'Malley and Ecker 2010, O'Malley et al. 2015);
- The Open Reading Frame (ORF) clone collections for functional genomics of *Arabidopsis*, maize, and other species (Burdo et al. 2014, Gaudinier et al. 2011, Pruneda-Paz et al. 2014);
- CRISPR/Cas9 genome editing vectors and mutant stocks (Fauser et al. 2014).

Next generation *Arabidopsis* resources are continuously added to the ABRC's holdings. The center has acquired a number of cutting-edge resources such as a set of HALO-tagged transcription factors (TFs) for genome-wide identification of TF binding sites (O'Malley et al. 2016), vectors and host strains for high throughput yeast two hybrid assays by deep sequencing (Trigg et al. 2017), and a comprehensive toolkit for advanced genome engineering, including CRISPR/Cas9 mutagenesis (Cermak et al. 2017).

The ABRC is actively pursuing novel research trends and technologies to expand the utility and improve the quality of the collection (Lindsey et al. 2017), as well as to make access to stocks more user-friendly. In support of the recent acquisition policy to diversify beyond *Arabidopsis* stocks, 12 deep-sequenced *Brassica* seed stocks have been added to the collection (Figure 2.1). The acquisition and propagation of more bioenergy-related stocks are planned for the near future. These include well-characterized/sequenced individual lines or small collections of crops such as *Brassica rapa* and bioenergy crops such as field pennycress or *Camelina.* Also included are large populations for genome wide association studies (GWAS), e.g., *Boechera stricta.* The ABRC database includes a robust search tool that fully supports the expansion of the collection to other species, making these resources accessible to a wider research audience.

The ABRC is currently implementing deep sequencing technology for stock identification and QC purposes by taking advantage of the relatively low cost to prepare and sequence genomic libraries (Pisupati et al. 2017). Deep sequencing can be used to identify, or to validate the identity of, donated plant lines and plasmid vectors and clones, as well to assess genetic relationship between stocks. Even at low coverage, sequencing has been recommended as one of the best practices for maintaining the identity of genetic stocks (Bergelson et al. 2016). By performing large-scale identification of preserved stocks, stock centers are in a historically unique position to take a leading role in addressing the societal and ethical aspects of research such as resource conservation and authentication, along with the data sharing and reproducibility (Halewood et al. 2018, McCluskey et al. 2017). Apart from advocating sharing of stocks and data within the *Arabidopsis* community, the ABRC has also been open to providing the knowledge and experience to other research communities (Brkljacic et al. 2011, McCluskey et al. 2017).

ANTICIPATED TECHNOLOGICAL APPROACHES TO IMPROVE COLLECTION MANAGEMENT PRACTICES

Personnel from the ABRC have completed the Six Sigma Green Belt Kaizen Blitz certification program. Using this training, the ABRC has focused on reducing waste and maximizing the efficiency of operations. As a result, a number of improvements have been made such as automation, barcoding, and improved inventory management techniques. Significant effort has also been devoted to documenting operational processes through the development of 75 Standardized Operating Procedures. These measures have resulted in an overall increase in the operational efficiency of the ABRC, effectively tripling the harvesting efficiency in just three years and reducing seed order processing time to 2–3 days. Current and planned improvements are listed separately below for each of the approaches.

Automation

The use of robotics for distribution and stock amplification purposes has increased the efficiency of the ABRC operation. The distribution of DNA stocks (clones and vectors) is accomplished using a Biomek 3000 robotic system (Beckman) that is used to prepare sets and individual stocks. For seed distribution, a custom-made automated seed dispensing system (Seed Aliquoting Machine, SAM, Figure 2.3) is used to prepare orders and to amplify seed sets upon depletion. SAM has helped reduce seed order processing time to 2–3 days, representing a 3-fold reduction. Future plans include the design and production of a robotic system that will assist with the harvesting stage of the seed reproduction process. This stage has been identified as a bottleneck of the seed reproduction process.

Barcoding

All ABRC DNA stocks are barcoded to prevent cross-contamination and increase the efficiency of stock handling. Planted seed stocks are also barcoded, streamlining data collection and increasing the efficiency of the QC process. Planned improvements include expanding barcoding to all donated and ordered seed stocks, with a long-term goal to expand to a vial inventory system. Each vial of a single stock will be barcoded separately, enabling the tracking of ordered stock vials. This tracking system will significantly improve the efficiency of addressing user inquiries by enabling the tracking of all ordered vials associated with each inquiry.

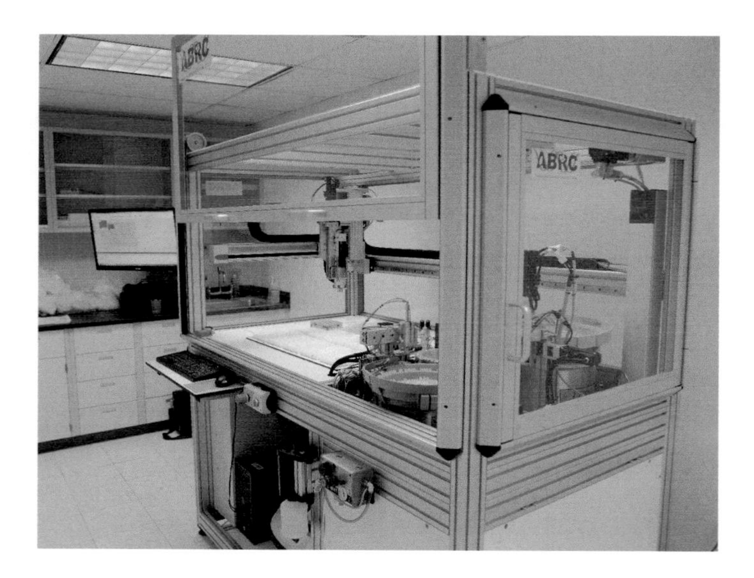

FIGURE 2.3 Seed Aliquoting Machine developed in collaboration with Labman Automation Ltd., UK.

Inventory Management

With over half a million seed stocks in the collection, individual vials are occasionally damaged or misplaced. To minimize the impact of this issue, a modification of MIPAR, an image analysis software tool developed at The Ohio State University (https://link.springer.com/article/10.1186/2193-9772-3-10), is currently being tested in combination with a new optical character recognition module. The MIPAR-OCR will translate images of inventory boxes into lists of stock numbers before and after a vial is pulled out of the box. The implementation of this system will further increase distribution efficiency by greatly reducing the number of inventory errors and the amount of time wasted searching for vials.

DATABASE MANAGEMENT AND COLLECTION ACCESS

Providing access to stock data continues to be a critical component of the ABRC's mission. From 1992 until 1994, stock information was made available through printed catalogs and was also accessible through a terminal login and email-based query system (*Arabidopsis* Information Management System, AIMS). In 1994 an online version of AIMS was launched and in 1995, a world wide web server that linked to the database and ordering system provided broader access to the community. Production of printed catalogs was discontinued in 1996, as the size of the collection and the volume of related data became too large to disseminate in this format. In 2001, stock data was incorporated into the TAIR database (Garcia-Hernandez et al. 2002), allowing online access to stock information and additional genomic and phenotypic data generated by the research community. This was essential for providing access to the large quantities of flank sequence data being generated for the T-DNA insertion mutant collection, and the full sequences of ORF clones representing the majority of the *Arabidopsis* genome.

In order to provide better access to ABRC stock information via associated gene and polymorphism data, and to allow for the incorporation of new resources into the stock collection, the ABRC has developed a stand-alone stock database and online ordering system separate from TAIR, available at https://abrc.osu.edu. The new system greatly improves the accessibility of the collection and makes stocks other than *Arabidopsis* seeds and clones (such as cell cultures, antibodies, vectors) easier to locate and order. Identifying non-*Arabidopsis* resources is also simplified due to an integrated species-specific search function. A simplified registration system, and streamlined ordering system that automates the process for submitting required shipping documentation, further simplifies access to the collection. Data sharing with NASC (http://arabidopsis.info), TAIR (www.arabidopsis.org), and others will be maintained to allow researchers to continue to use the platforms with which they are familiar to access stock information, while providing enhanced access to new resources and a streamlined ordering experience.

ABRC OUTREACH ACTIVITIES

In addition to serving as a valuable resource for the research community, the ABRC has expanded its mission to include promoting the use of plants in K-12 and college level instruction, and raising awareness about the importance of plants for the general public. This goal has been accomplished through the development and curation of a variety of education resources and programs. The following section outlines the challenges facing science education, highlights how plants, and specifically *Arabidopsis*, can be used to address current challenges, and provides an overview of the ABRC's outreach initiatives.

Challenges and opportunities facing science education today include declining student interest in science, technology, engineering, and math subjects, increased expectations of the depth of teacher domain knowledge, a focus on standardized testing, and the increased availability of high quality, low cost educational resources (Davis et al. 2006, Gilbert 2004). Research demonstrates the benefits of moving from a lecture-based science curriculum to more authentic science experiences that incorporate active learning opportunities (Freeman et al. 2014, Kontra et al. 2015, Michael 2006, Price et al. 2018). This includes making direct connections to current research and demonstrating the benefits of science to society (Braund and Reiss 2006). However, with the continued focus on standardized testing and the requirement to meet set curriculum standards, many teachers may feel limited in their ability to develop creative and engaging science lessons.

In view of the important role that model systems play in research, it follows that the use of such systems should be mirrored in educational endeavors (Coll et al. 2005, Gilbert 2004, Gobert and Buckley 2000, Hubbs et al. 2017, Keskin and Cam 2017, Treagust et al. 2002). *Arabidopsis* is uniquely positioned to serve as a model system while also addressing a concern well recognized in the plant science community—*plant blindness*. This term, first coined in the 1990s, refers to people's inability to notice plants in the environment, recognize the important role they play in our daily lives, and the disproportionate interest students have in studying animals rather than plants (Allen 2003, Mann et al. 2017, Wandersee and Schussler 1999). Integrating plants into their science curricula provides teachers with multiple opportunities to facilitate student-driven active learning experiences that have the potential to foster a greater interest in plants, and to provide a deeper understanding of the impact of plants and plant science-related concepts (Allen 2003, Wandersee and Schussler 1999).

Many characteristics of *Arabidopsis* make it an ideal organism for science education. The use of plants relieves many of the ethical and logistical concerns that arise with the use of animals (Cross and Cross 2004) in instances where teachers are looking to incorporate the study of live organisms in the classroom. The short life cycle, small size, and relatively low-maintenance growth requirements of *Arabidopsis* make it a viable option for modestly equipped science classrooms with limited space (Mann et al. 2017). *Arabidopsis* provides the opportunity to demonstrate a variety of science concepts including inheritance, variation, response to environment, growth and development, hormone physiology, and more (Mann et al. 2017, Price et al. 2018, Wyatt and Ballard 2007, Zheng 2006). In addition, as an important model system for research, and a relative of many economically important food crops, *Arabidopsis* provides opportunities for students to make relevant connections to real-world science.

In 2011, the ABRC began its outreach program thanks in part to funding from the American Society of Plant Biologists (ASPB) and the National Science Foundation.

Initial efforts centered on the establishment of a central hub for *Arabidopsis* teaching resources, and the release of a set of education kits. A kit consists of experimental protocols, supporting material, and *Arabidopsis* seeds—everything an educator needs to introduce *Arabidopsis* into his or her science curriculum. Many of the kits available through ABRC were developed by members of the research and education community. These kits are collectively known as "Translating Research on *Arabidopsis* Into a Network of Education Resources." Six of the kits, collectively known as "Greening the Classroom," were developed and tested by ABRC staff. All ABRC education resources are available to a global audience, with education kits having been shipped to 30 countries around the world. Currently, 22 kits are available via the ABRC's outreach website, with a number of additional kits under development. One example includes a partnership with the ASPB to develop kits corresponding to their "Teaching Tools in Plant Biology" modules (http://www.plantcell.org/content/teaching-tools-plant-biology). With this partnership, the ABRC seeks to add value to the existing modules by providing a visual demonstration of the concepts presented within a teaching tool (e.g., stocks presented in Figure 2.2 are part of the "Genetics of Floral Development Kit").

The ABRC's education kits are in high demand, with 300–350 kits ordered annually. The ABRC's outreach initiatives have grown to include regular presentations at education conferences, teacher professional development workshops, visits to local schools, school field trips to the ABRC facilities, remote support for teachers growing and using *Arabidopsis* in the classroom, and a number of partnerships with community organizations. In 2017, the ABRC served more than 4,000 individuals that included K-12 students, teachers, and pre-service teachers, as well as the general public.

Looking forward, ABRC has identified a number of opportunities to expand and diversify its impact on education. Future goals include engaging a wider plant community in outreach efforts, increasing the awareness of plants by the general public, creating interactive activities, and providing a more user-friendly online source of information for non-specialists. As long as model systems play a role in science, so too should they play a role in science education. The ABRC will continue to lead efforts to forge this connection.

CONCLUSIONS

Throughout the history of *Arabidopsis* research, the ABRC and other stock centers have played a key role in driving and supporting the efforts of the *Arabidopsis* research community. While the ABRC has adapted to the changes propelled by new technologies and cutting-edge research, the role of the ABRC as a critical resource for both basic and applied plant biology research has remained the same: to continue to inexpensively distribute seed and DNA resources to the scientific community, enabling discoveries across all plant biology and beyond.

ACKNOWLEDGMENTS

We would like to thank David Somers for critical reading of the manuscript. This work was supported by the NSF grants DBI-1049341, DBI-1561210, and MCB-1143813.

REFERENCES

Allen, W. 2003. Plant blindness. *Bioscience* 53:926–926.

Alonso, J.M., A.N. Stepanova, T.J. Leisse et al. 2003. Genome-wide insertional mutagenesis of *Arabidopsis thaliana*. *Science* 301:653–657.

Alonso, J.M. and J.R. Ecker. 2006. Moving forward in reverse: Genetic technologies to enable genomewide phenomic screens in *Arabidopsis*. *Nature Reviews Genetics* 7:524–536.

Al-Shehbaz, I.A. and S.L. O'Kane Jr. 2002. Taxonomy and phylogeny of *Arabidopsis* (Brassicaceae). *Arabidopsis Book* 1:e0001. doi:10.1199/tab.0001.

Anastasio, A.E., A. Platt, M. Horton et al. 2011. Source verification of mis-identified *Arabidopsis thaliana* accessions. *Plant Journal* 67:554–566.

Austin, R.S., D. Vidaurre, G. Stamatiou et al. 2011. Next-generation mapping of *Arabidopsis* genes. *Plant Journal* 67:715–725.

Beck, J.B., H. Schmuths and B.A. Schaal. 2008. Native range genetic variation in *Arabidopsis thaliana* is strongly geographically structured and reflects Pleistocene glacial dynamics. *Molecular Ecology* 17:902–915.

Bergelson, J., E.S. Buckler, J.R. Ecker, M. Nordborg and D. Weigel. 2016. A proposal regarding best practices for validating the identity of genetic stocks and the effects of genetic variants. *Plant Cell* 28:606–609.

Bosetti, C., M. Filomeno, P. Riso et al. 2012. Cruciferous vegetables and cancer risk in a network of case-control studies. *Annals of Oncology* 23:2198–2203.

Braund, M. and M. Reiss. 2006. Towards a more authentic science curriculum: The contribution of out-of-school learning. *International Journal of Science Education* 28:1373–1388.

Amtmann, A. 2009. Learning from evolution: *Thellungiella* generates new knowledge on essential and critical components of abiotic stress tolerance in plants. *Molecular Plant* 2:3–12.

Brennan, A.C., B. Méndez-Vigo, A. Haddioui, J.M. Martínez-Zapater, F.X. Picó and C. Alonso-Blancoet. 2014. The genetic structure of *Arabidopsis thaliana* in the south-western Mediterranean range reveals a shared history between North Africa and southern Europe. *BMC Plant Biology* 14:17. doi:10.1186/1471-2229-14-17.

Bowman, J.L., D.R. Smyth and E.M. Meyerowitz. 2012. The ABC model of flower development: Then and now. *Development* 139:4095–4098.

Brkljacic, J., E. Grotewold, R. Scholl et al. 2011. *Brachypodium* as a model for the grasses: Today and the future. *Plant Physiology* 157:3–13.

Burdo, B., J. Gray, M.P. Goetting-Minesky et al. 2014. The Maize TFome--development of a transcription factor open reading frame collection for functional genomics. *Plant Journal* 80:356–366.

Burow, M. and B.A. Halkier. 2017. How does a plant orchestrate defense in time and space? Using glucosinolates in *Arabidopsis* as case study. *Current Opinion in Plant Biology* 38:142–147.

Cermak, T., S.J. Curtin, J. Gil-Humanes et al. 2017. A multipurpose toolkit to enable advanced genome engineering in plants. *Plant Cell* 29:1196–1217.

Chanderbali, A.S., B.A. Berger, D.G. Howarth, P.S. Soltis and D.E. Soltis. 2016. Evolving ideas on the origin and evolution of flowers: New perspectives in the genomic era. *Genetics* 202:1255–1265.

Chhikara, S., H.M. Abdullah, P. Akbari, D. Schnell and O.P. Dhankher. 2018. Engineering *Camelina sativa* (L.) Crantz for enhanced oil and seed yields by combining diacylglycerol acyltransferase1 and glycerol-3-phosphate dehydrogenase expression. *Plant Biotechnology Journal* 16:1034–1045.

Chory, J., J.R. Ecker, S. Briggs et al. 2000. National Science Foundation-sponsored workshop report: "The 2010 Project" functional genomics and the virtual plant. A blueprint for understanding how plants are built and how to improve them. *Plant Physiology* 123:423–426.

Clouse, S.D., M. Langford and T.C. McMorris 1996. A brassinosteroid-insensitive mutant in *Arabidopsis thaliana* exhibits multiple defects in growth and development. *Plant Physiology* 111:671–678.

Coll, R.K., B. France and I. Taylor. 2005. The role of models/and analogies in science education: Implications from research. *International Journal of Science Education* 27:183–198.

Cross, T.R. and V.E. Cross. 2004. Scalpel or mouse? A statistical comparison of real and virtual frog dissections. *American Biology Teacher* 66:408–411.

Davis, E.A., D. Petish and J. Smithey. 2006. Challenges new science teachers face. *Review of Educational Research* 76:607–651.

Durvasula, A., A. Fulgione, R.M. Gutaker et al. 2017. African genomes illuminate the early history and transition to selfing in *Arabidopsis thaliana. Proceedings of National Academy of Sciences* (USA) 114:5213–5218.

Fauser, F., S. Schiml and H. Puchta. 2014. Both CRISPR/Cas-based nucleases and nickases can be used efficiently for genome engineering in *Arabidopsis thaliana. Plant Journal* 79:348–359.

Feldmann, K. and M. Marks. 1987. *Agrobacterium*-mediated transformation of germinating seeds of *Arabidopsis thaliana*: A non-tissue culture approach. *Molecular and General Genetics* 208:1–9.

Feldmann, K.A., M.D. Marks, M.L. Christianson and R.S. Quatrano. 1989. A dwarf mutant of *Arabidopsis* generated by T-DNA insertion mutagenesis. *Science* 243:1351–1354.

Francois, O., M.G.B. Blum, M. Jakobsson and N.A. Rosenberg. 2008. Demographic history of European populations of *Arabidopsis thaliana. PLoS Genetics* 4:e1000075. doi:10.1371/journal.pgen.1000075.

Freeman, S., S.L. Eddy, M. McDonough et al. 2014. Active learning increases student performance in science, engineering, and mathematics. *Proceedings of the National Academy of Sciences* (USA) 111:8410–8415.

Garcia-Hernandez, M., T.Z. Berardini, G. Chen et al. 2002. TAIR: A resource for integrated *Arabidopsis* data. *Functional and Integrative Genomics* 2:239–253.

Gaudinier, A., L. Zhang, J.S. Reece-Hoyes et al. 2011. Enhanced Y1H assays for *Arabidopsis. Nature Methods* 8:1053–1055.

Genomes Consortium. 2016. 1,135 genomes reveal the global pattern of polymorphism in *Arabidopsis thaliana. Cell* 166:481–491.

Gesch, R.W., T.A. Isbell, E.A. Oblath et al. 2015. Comparison of several *Brassica* species in the north central US for potential jet fuel feedstock. *Industrial Crops and Products* 75:2–7.

Gilbert, J.K. 2004. Models and modeling: Routes to more authentic science education. *International Journal of Science and Mathematics Education* 2:115–130.

Gobert, J.D. and B.C. Buckley. 2000. Introduction to model-based teaching and learning in science education. *International Journal of Science Education* 22:891–894.

Halewood, M., T. Chiurugwi, R.S. Hamilton et al. 2018. Plant genetic resources for food and agriculture: Opportunities and challenges emerging from the science and information technology revolution. *New Phytologist* 217:1407–1419.

Hauser, F., W. Chen, U. Deinlein et al. 2013. A genomic-scale artificial microRNA library as a tool to investigate the functionally redundant gene space in *Arabidopsis. Plant Cell* 25:2848–2863.

Hou, X., L. Li, Z. Peng et al. 2010. A platform of high-density INDEL/CAPS markers for map-based cloning in *Arabidopsis. Plant Journal* 63:880–888.

Hubbs, N.B., K.N. Parent and J.R. Stoltzfus. 2017. Models in the biology classroom: An in-class modeling activity on meiosis. *American Biology Teacher* 79:482–491.

James, G.V., V. Patel, K.J. Nordström et al. 2013. User guide for mapping-by-sequencing in *Arabidopsis. Genome Biology* 14:R61. doi:10.1186/gb-2013-14-6-r61.

Kawakatsu, T., S.C. Huang, F. Jupe et al. 2016. Epigenomic diversity in a global collection of *Arabidopsis thaliana* accessions. *Cell* 166:492–505.

Keskin, F. and A. Cam. 2017. Using a model to teach crossing over. *American Biology Teacher* 79:305–308.

Kontra, C., D.J. Lyons, S.N. Fischer and S.L. Beilock. 2015. Physical experience enhances science learning. *Psychological Science* 26:737–749.

Koornneef, M. and J.H. van der Veen. 1980. Induction and analysis of gibberellin sensitive mutants in *Arabidopsis thaliana* (L.) Heynh. *Theoretical and Applied Genetics* 58:257–263.

Koornneef, M. and D. Meinke. 2010. The development of *Arabidopsis* as a model plant. *Plant Journal* 61:909–921.

Koornneef, M., M.L. Jorna, D.L.C. Brinkhorst-van der Swan and C.M. Karssenet. 1982. The isolation of abscisic acid (ABA) deficient mutants by selection of induced revertants in non-germinating gibberellin sensitive lines of *Arabidopsis thaliana* (L.) Heynh. *Theoretical and Applied Genetics* 61:385–393.

Laibach, F. 1943. *Arabidopsis thaliana* (L.) Heynh. als objekt fur genetische und entwicklungsphysiologische untersuchungen. *Botanisches Archiv. Zeitschrift für die Gesamte Botanik* 44:439–455.

Langridge, J. 1955. Biochemical mutations in the crucifer *Arabidopsis thaliana* (L.) Heynh. *Nature* 176:260–261.

Lau, O.S. and X.W. Deng. 2010. Plant hormone signaling lightens up: Integrators of light and hormones. *Current Opinion in Plant Biology* 13:571–577.

Lepiniec, L., I. Debeaujon, J.M. Routaboul et al. 2006. Genetics and biochemistry of seed flavonoids. *Annual Review of Plant Biology* 57:405–430.

Lincoln, C., J.H. Britton and M. Estelle 1990. Growth and development of the axr1 mutants of *Arabidopsis. Plant Cell* 2:1071–1080.

Lindsey, B.E., L. Rivero, C.S. Calhoun, E. Grotewold and J. Brkljacic. 2017. Standardized method for high-throughput sterilization of *Arabidopsis* seeds. *Journal of Visualized Experiments* 128. doi:10.3791/56587.

Liu, Y.G., N. Mitsukawa, T. Oosumi and R.F. Whittier. 1995. Efficient isolation and mapping of *Arabidopsis thaliana* T-DNA insert junctions by thermal asymmetric interlaced PCR. *Plant Journal* 8:457–463.

Lu, C., J.A. Napier, T.E. Clemente and E.B. Cahoon. 2011. New frontiers in oilseed biotechnology: Meeting the global demand for vegetable oils for food, feed, biofuel, and industrial applications. *Current Opinion in Biotechnology* 22:252–259.

Lukowitz, W., C.S. Gillmor and W.R. Scheible. 2000. Positional cloning in *Arabidopsis.* Why it feels good to have a genome initiative working for you. *Plant Physiology* 123:795–805.

Mann, J.W., J. Larson, M. Pomeranz et al. 2017. Linking genotype to phenotype: The effect of a mutation in gibberellic acid production on plant germination. *CourseSource.* doi:10.24918/cs. 2017.18.

McCluskey, K., K. Boundy-Mills, G. Dye et al. 2017. The challenges faced by living stock collections in the USA. *Elife* 13:6. doi:10.7554/eLife.24611.

Meinke, D. and R. Scholl. 2003. The preservation of plant genetic resources: Experiences with *Arabidopsis. Plant Physiology* 133:1046–1050.

Meyerowitz, E.M. 1987. *Arabidopsis thaliana. Annual Review of Genetics* 21:93–111.

Meyerowitz, E.M. 2001. Prehistory and history of *Arabidopsis* research. *Plant Physiology* 125:15–19.

Michael, J. 2006. Where's the evidence that active learning works? *Advances in Physiology Education* 30:159–167.

Moser, B.R., G. Knothe, S.F. Vaughn and T.A. Isbell. 2009. Production and evaluation of biodiesel from field pennycress (*Thlaspi arvense* L.) oil. *Energy & Fuels* 23:4149–4155.

Negrutiu, I., A. Cattoir-Reynaerts and M. Jacobs. 1978. Selection and characterization of cell lines of *Arabidopsis thaliana* resistant to amino acid analogs. *Archives Internationales de Physiologie et de Biochimie* 86:442–443.

Ng, M. and M.F. Yanofsky. 2000. Three ways to learn the ABCs. *Current Opinion in Plant Biology* 3:47–52.

Nohales, M.A. and S.A. Kay. 2016. Molecular mechanisms at the core of the plant circadian oscillator. *Nature Structural and Molecular Biology* 23:1061–1069.

Novikova, P.Y., N. Hohmann, V. Nizhynska et al. 2016. Sequencing of the genus *Arabidopsis* identifies a complex history of nonbifurcating speciation and abundant trans-specific polymorphism. *Nature Genetics* 48:1077–1082.

O'Malley, R.C. and J.R. Ecker. 2010. Linking genotype to phenotype using the *Arabidopsis* unimutant collection. *Plant Journal* 61:928–940.

O'Malley, R.C., J.M. Alonso, C.J. Kim, T.J. Leisse and J.R. Ecker. 2007. An adapter ligation-mediated PCR method for high-throughput mapping of T-DNA inserts in the *Arabidopsis* genome. *Nature Protocols* 2:2910–2917.

O'Malley, R.C., C.C. Barragan and J.R. Ecker. 2015. A user's guide to the *Arabidopsis* T-DNA insertion mutant collections. *Methods in Molecular Biology* 1284:323–342.

O'Malley, R.C., S.C. Huang, L. Song et al. 2016. Cistrome and epicistrome features shape the regulatory DNA landscape. *Cell* 165:1280–1292.

Ostergaard, L. and M.F. Yanofsky. 2004. Establishing gene function by mutagenesis in *Arabidopsis thaliana. Plant Journal* 39:682–696.

Pandolfi, V., J.R.C.F. Neto, M.D. da Silva et al. 2017. Resistance (R) genes: Applications and prospects for plant biotechnology and breeding. *Current Protein and Peptide Science* 18:323–334.

Pico, F.X., B. Méndez-Vigo, J.M. Martínez-Zapater and C. Alonso-Blanco. 2008. Natural genetic variation of *Arabidopsis thaliana* is geographically structured in the Iberian Peninsula. *Genetics* 180:1009–1021.

Pisupati, R., I. Reichardt, Ü. Seren et al. 2017. Verification of *Arabidopsis* stock collections using SNPmatch, a tool for genotyping high-plexed samples. *Scientific Data* 4:170184. doi:10.1038/sdata.2017.184.

Platt, A., M. Horton, Y.S. Huang et al. 2010. The scale of population structure in *Arabidopsis thaliana. PLoS Genetics* 6:e1000843.

Price, C.G., E.M. Knee, J.A. Miller et al. 2018. Following phenotypes: An exploration of Mendelian genetics using *Arabidopsis* plants. *American Biology Teacher* 80:291–300.

Provart, N.J., J. Alonso, S.M. Assmann et al. 2016. 50 years of *Arabidopsis* research: Highlights and future directions. *New Phytologist* 209:921–944.

Pruitt, R.E. and E.M. Meyerowitz. 1986. Characterization of the genome of *Arabidopsis thaliana. Journal of Molecular Biology* 187:169–183.

Pruneda-Paz, J.L., G. Breton, D.H. Nagel et al. 2014. A genome-scale resource for the functional characterization of *Arabidopsis* transcription factors. *Cell Reports* 8:622–632.

Rédei, G. 1975. *Arabidopsis* as a genetic tool. *Annual Review of Genetics* 9:111–127.

Rivero, L., R. Scholl, N. Holomuzki, D. Crist, E. Grotewold and J. Brkljacic. 2014. Handling *Arabidopsis* plants: Growth, preservation of seeds, transformation, and genetic crosses. *Methods in Molecular Biology* 1062:3–25.

Seki, M., M. Narusaka, A. Kamiya et al. 2002. Functional annotation of a full-length *Arabidopsis* cDNA collection. *Science* 296:141–145.

Sharbel, T.F., B. Haubold and T. Mitchell-Olds. 2000. Genetic isolation by distance in *Arabidopsis thaliana*: Biogeography and postglacial colonization of Europe. *Molecular Ecology* 9:2109–2118.

Somerville, C. and W. Ogren. 1980. Photo respiration mutants of *Arabidopsis thaliana* deficient in serine glyoxylate amino transferase Ec-2.6.1.45 activity. *Proceedings of the National Academy of Sciences* (USA) 77:2684–2687.

Somerville, C.R. and W.L. Ogren. 1981. Photorespiration-deficient mutants of *Arabidopsis thaliana* lacking mitochondrial serine transhydroxymethylase activity. *Plant Physiology* 67:666–671.

Somerville, C.R. and W.L. Ogren. 1982. Mutants of the cruciferous plant *Arabidopsis thaliana* lacking glycine decarboxylase activity. *Biochemical Journal* 202:373–380.

Somerville, S.C. and W.L. Ogren. 1983. An *Arabidopsis thaliana* mutant defective in chloroplast dicarboxylate transport. *Proceedings of the National Academy of Sciences* (USA) 80:1290–1294.

Somerville, C.R., A.R. Portis and W.L. Ogren. 1982. A mutant of *Arabidopsis thaliana* which lacks activation of RuBP carboxylase in vivo. *Plant Physiology* 70:381–387.

Song, J., J. Irwin and C. Dean. 2013. Remembering the prolonged cold of winter. *Current Biology* 23:R807–R811.

Tao, L., A. Milbrandt, Y. Zhang and W-C. Wang. 2017. Techno-economic and resource analysis of hydroprocessed renewable jet fuel. *Biotechnology for Biofuels* 10:261. doi:10.1186/s13068-017-0945-3.

Till, B.J., S.H. Reynolds, E.A. et al. 2003. Large-scale discovery of induced point mutations with high-throughput TILLING. *Genome Research* 13:524–530.

Treagust, D.F., G. Chittleborough and T.L. Mamiala. 2002. Students' understanding of the role of scientific models in learning science. *International Journal of Science Education* 24:357–368.

Trigg, S.A., R.M. Garza, A. MacWilliams et al. 2017. CrY2H-seq: A massively multiplexed assay for deep-coverage interactome mapping. *Nature Methods* 14:819–825.

Wandersee, J.H. and E.E. Schussler. 1999. Preventing plant blindness. *American Biology Teacher* 61:82–86.

Wang, H. 2005. Signaling mechanisms of higher plant photoreceptors: A structure-function perspective. *Current Topics in Developmental Biology* 68:227–261.

Wellmer, F. and J.L. Riechmann. 2010. Gene networks controlling the initiation of flower development. *Trends in Genetics* 26:519–527.

Wu, H.J., Z. Zhang, J.Y. Wang et al. 2012. Insights into salt tolerance from the genome of *Thellungiella salsuginea*. *Proceedings of the National Academy of Sciences* (USA) 109:12219–12224.

Wyatt, S. and H.E. Ballard. 2007. *Arabidopsis* ecotypes: A model for course projects in organismal plant biology and evolution. *American Biology Teacher* 69:477–481.

Yamamoto, M. and T. Nishio. 2014. Commonalities and differences between *Brassica* and *Arabidopsis* self-incompatibility. *Horticulture Research* 1:14054. doi:10.1038/hortres.2014.54.

Zhang, M. and S.S. Malhi. 2010. Perspectives of oilseed rape as a bioenergy crop. *Biofuels* 1:621–630.

Zheng, Z.L. 2006. Use of the gl1 mutant & the CA-rop2 transgenic plants of *Arabidopsis thaliana* in the biology laboratory course. *American Biology Teacher* 68:e148–e153.

3 The Bacillus Genetic Stock Center/*Bacillus subtilis*

Daniel R. Zeigler

CONTENTS

Abstract: *Bacillus*, a large and phylogenetically diverse group of rod-shaped bacteria, have been isolated from virtually every conceivable environment on this planet. Their unique life cycle and other characteristics have made them valuable in industrial, agricultural, and medical research. In response to their widespread use, the Bacillus Genetic Stock Center was initiated in the 1970s. Its main collection now contains more than 11,000 accessions that includes

67 species in 12 genera of *Bacillus*. Isolates can usually be easily propagated and stored under conditions that maintain their long-term viability. Strains of *Bacillus*, sets of phages and plasmid tools, and associated resources are available from the collection, on request.

INTRODUCTION

Bacillus sensu lato comprises an enormous and phylogenetically diverse group of organisms that have adopted a common lifestyle. When conditions are favorable for growth, cells propagate rapidly. When nutrient sources are depleted, *Bacillus* populations differentiate into subpopulations of different cell types, many of them showing a high degree of multicellularity. One specialized type, the endospore, is a metabolically inert resting cell that is highly resistant to most environmental insults. When conditions are once again favorable for growth, spores germinate and grow out to establish new populations. This developmental suite has been termed the Bacillus Lifestyle (Zeigler and Nicholson 2017) and is not only a highly successful adaptation but also has practical consequences, many of them potentially beneficial, for human beings. Modern sequence-based taxonomic methods have revealed that *Bacillus sensu lato* comprises at least three bacterial families and dozens of genera. One species (*Bacillus subtilis*) and one particular laboratory strain (*B. subtilis* 168) have emerged as an important model organism. Thousands of useful genetic mutants and strain libraries have been developed over the past 70 years of *B. subtilis* research, especially during the 20 years since the genome sequence of strain 168 was determined. The Bacillus Genetic Stock Center (BGSC) exists as a steward of this legacy. It collects, maintains, distributes, and assembles data regarding these mutants and genetic tools, so that they will remain available in perpetuity for members of the research community.

THE GENUS *BACILLUS* AND THE BACILLUS LIFESTYLE

"The genus *Bacillus*," wrote Ruth Gordon and colleagues in their authoritative handbook, "encompasses the rod-shaped bacteria capable of aerobically forming refractile endospores that are more resistant than vegetative cells to heat, drying, and other destructive agencies" (Gordon et al. 1973). More than four decades later, despite the revolution ushered in by recent technological and theoretical advances in biology, the definition remains essentially the same: *Bacillus* bacteria have cells that are "rod-shaped, straight or slightly curved, occurring singly and in pairs, some in chains, and occasionally as long filaments. Endospores are formed, no more than one to a cell; these spores are very resistant to many adverse conditions" (Logan and De Vos 2015). This definition, while accurate for the purposes of taxonomy, needs to be placed in a much larger framework. Endospores are far more than a morphological feature or taxonomic trait. They are the centerpiece of a diverse range of developmental options, many of them requiring a remarkable degree of multicellular cooperation, which have allowed these bacteria to survive and thrive on a global scale (Figure 3.1).

Most isolates of the genus *Bacillus* and its closest phylogenetic relatives have simple nutritional requirements. Under ideal conditions, they proliferate rapidly, dividing by binary fission. Generation times for many species may be as short as a half

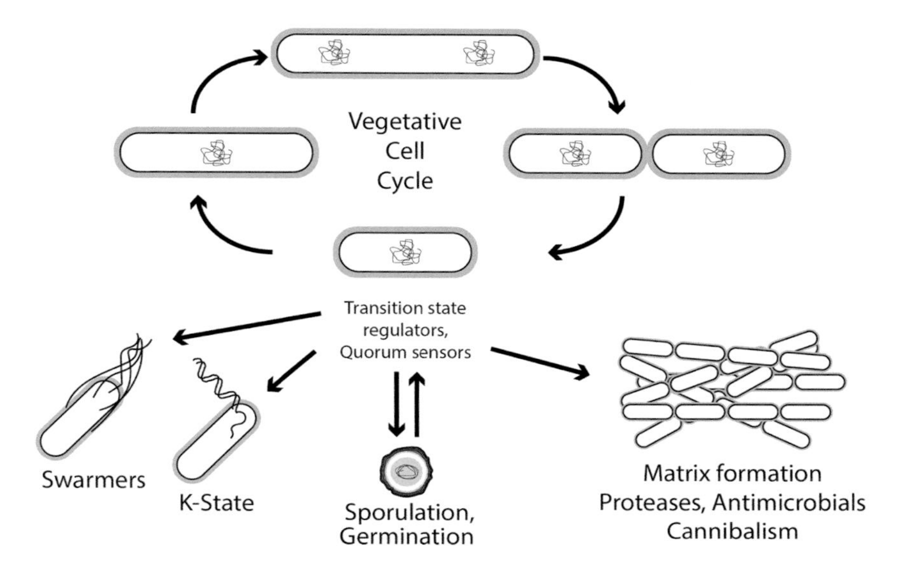

FIGURE 3.1 The *Bacillus* Lifestyle. *Bacillus* may proliferate rapidly when environmental conditions permit. Under unfavorable conditions, subpopulations differentiate into a variety of cell types.

hour or less in the laboratory (Zeigler, unpublished observations). In nature, however, conditions approximating the ideal occur only sporadically. When a population of *Bacillus* experiences nutrient depletion and enters stationary phase, several subpopulations emerge, each pursuing a different developmental path (Figure 3.1). This suite of options has been seen as a bet-hedging strategy, increasing the likelihood that at least part of the population survives the environmental challenge (Norman et al. 2015, Veening et al. 2008b). In *B. subtilis* cultures, for example, subpopulations of motile single cells and sessile chains quickly appear (Mukherjee and Kearns 2014). The motile cells may be able to locate microenvironments that are more favorable for growth. The sessile chains differentiate further, with one subpopulation secreting massive quantities of degradative enzymes and antimicrobials, perhaps converting potential nutritional sources in the environment into sugars, nucleotides, and amino acids for the benefit of the entire population (Veening et al. 2008a). Other sessile cells secrete a matrix that binds the population into a biofilm (Claessen et al. 2014).

Rather less altruistically, in certain species these matrix-builders may also secrete killing factors, allowing some cells to cannibalize others (Shank and Kolter 2011). Alternatively, *B. subtilis* and its relatives can form subpopulations that enter into the "K-state", a developmental stage in which growth and cell division are suppressed and machinery is assembled that allows cells to bind and import DNA from the environment (Berka et al. 2002, Yuksel et al. 2016). Finally, the Bacillus Lifestyle includes an option of last resort—endospore formation. These living time capsules include a cell membrane, a dehydrated cytoplasm with a starter set of gene expression machinery, and a condensed protein-bound chromosome (Huang and Hull 2017). They are resistant to a variety of environmental insults including UV radiation, abrasion, desiccation, lysozyme, and organic solvents. They can remain metabolically inert for long periods of time, years or perhaps centuries. However, when environmental conditions are once again favorable, spores can germinate and initiate outgrowth in a matter of hours (Moir and Cooper 2015). Rapid proliferation begins again.

Bacillus, then, encompasses an enormous, phylogenetically diverse, and ancient group of organisms that have adapted a common set of survival strategies. These bacteria have attracted intense interest as a model system for studying developmental biology. As described below, they also have many current and potential applications affecting the public health and economic development of human society.

A Taxonomic Explosion

When Ruth Gordon completed her handbook *The Genus Bacillus* (Gordon et al. 1973), she was able to classify every isolate then available to her using only 18 species designations. Her classification key made use of the limited technologies available before the dawn of the Recombinant DNA age: colony, cell, and spore morphology; carbohydrate utilization and acid production; enzymatic activity; and growth on discriminatory media. This was the paradigm into which the BGSC was born. It was to focus on an important but limited scope of bacteria, the handful of species within the genus *Bacillus*.

Over the next decade, there occurred one of the most important and far-reaching scientific revolutions of the twentieth century, the development of a DNA sequence-based phylogenetic tree based on the universally conserved 16S rRNA gene (Fox et al. 1980, Woese and Fox 1977). When conventional classification systems were superseded, it was quickly realized that the bacteria—now elevated to one of the three primary Domains on the tree of life—were far more phylogenetically diverse than previously imagined. This revolution had an especially dramatic impact on the classical genus *Bacillus*. When Carol Ash and colleagues applied the new technology to existing *Bacillus* species, they suggested that the high degree of DNA sequence diversity would support the creation of at least five genera, including *Bacillus sensu stricto* for *B. subtilis* and its closest relatives. It was quickly realized that many additional genera were actually required (Priest 1993) to accommodate known diversity, and that these genera clustered into several distinct taxonomic families besides the classical *Bacillaceae*. At the time of writing, there were four genera in the family *Alicyclobacillaceae*, ten in the family *Paenibacillaceae*, 13 in the family *Planococcaceae*, and 57 in the family *Bacilliaceae* alone. Those 74 genera comprise many hundreds of species. The classification of *Bacillus sensu lato*—the old genus *Bacillus* of Gordon—continues to evolve rapidly, so the interested reader is encouraged to access more current information on the relevant pages on the List of Prokaryotes with Standing in Nomenclature (LPSN) website (http://www.bacterio.net/-classifphyla.html#firmibacteria) (Parte 2014).

A Taxon with Many Dimensions

Isolates from *Bacillus sensu lato*, each featuring some variant of the Bacillus Lifestyle have been isolated from nearly every conceivable environment from around the globe. It is beyond the scope of this (and perhaps any) review to adequately summarize all of the phylogenetic and ecological diversity that they represent. However, a much smaller number of species have proven to be especially important for industrial, agricultural, and medical applications. They can only be introduced briefly here. A more thorough review of these organisms has been published elsewhere (Zeigler and Perkins 2015).

The *Bacillus subtilis* Group

Practically useful taxa are epitomized by the industrial workhorse, *B. subtilis*. It plays a key role as a production platform in many sustainable fermentation applications, including the production of enzymes, vitamins and other nutritional supplements, and biofuel components. The *Bacillus subtilis* Group (BSG), an informal grouping of about a dozen closely related species, contains several other important members. Like *B. subtilis*, *B. licheniformis* which is an industrially important producer of enzymes. *Bacillus atrophaeus*, which produces a black pigment on suitable media, is an important bioindicator strain for the sterilization/decontamination industry and is a safe simulant for studying dispersal of anthrax spores (Sella et al. 2015). *Bacillus amyloliquefaciens* and its relatives have been shown to have significant potential as plant growth-promoting rhizobacteria (Fan et al. 2017, Ye et al. 2018). Still other members of the BSG are fermentation organisms for traditional rice bran or soybean-based foods in Korea, Japan, Thailand, and Vietnam (Chen et al. 2012, Kanesaki et al. 2018, Kim et al. 2015). These observations have raised interest in the use of BSG isolates as probiotics for humans and livestock (Elshaghabee et al. 2017, Mingmongkolchai and Panbangred 2018).

The *Bacillus cereus* Group

The *Bacillus cereus* Group (BCG) includes several other species with significant applications in an array of fields (Ceuppens et al. 2013, Patino-Navarrete and Sanchis 2017). *Bacillus cereus* itself is an important food spoilage organism that can produce diarrheal or emetic toxins leading to food poisoning (Duport et al. 2016). *Bacillus anthracis* is the causative agent of the serious vertebrate disease anthrax, its virulence due to a three-component toxin and a protective capsule (Friebe et al. 2016). In contrast, most *B. thuringiensis* isolates have no demonstrated pathogenicity to humans or other vertebrates, but instead make one or more proteinaceous toxins that are pathogenic to a narrow range of invertebrates, including insects and nematodes. They have found wide application in agriculture, whether formulated as sprays or as sources of insecticidal genes for transgenic crops (Bravo et al. 2011, Roh et al. 2007).

Geobacillus

A third group of note are the members of the thermophilic genus *Geobacillus* which, despite their requirement for elevated growth temperatures, have a worldwide distribution and are easily isolated from nature (Zeigler 2014). *Geobacillus stearothermophilus* spores are already the industry standard as a biological indicator for steam sterilization (Huesca-Espitia et al. 2016). More recently, members of the genus have garnered attention as potential platforms for second-generation biofuel production using lignocellulosic feedstocks (Hussein et al. 2015). Further exploitation of the genus will require an expansion of its genetic toolkit.

Many other species of *Bacillus sensu lato* could be singled out for their practical applications, but these examples are sufficient to demonstrate the multidimensional importance of this taxon.

BACILLUS SUBTILIS, A MODEL ORGANISM

Working in the 1870s, Ferdinand Cohn, a bacteriologist at Cornell (Geneva), allowed a bacterial culture to multiply in boiled hay infusions. He reported seeing motile cells that transitioned into stationary filaments and then developed endospores. Eventually a film formed on the surface of the culture, which grew into a thick, flocculent pellicle (Cohn 1872). By studying what was probably *Bacillus subtilis*, Cohn had observed the Bacillus Lifestyle as described above, that is, a period of robust growth fueled by opportunistic decomposition of biomass until key nutrients were depleted, followed by a range of developmental pathways including motility, spore production, and biofilm formation (Zeigler and Nicholson 2017). When Cohn's contemporary, Robert Koch (then a physician working in Poland) showed that a morphologically similar organism was the causative agent for anthrax (Koch 1876), he firmly established *Bacillus* as an important topic for microbiological research. Decades of research were to follow.

THE ERAS OF *BACILLUS* BIOLOGY

THE "CLASSICAL ERA"

Aerobic, endospore-forming bacteria could be isolated quickly and easily from a wide variety of natural sources. However, this required tedious, methodical characterization to develop a taxonomic system to categorize them, creating order from chaos (Gordon and Smith 1949, Smith et al. 1946). Ferdinand Cohn's original *B. subtilis* isolate had been lost. In 1930, Harold J. Conn carefully compared F. Cohn's published description with a collection of laboratory stocks and selected a strain from the University of Marburg as the closest match (Conn 1930). The Marburg isolate became the nomenclatural type strain, not only for the species but also for the genus *Bacillus*. It was deposited in the American Type Culture Collection (ATCC) as ATCC 6051^T and later in the British National Collection of Industrial Bacteria as NCIB 3610^T. In the late 1940s, two Yale microbiologists, Norman Giles and Paul Burkholder, chose ATCC 6051^T as the subject of a series of mutagenesis experiments designed to test the recently developed "One Gene One Enzyme" hypothesis (Burkholder and Giles 1947). One of their mutants, a tryptophan auxotroph they called 168, was destined to play a large role in *Bacillus* research in the decades to come (Zeigler et al. 2008).

THE "GENETICS ERA"

Several years later, a small set of the Yale mutants were passed on to John Spizizen, then at Case Western Reserve University. His subsequent discovery that strain 168 could be genetically transformed by purified DNA (Spizizen 1958) ushered in the

"Genetics Era" when biochemical exploration was combined with genetic analysis to elucidate the processes underlying *Bacillus* physiology, metabolism, and development. Soon bacteriophage transduction was added as a genetic tool (Takahashi 1961). A picture of the *B. subtilis* genome gradually began to emerge from hundreds of studies employing mutagenesis, phenotypic characterization, and genetic mapping. Some of these experiments used asporogenous mutants, providing increasing insights into the complexities of the Bacillus Lifestyle (Schaeffer 1967). This culminated with the publication of an un-gapped circular genetic map for *B. subtilis* 168 (Lepesant-Kejzlarovà et al. 1975).

The "Recombinant DNA Era"

The "Recombinant DNA Era" of *Bacillus subtilis* research began with the first reports of cloning genes in and out of the organism (Chi et al. 1978, Duncan et al. 1977, Ehrlich 1978, Nagahari and Sakaguchi 1978, Segall and Losick 1977). The small, rolling circle replicons of *Staphylococcus aureus* were found to function in *B. subtilis*, forming the basis of the first generation of cloning vectors (Ehrlich 1977). A major breakthrough was the development of the integration vector, which allowed cloned genes to be tagged with a selective marker and mapped or subjected to knockout mutagenesis (Haldenwang et al. 1980). Other vectors were developed that could generate fusions to reporter genes, facilitating gene expression studies in the organism (Zuber and Losick 1983). Still other vectors were constructed that could deliver transposons to *B. subtilis*, allowing saturation mutagenesis with selectable markers and fusions (Youngman et al. 1983). As recombinant DNA techniques grew in sophistication, they provided ever-deepening insights into the structure, function, and development of *B. subtilis* cells.

The "Post-Genome Era"

As Sanger DNA sequencing became less expensive and more widely available, cloned *B. subtilis* genes of interest began to be sequenced routinely. In September 1989, an international consortium of *B. subtilis* geneticists began working systematically toward determining the complete genome sequence for the organism. Given the technology available at the time, it was a massive undertaking that required eight years for completion. In November 1989, the complete annotated sequence was released. This was a major milestone in *B. subtilis* studies (Kunst et al. 1997).

The ensuing two decades can be termed the "Post-Genome Era." The genome sequence facilitated the continued study of isolated genes or proteins, and it also enabled the analysis of *B. subtilis* as a biological system. Ongoing efforts, generally also accomplished by a consortia of laboratories, have measured the organism's transcriptome under dozens of physiological states in a variety of conditions (Nicolas et al. 2012), have generated single-gene knockout libraries for every non-essential gene (Koo et al. 2017) and a tunable "knockdown" library for every essential gene (Peters et al. 2016), and have created the possibility of rational metabolic engineering in the organism (Liu et al. 2017).

Each of these eras in *B. subtilis* research—the Classical Era (1930–1957), the Genetics Era (1958–1976), the Recombinant DNA Era (1977–1996), and the Post-Genome Era (1997—present)—has bequeathed a legacy of strains and fresh opportunities to the present generation.

ADVANTAGES OF *BACILLUS SUBTILIS* AS A MODEL ORGANISM

Model organisms are those that have been studied so frequently, for so long, and by so many dedicated scientists, that they can now "be drawn upon for formulating incisive experiments to illuminate" obscure biological processes and can serve as a "proving ground for developing new technologies, which typically spread quickly throughout the research community" (Fields and Johnston 2005). For the study of an enormous group of bacteria with high economic, environmental, and medical importance—the Gram-positive phylum Firmicutes—that model organism is *Bacillus subtilis*. As the historical survey above demonstrates, the most important strain for this purpose is *B. subtilis* 168. This organism has many advantages over most of its Gram-positive relatives. It is easily grown and maintained, genetically tractable and non-pathogenic and completely innocuous (Schallmey et al. 2004). Liu and Deutschbauer (2018) have recently argued that for an organism to be considered a modern model system, it must have at least four qualifications all of which *B. subtilis* 168 very clearly possesses.

When working with the *B. subtilis* system, researchers have access to:

1. An inventoried parts list of genes, along with their products and promoters. As discussed above, the complete genome sequence of *B. subtilis* 168 was determined over twenty years ago (Kunst et al. 1997). A more recent resequencing effort, making full use of next generation sequencing technologies, refined the accuracy of the genome sequence (Barbe et al. 2009). Components of the transcriptome have been identified and studied under a variety of conditions (Arrieta-Ortiz et al. 2015, Nicolas et al. 2012). Regulatory RNAs have also been similarly identified and studied (Mars et al. 2016). For *B. subtilis* 168, then, the parts list is completely known.
2. A suite of genetic tools for manipulation of the organism. For *B. subtilis* 168, one component of the Bacillus Lifestyle is the developmental process termed the "K-state" during which cells become competent for natural transformation with exogenous DNA, accompanied by an activation of recombination enzymes for integrating this DNA into the chromosome (Berka et al. 2002). Natural competence can be supplemented with other tools, most notably electroporation (Meddeb-Mouelhi et al. 2012) and bacteriophage SPP1-mediated transduction (Valero-Rello et al. 2017, Yasbin and Young 1974). A large variety of multipurpose plasmid vectors designed for cloning, regulated expression, epitope tagging, marker gene fusion, and chromosomal integration are readily available. Many of these tools, in addition to the model organism itself and ordered knockout collections derived from it, are available from the BGSC.
3. Robust, curated, data-driven gene annotations. Most genome projects make use of a genome annotation pipeline using sequence searching and alignment algorithms that essentially import annotations from other organisms.

Machine annotation, although rapid, is prone to error and is sometimes responsible for old errors percolating through a database (Gilks et al. 2002). The genome sequence of *B. subtilis* 168, in contrast, has been manually curated for over 20 years, and annotations are based on published data (Borriss et al. 2018). SubtiWiki 2.0, a community curated genome encyclopedia for the organism, is also available (Michna et al. 2016).

4. Computational platforms for systems-level analysis. SubtiWiki 2.0, mentioned above, is much more than an alternative annotation platform. It also incorporates and graphically presents data on protein-protein interactions, regulatory pathways, and gene expression for each locus in the genome. Through outlinks, SubtiWiki 2.0 serves as a central hub for complementary platforms, including the BsubCyc database of metabolic pathways (Caspi et al. 2018), the GenoList platform for genome searching and comparison (Lechat et al. 2008), and the *B. subtilis* Expression Data Browser maintained by the Institut National de la Recherche Agronomique (INRA) research center at Jouy-en-Josas, France.

THE BACILLUS GENETIC STOCK CENTER

Establishment

The BGSC traces its origins to the 1974 and 1975 annual meetings of the American Society for Microbiology. The "Genetics Era" of *B. subtilis* research was coming to full fruition, and hundreds of well-characterized mutants were available. Other species from the genus *Bacillus* were already proving useful in industrial, agricultural, and medical research. Several geneticists, including Frank Young, Manley Mandel, Arnold Demain, James Copeland, Harlyn Halvorson, and Donald Dean, perceived the need for a centralized organization to collect genetically characterized strains from key species, document the stability and validity of these cultures, and distribute them to interested persons. Participants at the First Conference on *Bacillus*, held August 6–9, 1975 at Cornell University, formed an organizing committee to explore sources of funding for the culture collection. This committee, chaired by Copeland, consisted of the original group of academic scientists, joined by Robert Erickson (Miles Laboratories) and Charles A. Claridge (Bristol Laboratories) to represent industrial firms interested in the fermentation biology and genetics of *Bacillus*. The committee chose Dean to direct the BGSC. In 1997, Daniel Zeigler, who first joined the BGSC staff in 1985, assumed the position of Director. The BGSC has received generous support from the National Science Foundation since its inception.

Mission Statement

The primarily goal of the BGSC is to "maintain genetically characterized strains, cloning vectors, and bacteriophage for the genus *Bacillus* and related organisms and to distribute these materials without prejudice to qualified scientists and educators throughout the world." The BGSC serves the scientific research and education communities as the steward of the rich legacy of strains that have been produced by previous generations and the exciting new tools that continue to be created in our day.

Scope

The BGSC comprises a main collection and a strain warehouse. The main collection is fully searchable from the BGSC website (http://www.bgsc.org) and is publicized in various special focus catalogs and news blogs on the site. The warehouse contains a select group of personal strain collections from retired scientists including Stanley Zahler (Cornell), Ernst Freese (National Institutes of Health), Bernard Reilly (Case Western and University of Minnesota), and Joshua Lederberg (Stanford). These collections are maintained because of their historical interest, but they are not documented in the public database. Individual strains are available on request.

As of July 2017, the main collection was comprised of 11,498 accessions (Table 3.1), an increase of over 400% since July 2013. Represented are 67 species in 12 genera of *Bacillus sensu lato*, along with sets of phages and plasmid tools. Over 80% of the accessions are derived from the Gram-positive model organism, *B. subtilis* 168. Among them are three systems-biology libraries: the BKE and BKK knockout libraries, in which every non-essential gene is individually inactivated with erythromycin or kanamycin resistance cassettes, respectively (Koo 2017), and the BEC knockdown library, in which every essential gene has been placed under the control of a tunable promoter (Peters et al. 2016). Other well-represented groups are:

- The insect pathogen *B. thuringiensis* and other members of the *B. cereus* species cluster,
- *B. megaterium*, a species with large cells useful for direct visualization methods,
- *Geobacillus*, a group of industrially important obligate thermophiles, and
- *Lysinibacillus sphaericus*, a species pathogenic to mosquitos and blackflies.

The smaller taxonomic groups are interesting in their own right, with extremophiles, marine bacteria, plant-associated microbes, and antibiotic producers

TABLE 3.1
Classification and Number of Components in the Main Bacillus Genetic Stock Center Collection

Classification	Number	Classification	Number
Bacillus subtilis 168 mutants	9596	*Aneurinibacillus* sp.	2
Gene knockout	*7935*	*Brevibacillus* sp.	18
Gene knockdown	*314*	*Geobacillus* sp.	161
Other *B. subtilis* isolates	80	*Lysinibacillus* sp.	155
B. thuringiensis	198	*Paenibacillus* sp.	15
B. cereus species cluster	108	*Rummelibacillus* sp.	4
B. megaterium	244	*Sporosarcina* sp	1
B. licheniformis	43	*Ureibacillus* sp.	3
B. amyloliquefaciens cluster	23	Other *Bacillus sensu lato*	47
B. pumilus species cluster	10	Plasmid tools	741
B. atrophaeus	9	*Bacillus* phages and lysogens	40

represented. Currently, over 200 strains in the collection, most of them wild type isolates, have publicly available genome sequences.

The BGSC also maintains and distributes cloning vectors and other genetic tools for manipulating *B. subtilis* and other important species. Currently there are over 700 such tools available. Included are the "Bacillus BioBrick Box" 1.0 and 2.0 sets of parts for synthetic biology (Popp et al. 2017, Radeck et al. 2013); the "Sporobeads" set for surface display of proteins on *B. subtilis* spore coats (Bartels et al. 2018); a toolbox for creating "personalized" integration vectors (Radeck et al. 2017); a tunable gene expression toolkit (Guiziou et al. 2016); and many more. Most of the BGSC vector collection are replicate in *Escherichia coli* and are distributed in that form.

Maintenance and Distribution

As a consequence of their Bacillus Lifestyle, *Bacillus* isolates can usually be propagated and stored very easily, with excellent long-term viability. There has been little motivation to modify the methods that were described in detail nearly two decades ago (Sanderson and Zeigler 1991). The main innovation has been the increasing tendency to store larger libraries in 96-well microplates or compatible microvials, both to save space and to facilitate further robotic manipulation and high-throughput analysis. Very few members of *Bacillus sensu lato* are pathogenic; the large majority, including all *B. subtilis* isolates, can be treated as innocuous Biosafety Level 1 organisms. Some members of the *B. cereus* group should be treated as Biosafety Level 2. *Bacillus anthracis* cultures are not stocked at the BGSC.

Isolates from the most commonly studied species, including *B. subtilis*, *B. cereus*, and their closest relatives, grow rapidly at mesophilic temperatures (30°C–37°C) under aerobic conditions. Most general-purpose complex microbiological media are adequate. The BGSC typically uses Nutrient Broth No. 2 (NB; Oxoid) for propagating these species. Anecdotal evidence, together with a perusal of the published literature, suggests that many labs use Lysogeny Broth (LB), the growth medium found ubiquitously in *Escherichia coli* labs and molecular biology protocols. However, NB allows for higher yields, healthier stationary phase cells, and higher sporulation frequencies than does LB (Zeigler, unpublished results). Isolates from other species of *Bacillus sensu lato* may prefer other media and may require different growth conditions.

Members of the industrially important genus *Geobacillus* are obligate thermophiles and should be propagated at higher temperatures; the BGSC routinely uses 50°C. Most *Geobacillus* isolates in the BGSC collection grow well on Tryptose Blood Agar Base (without blood), although some prefer Trypticase Soy Agar. For many isolates of *Bacillus sensu lato*, incubation for 3–4 days in NB is sufficient to produce a good yield of spores. A very effective sporulation medium for *B. subtilis* and its relatives, routinely used at the BGSC, is 2xSG (Leighton and Doi 1971). Schaeffer's complex sporulation medium (abbreviated SSM or less accurately DSM) (Schaeffer et al. 1965) is perhaps more commonly used in *B. subtilis* labs, although its spore yields are lower than those of 2xSG (Zeigler, unpublished results). Other sporulation media are preferable for *B. cereus* and its relatives or for the bioindicator strain *B. atrophaeus* (Buhr et al. 2008).

For any *Bacillus* isolate capable of producing them, spores are the preferred form for long term maintenance. After liquid cultures have undergone bulk sporulation, sterile glycerol can be added to a final concentration of 10% (v/v) and the spore suspension transferred to cryovials. Alternatively, sporulated cultures can be scraped from the surface of a sporulation agar plate with a sterile microbiological loop and suspended directly in a sterile 10% glycerol solution and transferred to vials. *Bacillus* isolates that do not sporulate under standard laboratory conditions and asporogenous mutants of *B. subtilis* can also be stored in glycerol solutions, but these cultures should be harvested during the exponential phase of growth. The cryovials can be stored by freezing them in mechanical −80°C freezers or in liquid nitrogen vapor phase, without any need for specific cooling protocols. Spore suspensions will survive indefinitely (at least for decades at the BGSC) under these conditions.

If stocks have been stored as frozen spore suspensions, their distribution is simple and inexpensive. A loopful of spore-laden ice can be scraped from the surface of a frozen vial. The frozen spore suspension is transferred to a small stack of filter paper disks so that the spores can adsorb into the paper as the ice melts. The disks are then wrapped either in sterile aluminum foil packets or in plastic wrap. The packets are mailed or shipped inside a business envelope. When they arrive at their destination, the packets can be opened and the disks transferred to a suitable growth medium. Under typical ambient conditions, spores will last at least several weeks, if not longer. Refrigeration of disks or of *Bacillus* cultures is not recommended. Early blocked sporulation mutants die quickly on filter paper under normal ambient conditions, with half-lives on the order of one day at 22°C (Zeigler, unpublished results). These mutants may be shipped as cultures growing on agar slants or in small Petri dishes. Alternatively, asporogenous mutants may be transported in freeze-dried lyophiles, although the BGSC does not make use of this method.

Use of the Collection

The breadth of the taxon *Bacillus sensu lato* is seen in the diverse community of researchers who make use of it. During the last five years, cultures were distributed to scientists and educators in 57 nations. One measure of the BGSC's impact is the collection's citation in peer-reviewed publications. However, this metric fails to account for strains used in teaching laboratories. Since April 2013, at least 522 publications have acknowledged the BGSC as a source of research materials. This is likely a large underestimate as frequently, if a series of publications makes use of a set of strains, their original source is only cited in the initial reference. Even an incomplete tally reveals a broad impact on a variety of important fields. These publications have been grouped into 32 research fronts in BGSC analyses. Figure 3.2 depicts those fronts with the greatest use of BGSC strains in published papers. Although omitted from the figure for clarity, some of the smaller research fronts are exciting, fast moving fields, such as the investigation of *B. thuringiensis* toxins that target cancer cells, the use of *B. subtilis* in arsenic bioremediation, and the development of *Bacillus* spores as oral vaccines. During this same period, there have also been at least 69 US patents issued disclosing inventions based in part on BGSC strains. In short, the BGSC has played a measurably valuable part in the scientific infrastructure by serving as a toolbox for innovation.

Well over half of the strains in the BGSC collection have been distributed at least once during the past five-year period. However, certain accessions can be singled out as especially popular (Table 3.2). The most commonly requested stock, by a

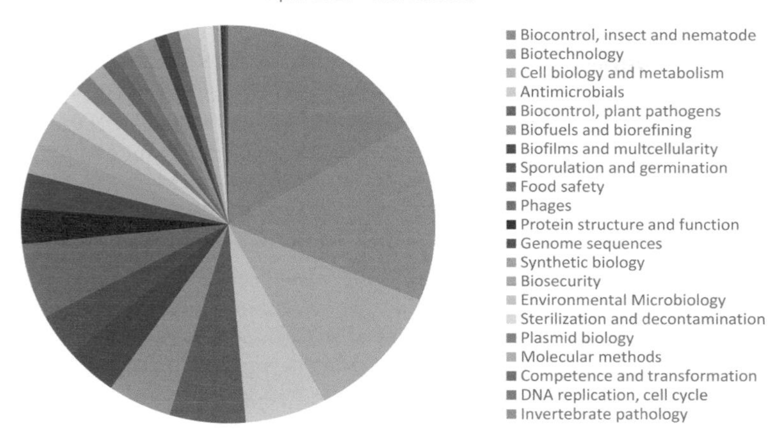

FIGURE 3.2 **(See color insert.)** References citing use of Bacillus Genetic Stock Center strains from April 2013 to March 2018. Pie chart indicates the proportion of references belonging to an identified research front.

TABLE 3.2
Most Frequently Distributed BGSC Strains and Plasmids (Cumulative, July 1, 2012–June 30, 2017)

Strain	Species[a]	Total	Plasmid	Type	Total
1A1	*Bacillus subtilis*	265	pDR244	Cre-production	89
1A976	*Bacillus subtilis*	85	pMUTIN4	Conditional knockout	78
3A1^T	*Bacillus subtilis*	76	pMutin-GFP+	Fluorescent tagging	73
1A747	*Bacillus subtilis*	74	pAX01	Inducible expression	68
10A6	*Bacillus velezensis*	60	pNW33N	Thermophilic cloning	67
3A38	*Bacillus subtilis*	55	pSG1156	Fluorescent tagging	59
4D1	*Bt kurstaki*	45	pBS4S	Simple integration	55
4D4	*Bt kurstaki*	45	pSG1154	Fluorescent tagging	53
1A1133	*Bacillus subtilis*	44	P*veg*	Constitutive expression	52
4AA1	*Bt morrisoni*	40	pDG1662	Ectopic integration	50
1P7	Bacteriophage SPP1	36	pA-spac	Inducible expression	50
4Q7	*Bt israelensis*	36	pLIKE-rep	Inducible expression	49
7A16	*Bacillus megaterium*	36	pBS2E	Ectopic integration	49
9A5	*Geobacillus* sp.	33	pBGSC6	Simple integration	45
25A2	*Paenibacillus polymyxa*	31	pBS3C-lux	Expression reporter	45
1A751	*Bacillus subtilis*	30	pDG1730	Ectopic integration	40
1A96	*Bacillus subtilis*	30	pLIKE-int	Inducible expression	40
3A36	*Bacillus subtilis*	30	P*liaG*	Constitutive expression	38

[a] *Bt*, abbreviation for *B. thuringiensis*.

large margin, remains the classic model organism, *B. subtilis* 168 (BGSC accession 1A1). Its wild type parent, *B. subtilis* NCIB 3610^{T} (BGSC 3A1^{T}), is also in high demand. Other *B. subtilis* laboratory strains, such as PY79 (BGSC 1A747) and JH642 (BGSC 1A96) continue to be useful to the research community. Also frequently requested were two more recently developed *B. subtilis* derivatives. Strain KO7 (BGSC 1A1133), developed in 2016 at the BGSC, is a marker-free, prototrophic strain with knockouts in all seven secreted proteases, an important feature for production of foreign proteins (Zeigler, unpublished). *Bacillus subtilis* SCK6 is a strain engineered to have xylose-inducible genetic competence (Zhang and Zhang 2011). Standard strains from certain other species have also been popular requests, including *B. thuringiensis* isolates that are toxic to lepidopteran pests (BGSC 4D1, 4D4); *Bacillus velezensis* (BGSC 10A6) and *Paenibacillus polymyxa* (BGSC 25A2) isolates that have plant growth-promoting characteristics; and *B. thuringiensis* (BGSC 4Q7) and thermophilic *Geobacillus* (BGSC 9A5) derivatives that have been modified to serve as cloning hosts. Popular vectors and genetic tools include the transducing bacteriophage SPP1 (BGSC 1P7) and a variety of expression vectors and fluorescent protein fusion vectors.

Databases

Digitization of strain accession data has long been a priority at the BGSC. In the late 1970s, a major in-house programming effort generated a proprietary database, housed on a mainframe computer, which linked accession numbers, strain names, provenance, genotypes, and phenotypes for every item in the collection. It was used to generate the first edition of the BGSC strain catalog in 1980. By 1990, these data, together with similar records for newer accessions, were incorporated into a commercially available relational database package to generate a "catalog on diskette" that was physically distributed to thousands of users in the *Bacillus* genetics community (Sanderson and Zeigler 1991). The format of the strain database has long since changed to the ISO- and ANSI-compliant SQL programming language, which is used to power the online search engines at the BGSC website. The database still includes key information for each stock (BGSC accession number, taxonomic identity or product class, original codes, strain history, and genotypic description), but now also includes key published references, recommended growth media, and accession numbers for genomic or 16S rRNA gene sequences.

The online search engine allows the user to locate a strain by any word or phrase that might exist in its record, for example a species name, a mutant allele, an author who has published on the strain, or a phrase in the title of a key reference. Orders can be placed directly on the website with a convenient online form; this form generates a document that is automatically emailed to the BGSC director for documentation purposes and for transfer of data to the private strain usage spreadsheet and the user database. Orders are also accepted by e-mail, telephone, or fax, using the contact information displayed on the website. When orders are filled, all data in the strain database for the requested stocks, with the addition of a "notes" field, are outputted to a form letter accompanying the shipment. The public strain database is updated immediately whenever a new strain or plasmid is accessioned into the BGSC collection. The community-wide platforms SubtiWiki 2.0 and BSubCyc include information about BGSC mutant strains, such as knockouts, associated with particular gene loci.

CONCLUSIONS

For well over a century, endospore-forming bacteria capable of aerobic metabolism have generated intense scientific interest. Research efforts have focused on understanding the surprisingly complex developmental changes and multicellular interactions that characterize these organisms, on understanding and managing the public health risks of the few pathogenic isolates, and on harnessing the enormous biotechnological and agricultural potential of the group. One species, *Bacillus subtilis*, emerged as an important model organism, not only for the endospore-formers, but for all of the Gram-positive bacteria of the phylum Firmicutes. As technology developed over the course of the twentieth and twenty-first centuries, novel mutant strains and genetic tools were constructed for these research efforts. The BGSC exists to collect, maintain, and distribute these strains. All indications are that *Bacillus* isolates, including those of the model organism *B. subtilis*, will continue to be in high demand for years to come.

REFERENCES

Arrieta-Ortiz, M.L., C. Hafemeister, A.R. Bate et al. 2015. An experimentally supported model of the *Bacillus subtilis* global transcriptional regulatory network. *Molecular Systems Biology* 11:839. doi:10.15252/msb.20156236.

Barbe, V., S. Cruveiller, F. Kunst et al. 2009. From a consortium sequence to a unified sequence: The *Bacillus subtilis* 168 reference genome a decade later. *Microbiology* 155:1758–1775.

Bartels, J., S. Lopez Castellanos, J. Radeck and T. Mascher. 2018. Sporobeads: The utilization of the *Bacillus subtilis* endospore crust as a protein display platform. *ACS Synthetic Biology* 7:452–461.

Berka, R.M., J. Hahn, M. Albano. et al. 2002. Microarray analysis of the *Bacillus subtilis* K-state: Genome-wide expression changes dependent on ComK. *Molecular Microbiology* 43:1331–1345.

Borriss, R., A. Danchin, C.R. Harwood. et al. 2018. *Bacillus subtilis*, the model Gram-positive bacterium: 20 years of annotation refinement. *Microbial Biotechnology* 11:3–17.

Bravo, A., S. Likitvivatanavong, S.S. Gill and M. Soberon. 2011. *Bacillus thuringiensis*: A story of a successful bioinsecticide. *Insect Biochemistry and Molecular Biology* 41:423–431.

Buhr, T.L., D.C. McPherson and B.W. Gutting. 2008. Analysis of broth-cultured *Bacillus atrophaeus* and *Bacillus cereus* spores. *Journal of Applied Microbiology* 105:1604–1613.

Burkholder, P.R. and N.H. Giles. 1947. Induced biochemical mutations in *Bacillus subtilis*. *American Journal of Botany* 34:345–348.

Caspi, R., R. Billington, C.A. Fulcher et al. 2018. The MetaCyc database of metabolic pathways and enzymes. *Nucleic Acids Research* 46:D633–D639.

Ceuppens, S., N. Boon and M. Uyttendaele. 2013. Diversity of *Bacillus cereus* group strains is reflected in their broad range of pathogenicity and diverse ecological lifestyles. *FEMS Microbiology Ecology* 84:433–450.

Chen, K.I., M.H. Erh, N.W. Su, W.H. Liu, C.C. Chou and K.C. Cheng. 2012. Soyfoods and soybean products: From traditional use to modern applications. *Applied Microbiology and Biotechnology* 96:9–22.

Chi, N.Y., S.D. Ehrlich and J. Lederberg. 1978. Functional expression of two *Bacillus subtilis* chromosomal genes in *Escherichia coli. Journal of Bacteriology* 133:816–821.

Claessen, D., D.E. Rozen, O.P. Kuipers, L. Sogaard-Andersen and G.P. van Wezel. 2014. Bacterial solutions to multicellularity: A tale of biofilms, filaments and fruiting bodies. *Nature Reviews Microbiology* 12:115–124.

Cohn, F. 1872. Untersuchen über Bacterien. *Beiträge zur Biologie der Pflanzen* 1 (Heft 2):127–224.

Conn, H.J. 1930. The identity of *Bacillus subtilis. Journal of Infectious Diseases* 46:341–350.

Duncan, C.H., G.A. Wilson and F.E. Young. 1977. Transformation of *Bacillus subtilis* and *Escherichia coli* by a hybrid plasmid pCD1. *Gene* 1:153–167.

Duport, C., M. Jobin and P. Schmitt. 2016. Adaptation in *Bacillus cereus*: From stress to disease. *Frontiers in Microbiology* 7:1550. doi:10.3389/fmicb.2016.01550.

Ehrlich, S.D. 1977. Replication and expression of plasmids from *Staphylococcus aureus* in *Bacillus subtilis. Proceedings of the National Academy of Sciences* (USA) 74:1680–1682.

Ehrlich, S.D. 1978. DNA cloning in *Bacillus subtilis. Proceedings of the National Academy of Sciences* (USA) 75:1433–1436.

Elshaghabee, F.M.F., N. Rokana, R.D. Gulhane, C. Sharma and H. Panwar. 2017. *Bacillus as* potential probiotics: Status, concerns, and future perspectives. *Frontiers in Microbiology* 8:1490. doi:10.3389/fmicb.2017.01490.

Fan, B., J. Blom, H.P. Klenk and R. Borriss. 2017. *Bacillus amyloliquefaciens*, *Bacillus velezensis*, and *Bacillus siamensis* form an "operational group *B. amyloliquefaciens*" within the *B. subtilis* species complex. *Frontiers in Microbiology* 8:22. doi:10.3389/fmicb.2017.00022.

Fields, S. and M. Johnston. 2005. Cell biology. Whither model organism research? *Science* 307:1885–1886.

Fox, G.E., E. Stackebrandt, R.B. Hespell et al. 1980. The phylogeny of prokaryotes. *Science* 209:457–463.

Friebe, S., F.G. van der Goot and J. Burgi. 2016. The ins and outs of Anthrax toxin. *Toxins* (Basel) 8:69. doi:10.3390/toxins8030069.

Gilks, W.R., B. Audit, D. De Angelis, S. Tsoka and C.A. Ouzounis. 2002. Modeling the percolation of annotation errors in a database of protein sequences. *Bioinformatics* 18:1641–1649.

Gordon, R.E. and N.R. Smith. 1949. Aerobic sporeforming bacteria capable of growth at high temperatures. *Journal of Bacteriology* 58:327–341.

Gordon, R.E., W.C. Haynes and C.H.-N. Pang. 1973. *The genus Bacillus*. Washington, DC: United States Department of Agriculture.

Guiziou, S., V. Sauveplane, H.J. Chang, C. Clerte, N. Declerck, M. Jules and J. Bonnet. 2016. A part toolbox to tune genetic expression in *Bacillus subtilis. Nucleic Acids Research* 44:7495–7508.

Haldenwang, W.G., C.D.B. Banner, J.F. Ollington, R. Losick, J.A. Hoch, M.B. O'Connor and A.L. Sonenshein. 1980. Mapping a cloned gene under sporulation control by insertion of a drug-resistance marker into the *Bacillus subtilis* chromosome. *Journal of Bacteriology* 142:90–98.

Huang, M. and C.M. Hull. 2017. Sporulation: How to survive on planet Earth (and beyond). *Current Genetics* 63:831–838.

Huesca-Espitia, L.C., M. Suvira, K. Rosenbeck, G. Korza, B. Setlow, W. Li, S. Wang, Y.Q. Li and P. Setlow. 2016. Effects of steam autoclave treatment on *Geobacillus stearothermophilus* spores. *Journal of Applied Microbiology* 121:1300–1311.

Hussein, A.H., B.K. Lisowska and D.J. Leak. 2015. The genus *Geobacillus* and their biotechnological potential. *Advances in Applied Microbiology* 92:1–48.

Kanesaki, Y., E. Kubota, R. Ohtake, Y. Higashi, J. Nagaoka, T. Suzuki and S. Akuzawa. 2018. Draft genome sequence of *Bacillus licheniformis* Heshi-B2, isolated from fermented rice bran in a Japanese fermented seafood dish. *Genome Announcements* 6:1. doi:10.1128/genomeA.00118-18.

Kim, S.J., C.A. Dunlap, S.W. Kwon and A.P. Rooney. 2015. *Bacillus glycinifermentans* sp. nov., isolated from fermented soybean paste. *International Journal of Systematic and Evolutionary Microbiology* 65:3586–3590.

Koch, R. 1876. Die Ätiologie der Milbrandkrankheit, begrüdet die Entwicklunsgesicht des Bacillus Anthracis. *Beiträge zur Biologie der Pflanzen* 2:277–310.

Koo, B.M., G. Kritikos, J.D. Farelli et al. 2017. Construction and analysis of two genome-scale deletion libraries for *Bacillus subtilis*. *Cell Systems* 4:291–305.

Kunst, F., N. Ogasawara, I. Moszer et al. 1997. The complete genome sequence of the gram-positive bacterium *Bacillus subtilis*. *Nature* 390:249–256.

Lechat, P., L. Hummel, S. Rousseau and I. Moszer. 2008. GenoList: An integrated environment for comparative analysis of microbial genomes. *Nucleic Acids Research* 36:D469–D474.

Leighton, T.J. and R.H. Doi. 1971. The stability of messenger ribonucleic acid during sporulation in *Bacillus subtilis*. *Journal of Biological Chemistry* 246:3189–3195.

Lepesant-Kejzlarovà, J., J.-A. Lepesant, J. Walle, A. Billault and R. Dedonder. 1975. Revision of the linkage map of *Bacillus subtilis* 168: Indications for circularity of the chromosome. *Journal of Bacteriology* 121:823–834.

Liu, H. and A.M. Deutschbauer. 2018. Rapidly moving new bacteria to model-organism status. *Curremt Opinion in Biotechnology* 51:116–122.

Liu, Y., J. Li, G. Du, J. Chen and L. Liu. 2017. Metabolic engineering of *Bacillus subtilis* fueled by systems biology: Recent advances and future directions. *Biotechnology Advances* 35:20–30.

Logan, N.A. and P. De Vos. 2015. *Bacillus*. In *Bergey's Manual of Systematics of Archaea and Bacteria*, W.B. Whitman, F. Rainey, P. Kämpfer, M. Trujillo, J. Chun, P. DeVos, B. Hedlund and S. Dedysh (Eds.). Hoboken, NJ: Wiley.

Mars, R.A., P. Nicolas, E.L. Denham and J.M. van Dijl. 2016. Regulatory RNAs in *Bacillus subtilis*: A gram-positive perspective on bacterial RNA-mediated regulation of gene expression. *Microbiology and Molecular Biology Reviews* 80:1029–1057.

Meddeb-Mouelhi, F., C. Dulcey and M. Beauregard. 2012. High transformation efficiency of *Bacillus subtilis* with integrative DNA using glycine betaine as osmoprotectant. *Analytical Biochemistry* 424:127–129.

Michna, R.H., B. Zhu, U. Mader and J. Stulke. 2016. SubtiWiki 2.0—An integrated database for the model organism *Bacillus subtilis*. *Nucleic Acids Research* 44:D654–D662.

Mingmongkolchai, S. and W. Panbangred. 2018. *Bacillus* probiotics: An alternative to antibiotics for livestock production. *Journal of Applied Microbiology* 124:1334–1346.

Moir, A. and G. Cooper. 2015. Spore germination. *Microbiology Spectrum* 3.doi:10.1128/microbiolspec.

Mukherjee, S. and D.B. Kearns. 2014. The structure and regulation of flagella in *Bacillus subtilis*. *Annual Review of Genetics* 48:319–340.

Nagahari, K. and K. Sakaguchi. 1978. Cloning of *Bacillus subtilis* leucine A, B and C genes with *Escherichia coli* plasmids and expression of the *leuC* gene in *E. coli*. *Molecular and General Genetics* 158:263–270.

Nicolas, P., U. Mader, E. Dervyn et al. 2012. Condition-dependent transcriptome reveals high-level regulatory architecture in *Bacillus subtilis*. *Science* 335:1103–1106.

Norman, T.M., N.D. Lord, J. Paulsson and R. Losick. 2015. Stochastic switching of cell fate in microbes. *Annual Review of Microbiology* 69:381–403.

Parte, A.C. 2014. LPSN-list of prokaryotic names with standing in nomenclature. *Nucleic Acids Research* 42:D613–D616.

Patino-Navarrete, R. and V. Sanchis. 2017. Evolutionary processes and environmental factors underlying the genetic diversity and lifestyles of *Bacillus cereus* group bacteria. *Research in Microbiology* 168:309–318.

Peters, J.M., A. Colavin, H. Shi et al. 2016. A comprehensive, CRISPR-based functional analysis of essential genes in bacteria. *Cell* 165:1493–1506.

Popp, P.F., M. Dotzler, J. Radeck, J. Bartels and T. Mascher. 2017. The *Bacillus* BioBrick Box 2.0: Expanding the genetic toolbox for the standardized work with *Bacillus subtilis*. *Scientific Reports* 7:15058. doi:10.1038/s41598-017-15107-z.

Priest, F.G. 1993. Systematics and ecology of *Bacillus*. In Bacillus subtilis *and Other Gram-Positive Bacteria: Biochemistry, Physiology, and Molecular Genetics*, pp. 1–16, A.L. Sonenshein, J.A. Hoch & R. Losick (Eds.). Washington, DC: American Society for Microbiology.

Radeck, J., K. Kraft, J. Bartels et al. 2013. The *Bacillus* BioBrick Box: Generation and evaluation of essential genetic building blocks for standardized work with *Bacillus subtilis*. *Journal of Biological Engineering* 7:29. doi:10.1186/1754-1611-7-29.

Radeck, J., D. Meyer, N. Lautenschlager and T. Mascher. 2017. *Bacillus* SEVA siblings: A golden gate-based toolbox to create personalized integrative vectors for *Bacillus subtilis*. *Scientific Reports* 7:14134. doi:10.1038/s41598-017-14329-5.

Roh, J.Y., J.Y. Choi, M.S. Li, B.R. Jin and Y.H. Je. 2007. *Bacillus thuringiensis* as a specific, safe, and effective tool for insect pest control. *Journal of Microbiology and Biotechnology* 17:547–559.

Sanderson, K.E. and D.R. Zeigler. 1991. Storing, shipping, and maintaining records on bacterial strains. *Methods in Enzymology* 204:248–264.

Schaeffer, P., J. Millet and J.P. Aubert. 1965. Catabolic repression of bacterial sporulation. *Proceedings of the National Academy of Sciences* (USA) 54:704–711.

Schaeffer, P. 1967. Asporogenous mutants of *Bacillus subtilis* Marburg. *Folia Microbiologica (Praha)* 12:291–296.

Schallmey, M., A. Singh and O.P. Ward. 2004. Developments in the use of *Bacillus* species for industrial production. *Canadian Journal of Microbiology* 50:1–17.

Segall, J. and R. Losick. 1977. Cloned *Bacillus subtilis* DNA containing a gene that is activated early during sporulation. *Cell* 11:751–761.

Sella, S.R., L.P. Vandenberghe and C.R. Soccol. 2015. *Bacillus atrophaeus*: Main characteristics and biotechnological applications—A review. *Critical Reviews in Biotechnology* 35:533–545.

Shank, E.A. and R. Kolter. 2011. Extracellular signaling and multicellularity in *Bacillus subtilis*. *Current Opinion in Microbiology* 14:741–747.

Smith, N.R., R.E. Gordon and F.E. Clark. 1946. *Aerobic Mesophilic Sporeforming Bacteria*. Washington, DC: United States Department of Agriculture.

Spizizen, J. 1958. Transformation of biochemically deficient strains of *Bacillus subtilis* by deoxyribonucleate. *Proceedings of the National Academy of Sciences* (USA) 44:1072–1078.

Takahashi, I. 1961. Genetic transduction in *Bacillus subtilis*. *Biochemical and Biophysical Research Communications* 5:171–175.

Valero-Rello, A., M. Lopez-Sanz, A. Quevedo-Olmos, A. Sorokin and S. Ayora. 2017. Molecular mechanisms that contribute to horizontal transfer of plasmids by the bacteriophage SPP1. *Frontiers in Microbiology* 8:1816. doi:10.3389/fmicb.2017.01816.

Veening, J.W., O.A. Igoshin, R.T. Eijlander, R. Nijland, L.W. Hamoen and O.P. Kuipers. 2008a. Transient heterogeneity in extracellular protease production by *Bacillus subtilis*. *Molecular Systems Biology* 4:184. doi:10.1038/msb.2008.18.

Veening, J.W., W.K. Smits and O.P. Kuipers. 2008b. Bistability, epigenetics, and bet-hedging in bacteria. *Annual Review of Microbiology* 62:193–210.

Woese, C.R. and G.E. Fox. 1977. Phylogenetic structure of the prokaryotic domain: The primary kingdoms. *Proceedings of the National Academy of Sciences* (USA) 74:5088–5090.

Yasbin, R.E. and F.E. Young. 1974. Transduction in *Bacillus subtilis* by bacteriophage SPP1. *Journal of Virology* 14:1343–1348.

Ye, M., X. Tang, R. Yang, H. Zhang, F. Li, F. Tao, F. Li and Z. Wang. 2018. Characteristics and application of a novel species of *Bacillus*: *Bacillus velezensis*. *ACS Chemical Biology* 13:500–505.

Youngman, P.J., Perkins, J.B. and R. Losick. 1983. Genetic transposition and insertional mutagenesis in *Bacillus subtilis* with *Streptococcus faecalis* transposon Tn*917*. *Proceedings of the National Academy of Sciences* (USA) 80:2305–2309.

Yuksel, M., J.J. Power, J. Ribbe, T. Volkmann and B. Maier. 2016. Fitness trade-offs in competence differentiation of *Bacillus subtilis*. *Frontiers in Microbiology* 7:888. doi:10.3389/fmicb.2016.00888.

Zeigler, D.R., Z. Pragai, S. Rodriguez, B. Chevreux, A. Muffler, T. Albert, R. Bai, M. Wyss and J.B. Perkins. 2008. The origins of 168, W23, and other *Bacillus subtilis* legacy strains. *Journal of Bacteriology* 190:6983–6995.

Zeigler, D.R. 2014. The *Geobacillus* paradox: Why is a thermophilic bacterial genus so prevalent on a mesophilic planet? *Microbiology* 160:1–11.

Zeigler, D.R. and J.B. Perkins. 2015. The genus *Bacillus*. In *Practical Handbook of Microbiology*, pp. 429–466. E. Goldman and L.H. Green (Eds.). Boca Raton, FL: CRC Press.

Zeigler, D.R. and W.L. Nicholson. 2017. Experimental evolution of *Bacillus subtilis*. *Environmental Microbiology* 19:3415–3422.

Zhang, X.Z. and Y. Zhang. 2011. Simple, fast and high-efficiency transformation system for directed evolution of cellulase in *Bacillus subtilis*. *Microbial Biotechnology* 4:98–105.

Zuber, P. and R. Losick. 1983. Use of a *lacZ* fusion to study the role of the *spo0* genes of *Bacillus subtilis* in developmental regulation. *Cell* 35:275–283.

4 Genetic Resources of Rotifers in the Genus *Brachionus*

Terry W. Snell

CONTENTS

Abstract: A collection of *Brachionus* rotifer diapausing eggs is described. These eggs are desiccated and frozen, with a shelf life of many years. Diapausing eggs can be hatched by hydrating for 24 hours, yielding hatchlings suitable for many applications in comparative biology, aquaculture, and ecotoxicology. Rotifer diapause eggs can be inexpensively shipped globally, making live rotifers of these species available on demand to virtually all researchers. The ease of this "no culture" system is expected to enhance the attractiveness of rotifers as models for investigations in basic and applied biology. The *Brachionus* diapause egg collection makes a range of rotifer biodiversity easily accessible to aquaculturists in a form that is cost-effective, reliable, and free of contaminants.

INTRODUCTION TO ROTIFERS

Rotifers were first described by Leeuwenhoek in the late seventeenth century (Ford 1982) and, as such, have a rich history in research (Ratcliff 2000). The phylum Rotifera includes microscopic invertebrates in the supraphylum Lophotrochozoa (Dunn et al. 2008). With more than 2000 described species (Segers 2008), rotifers are common and important components of freshwater and coastal marine environments in both permanent and ephemeral water bodies (Wallace et al. 2005, Wallace and Snell 2010).

The genus *Brachionus* is the oldest valid genus name among monogonont rotifers (de Beauchamp 1952). This genus, containing 63 species (Segers 2008), exhibits great morphological variation in size, shape, and spines, and ecological variation in diet and habitat. Most species inhabit freshwater. Within the genus, the *Brachionus plicatilis* species complex is a large and important group of cryptic species that inhabit inland salt lakes and coastal marine environments (Mills et al. 2017). The main species in the diapause egg collection are members of the *B. plicatilis* species complex. Also included are the freshwater *B. calyciflorus* and the very small *Proales similis*, which are important in aquaculture. New species in the *B. plicatilis* complex are still being discovered and named. This is an active area of research in rotifer systematics. Another complex of cryptic species, *B. calyciflorus* has also been recognized, and new species have recently been named in it (Michaloudi et al. 2018).

ROTIFER GENERAL BIOLOGY AND LIFE CYCLE

Most rotifer species are filter feeders, living on bacteria, phytoplankton, and fungi. Other species are predatory and consume other zooplankton (Wallace et al. 2005). Rotifers range in size from 80 to 2000 μm and possess about 1000 cells with fully differentiated digestive, reproductive, nervous, sensory, excretory, and muscular systems (Figure 4.1). Most species develop directly with no larval stage and are eutelic, with no cell division after hatching. Two main groups are recognized: monogononts which have a single gonad, and bdelloids which are characterized by a wormlike crawling behaviour. Monogononts reproduce by cyclical parthenogenesis (Figure 4.2), which is dominated by asexual egg production via mitosis (Wallace and Snell 2010). During episodic bouts of sexual reproduction, some females produce haploid eggs via meiosis (Snell 2011). Sex in brachionids is triggered by environmental factors like crowding (Gilbert 1977, Kubanek and Snell 2008, Snell 2017, Snell and Boyer 1988). Sexual eggs hatch into either small haploid males, or diploid diapause (resting) eggs if they are fertilized. Diapause eggs (Figure 4.3) can withstand high temperatures, freezing, desiccation, and anoxic conditions and remain viable for decades.

USING ROTIFERS FOR COMPARATIVE BIOLOGY

Genetic analysis of traditional animal models typically uses screens of single gene mutants in highly inbred, laboratory-adapted strains. This approach provides limited insight into how fully adapted genetic systems behave in nature (Austad 2009, Jones et al. 2014, Nussey et al. 2012). An alternative strategy is to utilize natural variants that differ in their responses to a variety of variables such as diet, temperature, salinity, and predators. Comparing natural variation in phenotypes and genotypes between well-adapted, closely related species can be a potentially more powerful approach than comparing a small number of mutations in a few metabolic pathways (Mark Welch 2018). Characterization of natural variants allows the rigorous dissection of the overlap, differentiation, and trade-offs between genetic systems involved in adaptation to environmental variation. Investigating the comparative biology of natural variation in traits within and between rotifer clones and species is an especially powerful experimental design (Eldredge and Cracraft 1980). These and other

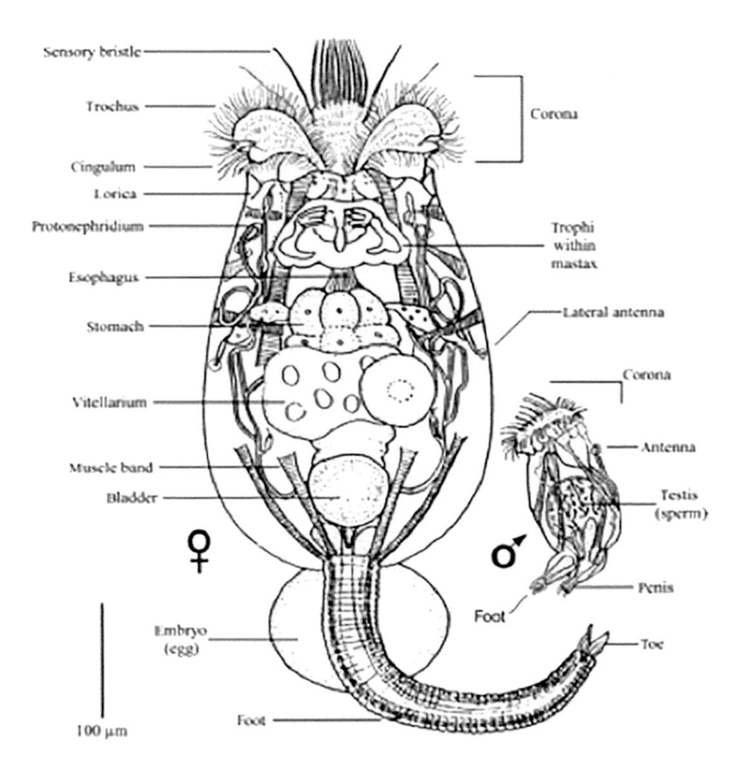

FIGURE 4.1 The anatomy of *Brachionus plicatilis* females and males. (From Wallace, R.L. and Snell, T.W., Rotifera, in *Ecology and Systematics of North American Freshwater Invertebrates*, J.H. Thorp and A.P. Covich (Eds.), pp. 173–235, Academic Press, New York, 2010.)

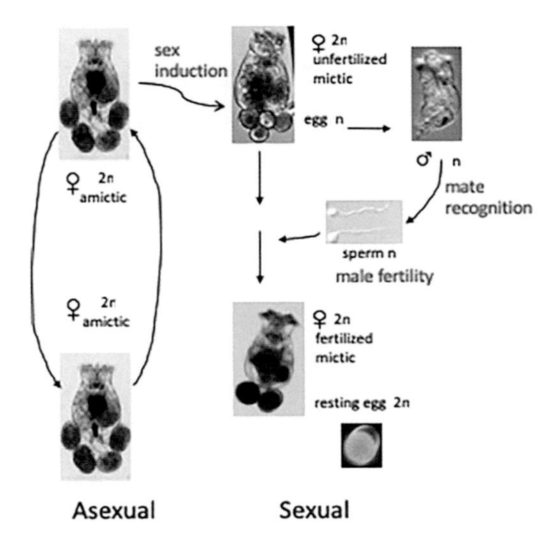

FIGURE 4.2 The cyclical parthenogenetic life cycle of *Brachionus*.

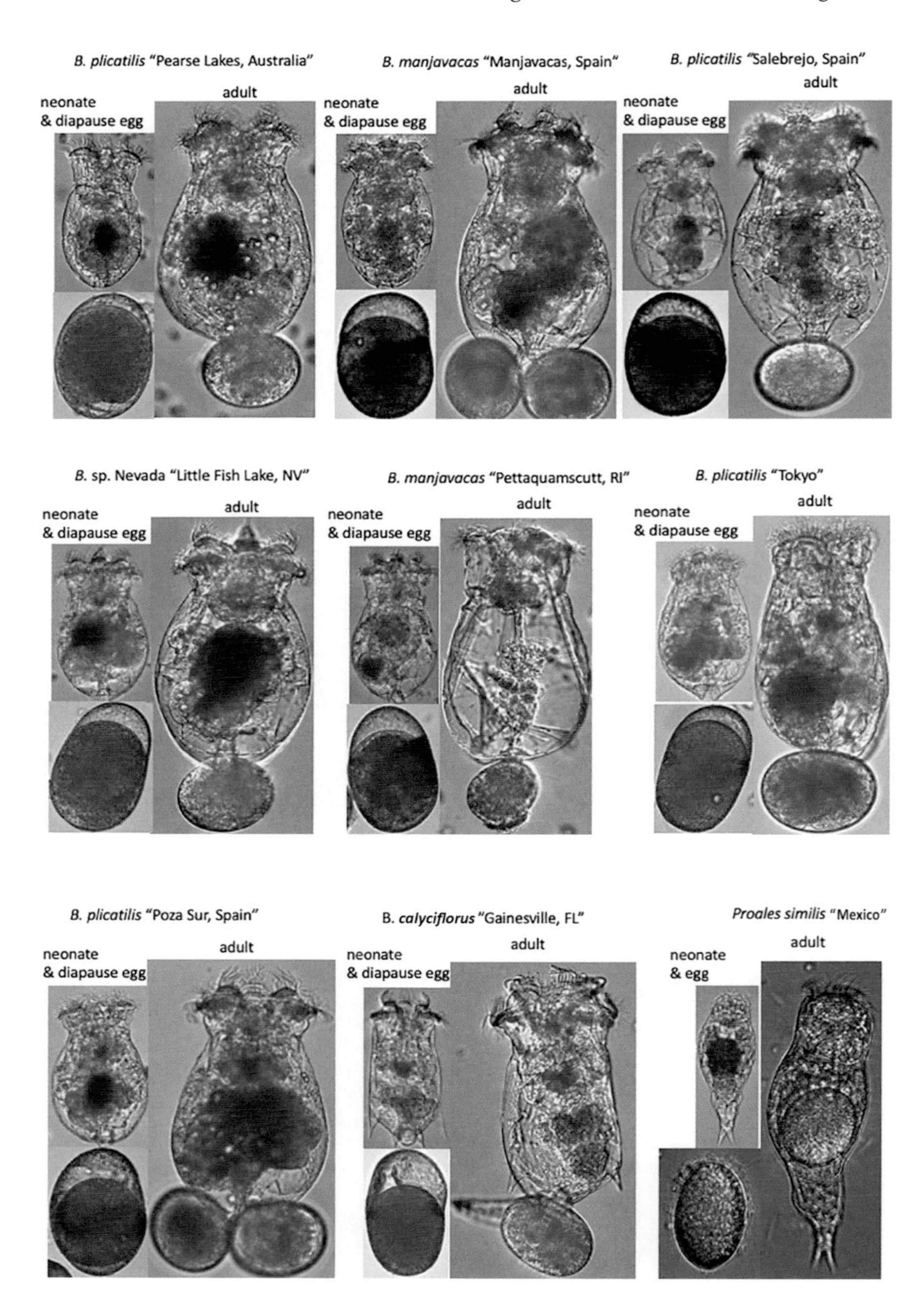

FIGURE 4.3 (See color insert.) Photomicrographs of representative members in the *Brachionus* diapause egg collection. For each species, the large photo on the right is an adult female, the small photo in the upper left is a neonate female, and a diapause egg is pictured in the lower left.

factors illustrate the value of a standardized collection of genetically distinct lineages which researchers can draw upon for their investigations.

Rotifers have been a valuable research tool for more than 100 years for studies in ecology, evolution, and applied biology. Rotifers are critical links in aquatic food webs and hundreds of papers have described their ecology (e.g., Declerck et al. 2005, Sarma et al. 2006). Due to their short life cycles, rotifers have frequently been the objects of studies in evolutionary ecology (e.g., Franch-Gras et al. 2018, Fussman et al. 2007) and experimental evolution (e.g., Becks and Agrawal 2010, Carmona et al. 2009).

In applied biology, rotifers are perhaps best known for their role as food for larval fish in aquaculture. Most marine aquaculture is heavily reliant on the mass culture of live rotifer feeds for larval rearing of many marine finfish (Conceicao et al. 2010, Hagiwara et al. 2001, 2017, Lubzens et al. 2001). Rotifers with different body sizes are utilized for larviculture of fish based on their corresponding mouth gape sizes (Hagiwara et al. 2001, Kotani 2017). Using appropriate size rotifer strains for marine finfish larviculture determines, to a large extent, its likelihood of success (Hagiwara et al. 2001, Kotani 2017, Lubzens et al. 2001). Paramount in efforts to supply sufficient rotifers of different sizes in hatcheries is the maintenance of separate, uncontaminated mass cultures of different species (Hagiwara et al. 2007).

A variety of *Brachionus* species are currently utilized in European hatcheries. Different European hatcheries deploy different *Brachionus* species because different species have markedly different optimal growing conditions. Over the geographic range from Norway to Greece, each fish hatchery operates under a unique set of environmental conditions and style of operation. Having a diversity of *Brachionus* species available increases the likelihood that hatcheries will be able to closely match the selected species with their rotifer production growing conditions and the needs of their larval predator (Conceicao et al. 2010, Kim et al. 2017, Kotani 2017).

Rotifers are popular model animals in ecotoxicology, as demonstrated by recent reviews (Rico-Martinez et al. 2017, Snell and Marcial 2017, Won et al. 2017). Rapid, cost-effective, and reproducible toxicity assessments, based on the diapause eggs of aquatic invertebrates, have become popular for a variety of reasons (Snell et al. 2018b, Wells et al. 1997). These include the technical simplicity of their use and the elimination of the need to master culture techniques in order to produce test animals and their food. Standard protocols have been validated and published for a rotifer acute (ASTM 1991) and a reproductive toxicity test (Standard Methods for the Examination of Water and Wastewater 1998). Different rotifer species have different sensitivities to toxicants (McDaniel and Snell 1999) and so employing greater rotifer biodiversity in toxicity tests could provide more ecologically realistic assessments of toxicity (Snell et al. 2018b). Diapause eggs, like those in the *Brachionus* collection, make it more feasible to incorporate several species into a battery of toxicity tests. Toxicity tests in freshwater have been almost exclusively based on the use of *B. calyciflorus* Pallas, and those in marine waters on *Brachionus plicatilis* Muller.

Rotifers are an emerging invertebrate model for studying the biology of aging. Contemporary reviews have summarized the recent contributions of rotifers to aging research (Gribble and Snell 2018, Snell 2014, Snell et al. 2015). These reviews have emphasized their use in quantifying aging rates, the effects of food quality and quantity, and the roles of diapause and inhibition of DNA synthesis on lifespan.

Rotifers offer many advantages as aging models, including their facile culture, a short 2- to 4-week lifespan, distinctive phenotypes of aging, and the ease of life table experiments enabling high levels of replication (King and Miracle 1980, Snell et al. 2012). Body transparency enables the microscopic examination of internal structures and the localization of gene expression and protein abundance. Asexual reproduction ensures clonal cultures, eliminating genetic variation as a confounding factor between treatments. Inducible sexual reproduction in monogononts permits the outcrossing of age-related traits and forward genetic analyses. Studies of rotifer aging are likely to benefit by comparing the responses of closely related species to caloric restriction (Gribble and Mark Welch 2013), metabolic shifting (Snell and Johnston 2014), and pharmaceutical interventions (Snell et al. 2018a).

COLLECTIONS OF ROTIFER BIODIVERSITY

Rotifer biodiversity is characterized by collections of photomicrographs in online databases such as the Rotifer World Catalog (http://www.rotifera.hausdernatur.at/) and the Encyclopedia of Life (http://eol.org/search?q=rotifers&search=Go). These resources play a vital role in providing information on the range of rotifer diversity, as it is currently understood. However, if researchers wish to utilize live rotifers, there are currently few options. Rotifers can be collected from natural populations, which is more challenging than one might think. Not all rotifer species can be found in local environments or during all seasons of the year. Rotifers collected from natural populations will require taxonomic classification. This could be challenging for non-specialists. Alternatively, live samples can be requested from the few research labs that maintain rotifer stock cultures. However, there are only a few such labs in the world and shipping live cultures internationally can be challenging. The use of rotifer diapause eggs is currently the best option as they are resistant to the stresses of shipping globally and can be made continuously available. Presently, mainly brachionid species are available as diapause eggs. However, as more monogononts are investigated and cultured, the controlled production of diapause eggs by additional species will become possible. Many bdelloid species are also tolerant of desiccation, and these can be readily shipped as anhydrobiotic xerosomes.

Several research labs around the world maintain a variety of rotifer species by serial dilution culture. These typically are maintained in controlled lab environments at constant temperature, under a fixed light regime, on artificial medium, and microalgae diets. The rotifers in these collections were usually collected as live animals from natural habitats, and isolated in single species cultures. Many researchers take advantage of the asexual reproduction of rotifers and clone their new samples. The collections in these labs represent species which are amenable to lab culture. In fact, little systematic investigation has been done to estimate what fraction of all rotifer species are culturable in a laboratory environment.

Specific laboratories have become known for culturing certain species of rotifers. These include:

- Atsushi Hagiwara at Nagasaki University—marine brachionids,
- Elizabeth Walsh at the University of Texas El Paso—monogonont species from temporary water bodies in arid environments,

- S.S.S. Sarma and S. Nandini of the National Autonomous University of Mexico (UNAM) Iztacala, Mexico City—many freshwater monogonont species,
- Roberto Rico-Martinez of the University of Aguascalientes (Mexico)—many freshwater monogononts,
- Manuel Serra of the University of Valencia (Spain)—many marine and inland salt-lake brachionids, and
- Karine Van Doninick of the University of Namur (Belgium)—several bdelloid species.
- These programs often share cultures with other researchers, upon request.

The most readily and reliably available source of specific rotifer species is samples of their diapause eggs. Large quantities of rotifer diapause eggs are produced in laboratories in Japan (Atsushi Hagiwara, Nagasaki University) and in the USA (Terry Snell, Georgia Institute of Technology). The main species typically available from these programs as diapause eggs are brachionids due to their importance in aquaculture and ecotoxicology. Initiating cultures from lab-produced diapause eggs ensures the correct taxonomic classification of the species and the initiation of cultures free of contamination by other rotifer species or ciliates.

Rotifer diapause eggs currently are commercially available from at least three sources: Florida Aqua Farms (http://floridaaquafarms.com/), MicroBioTests (http://www.microbiotests.be/), and Sustainable Aquatics (http://sustainableaquatics.com/).

GENERAL DESCRIPTION OF THE *BRACHIONUS* DIAPAUSE EGG COLLECTION AT THE GEORGIA INSTITUTE OF TECHNOLOGY (GEORGIA TECH)

The *Brachionus* diapause egg collection at Georgia Tech was assembled over many years by collecting from natural populations and as gifts received from other rotifer researchers. In order to become part of the collection, strains have to be able to grow well in the laboratory and have the capacity for sexual reproduction (Snell et al. 2018b). If strains meet these requirements, they are subjected to a standard diapause egg production protocol, with small adjustments for each species, to maximize egg production. It is well documented that some strains lose their ability to reproduce sexually after many generations in laboratory culture (Serra and Snell 2009). The *Brachionus* diapause egg collection at Georgia Tech currently consists of seven brachionid species and one *Proales* (Table 4.1). These include four *B. manjavacas* (Fontaneto et al. 2007) geographical strains, five *B. plicatilis* Muller strains, two *B. rotundiformis* Tschugunoff strains, and one strain each of four additional *Brachionus* species. These were all collected from a variety of habitats including inland salt lakes, coastal marine environments, and freshwater ponds. Adult females range in average length from 88 μm for *Proales similis* De Beauchamp to 305 μm for the *B. plicatilis* Tokyo strain.

Diapause eggs in the Georgia Tech collection are produced in a laboratory under the standard environmental conditions of; constant 25°C, 15 ppt salinity, a diet of *Tetraselmis suecica*, constant aeration, and constant illumination provided by two 20 W cool-white fluorescent lights. Supplemental feeding is provided using the commercial products S.parkle (Inve Technologies, Belgium) or Amplifeed

TABLE 4.1
Ecological and Morphological Characteristics of Rotifers in the *Brachionus* Diapause Egg Collection. Length and Width/Standard Deviation Measurements Were Obtained from Adult Females Without Eggs

Strain	Species	Collection Location	Habitat[a]	Length (µm)	Width (µm)
BM RUS	*B. manjavacas*	Azov Sea, Russia	CM	254.3/17.9	201.7/14.4
BmPetta	*B. manjavacas*	Pettaquamscutt, RI	CSP	244.6/11.1	200.6/7.0
Bm MAN	*B. manjavacas*	Manjavacas, Spain	IS	244.4/14.6	197.9/12.4
Bm Gaynor	*B. manjavacas*	Gaynor Pond, CO	ISP	245.2/19.1	202.4/22.6
Bp Posa Sur	*B. plicatilis*	Torreblanca, Spain	CSP	270.7/24.1	209.8/13.1
Bp China	*B. plicatilis*	Tiajin, China	CSP	260.7/19.0	208.7/11.6
BP Tokyo	*B. plicatilis*	Mie, Japan	ECP	3043.9/20.0	222.0/24.2
Bp SAL	*B. plicatilis*	Salebrejo, Spain	ISP	246.1/14.4	193.2/12.2
AUPEA006	*B. plicatilis*	Australia	CSP	224.1/17.3	174.1/18.3
Br HAW	*B. rotundiformis*	Hawaii	AF	163.5/13.8	134.7/8.9
Br Italy	*B. rotundiformis*	Adriatic, Italy	CM	136.9/15.5	113.0/9.1
JPN S-type	*Brachionus* sp.	Japan	FH	144.5/9.1	125.2/7.8
Nev2	*Brachionus* sp.	Nevada	ISL	246.4/10.3	195.4/14.8
LL1	*Brachionus* sp.	Lost Lake, CT	CSP	232.2/18.9	194.8/11.0
Bc GAINES	*B. calyciflorus*	Gainesville, FL	PD	216.0/24.3	148.0/12.5
Proales	*Proales similis*	Mazatlan, Mexico	SP	88.3/7.3	47.5/4.4

[a] Abbreviations: CM—coastal marine. PD—pond. ISP—inland salt pond. CSP—coastal salt pond. ECP—eel culture pond. AF—aquaculture facility. FH—fish hatchery. ISL—inland salt lake. SP—shrimp pond.

(Sustainable Aquatics, Jefferson City, TN). Rotifers are grown in batch cultures in 5 to 240 L plastic bags in about a two-week growth cycle. Cultures are inoculated at ~0.1 rotifer per mL using animals newly hatched from diapause eggs.

Once diapause egg production is complete, cultures are allowed to settle and the liquid is drained off. The eggs are concentrated, cleaned of debris, and stored at 4°C for one month. Diapause eggs are then filtered onto 53 µm plankton netting and air-dried. Dried diapause eggs are powdered, transferred to glass vials, and stored at −20°C. The eggs remain viable for decades when stored under these conditions. Maintaining rotifers as diapause eggs requires no further culturing and avoids the constant risk of cross-contamination that exists when stocks are maintained as serial dilution cultures. The taxonomic classification of all strains in the collection has been verified via gene sequence analysis of cytochrome c oxidase subunit I (COI) (Gomez et al. 2000).

CHARACTERIZATION AND EVALUATION OF THE GEORGIA TECH ROTIFER COLLECTION

The location of their original collection site of each strain and species is documented. The size of neonate and adult females of all strains and species in the *Brachionus* diapause egg collection has been characterized (Snell et al. 2018b). Each member of the collection has been characterized by the sequence of their COI gene.

Because of the extensive GenBank database of COI sequences for species in the *Brachionus plicatilis* species complex, reliable classification of populations can be achieved. The systematics of this species complex is probably the best studied of all rotifer species (Mills et al. 2017). The main description of the rotifer strains and species in the diapause egg collection has been presented by Snell et al. (2018b). Additional online resources that describe many of the species in the collection can be found on the Rotifer World Catalog website, the Encyclopedia of Life, http://www.micrographia.com/specbiol/rotife/homebdel/bdel0100.htm#bdellink, and http://www.ucmp.berkeley.edu/phyla/rotifera/rotifera.html.

COLLECTION DISTRIBUTION PRACTICES AND POLICIES

The international distribution of rotifer diapause eggs is non-problematic because they can be shipped dry in small vials using commercial delivery services. In a dried state, they are quite resistant to the stresses of shipping. It is the responsibility of requesting researchers to comply with all of the quarantine and movement of biodiversity regulations of their respective country. Rotifers generally are not considered pests nor has there been any case where they have been recognized as a problematic or invasive species. For several decades, fish hatcheries have exchanged starter cultures of brachionid rotifers from all over the world, with little regard for their potential impact on indigenous zooplankton. Throughout the history of these uncontrolled introductions, researchers have not recorded any ecological disturbances to native aquatic food webs. Nonetheless, researchers and hatcheries are advised to follow good laboratory practices to avoid accidental releases of cultured rotifers into natural environments.

LOOKING FORWARD

The *Brachionus* diapause egg collection at Georgia Tech will continue to expand as interesting new populations are identified. An immediate effort is being made to add representatives of the species *B. koreanus* (Hwang et al. 2013), *B. asplanchnoidis* (Michaloudi et al. 2017), and *B. ibericus* (Ciros-Perez et al. 2001). The intent is to incorporate as large a range of *Brachionus* biodiversity as possible. Traits such as exceptionally large or small size are desirable because of their potential importance in aquaculture. As new strains and species are collected and evaluated, attributes of their physiology or genetics may make them good candidates for addition to the collection.

Genomics resources for many of the species in the collection can be found in the publications of Jae-Seong Lee—Sungkyunkwan University, S. Korea (e.g., Hwang et al. 2014, Jeong et al. 2017, Lee et al. 2011), David Mark Welch—Marine Biological Lab, USA (Gribble and Mark Welch 2017, Suga et al. 2008), and Karine Van Doninick (Flot et al. 2013). Rotifer genomics is an active area of research, and it is expected that full genome sequences of many of the species in the *Brachionus* diapause egg collection will be available within the next few years in GenBank.

CONCLUSIONS

Research on the biology of rotifers over the past 50 years has enabled the assembly of a collection of rotifer diapause eggs that are a valuable resource for biological researchers, aquaculturists, and ecotoxicologists. Rotifer diapause eggs are commercially available from a few companies at reasonable cost. Diapause eggs enable live rotifers to be shipped globally quickly, cost-effectively, and reliably. The availability of rotifer diapause eggs makes feasible the comparative biology of closely related natural variants, for a variety of applications. Rotifer diapause eggs enable aquaculturists to supply larval predators with prey of the optimal size to maximize their survival and growth. Ecotoxicologists can develop more ecologically realistic models of toxicity by incorporating a broader range of rotifer biodiversity into toxicity tests.

REFERENCES

ASTM. 1991. A standard practice for performing acute toxicity tests using rotifers in the genus *Brachionus*. *American Society of Testing and Materials* 11:1210–1216.

Austad, S.N. 2009. Is there a role for new invertebrate models for aging research? *Journals of Gerontology Series A, Biological Sciences and Medical Sciences* 64:192–194.

Becks, L. and A.F. Agrawal. 2010. Higher rates of sex evolve in spatially heterogeneous environments. *Nature* 468:89–92.

Carmona, M.J., N. Dimas-Flores, E.M. Garcia-Roger and M. Serra. 2009. Selection of low investment in sex in a cyclically parthenogenetic rotifer. *Journal of Evolutionary Biology* 22:1975–1983.

Ciros-Perez, J., A. Gomez and M. Serra. 2001. On the taxonomy of three sympatric sibling species of the *Brachionus plicatilis* (Rotifera) complex from Spain, with the description of *B. ibericus* n. sp. *Journal of Plankton Research* 23:1311–1328.

Conceicao, L.E.C., M. Yufera, P. Makridis, S. Morais and M.T. Dinis. 2010. Live feeds for early stages of fish rearing. *Aquaculture Research* 41:613–640.

de Beauchamp, P. 1952. Variation chez les rotiferes du genre *Brachionus*. *Comptes Rendus* 235:1355–1357.

Declerck, S., J. Vandekerkhove, L. Johansson et al. 2005. Multi-group biodiversity in shallow lakes along gradients of phosphorus and water plant cover. *Ecology* 86:1905–1913.

Dunn, C.W., A. Hejnol, D.Q. Matus et al. 2008. Broad phylogenomic sampling improves resolution of the animal tree of life. *Nature* 452:745–749.

Eldredge, N. and J. Cracraft. 1980. *Phylogenetic Patterns and the Evolutionary Process: Method and Theory in Comparative Biology*. New York, Columbia University Press.

Flot, J-F., B. Hespeels, X. Li et al. 2013. Genomic evidence for ameiotic evolution in the bdelloid rotifer *Adineta vaga*. *Nature* 500:453–457.

Fontaneto, D., I. Giordani, G. Melone and M. Serra. 2007. Disentangling the morphological stasis in two rotifer species of the *Brachionus plicatilis* species complex. *Hydrobiologia* 583:297–307.

Ford, B.J. 1982. The Rotifera of Antony van Leeuwenhoek. *Quekett Journal of Microscopy* 34:362–373.

Franch-Gras, L., C. Hahn, E.M. Garcia-Roger, M.J. Carmona, M. Serra and A. Gómez. 2018. Genomic signatures of local adaptation to the degree of environmental predictability in rotifers. *Scientific Reports* 8:16051. doi:10.1038/s41598-018-34188-y.

Fussmann, G.F., M. Loreau and P.A. Abrams. 2007. Eco-evolutionary dynamics of communities and ecosystems. *Functional Ecology* 21:465–477.

Gilbert, J.J. 1977. Mictic-female production in monogonont rotifers. *Archiv fur Hydrobiologie Beih* 8:142–155.

Gomez, A., G.R. Carvalho and D.H. Lunt. 2000. Phylogeography and regional endemism of a passively dispersing zooplankter: Mitochondrial DNA variation in rotifer resting egg banks. *Proceedings of the Royal Society of London B* 267:2189–2197.

Gribble, K.E. and D.B. Mark Welch. 2013. Life-span extension by caloric restriction is determined by type and level of food reduction and by reproductive mode in *Brachionus manjavacas* (Rotifera). *Journals of Gerontology Series A. Biological Sciences* 68:349–358.

Gribble, K.E. and D.B. Mark Welch. 2017. Genome-wide transcriptomics of aging in the rotifer *Brachionus manjavacas*, an emerging model system. *BMC Genomics* 18:217. doi:10.1186/s12864-017-3540-x.

Gribble, K.E. and T.W. Snell. 2018. Rotifers as models for the biology of aging. In *Handbook of Models on Human Aging*, P.M. Conn and J. Ram (Eds.), pp. 483–495. New York, Elsevier.

Hagiwara, A., W.G. Gallardo, M. Assavaaree, T. Kotani and A.B. de Araujo. 2001. Live food production in Japan: Recent progress and future aspects. *Aquaculture* 200:111–127.

Hagiwara, A., K. Suga, A. Akazawa, T. Kotani and Y. Sakakura. 2007. Development of rotifer strains with useful traits for rearing fish larvae. *Aquaculture* 268:44–52.

Hagiwara, A., H.J. Kim and H. Marcial. 2017. Mass culture and preservation of *Brachionus plicatilis* sp. complex. In *Rotifers*, A. Hagiwara and T. Yoshinaga (Eds.), pp. 35–46. Singapore, Springer Nature.

Hwang, D.S., H.U. Dahms, H.G. Park and J.S. Lee. 2013. A new intertidal *Brachionus* and intrageneric phylogenetic relationships among *Brachionus* as revealed by allometry and CO1-ITS1 gene analysis. *Zoological Studies* 52:1–10.

Hwang, D.S., K. Suga, Y. Sakakura et al. 2014. Complete mitochondrial genome of the monogonont rotifer, *Brachionus koreanus* (Rotifera, Brachionidae). *Mitochondrial DNA Part B*. 2. doi:10.1080/23802359.2016.1202743.

Jeong, C.B., H.S. Kim, H.M. Kang et al. 2017. Genome-wide identification of ATP-binding cassette (*ABC*) transporters and conservation of their xenobiotic transporter function in the monogonont rotifer (*Brachionus koreanus*). *Comparative Biochemistry and Physiology Part D: Genomics and Proteomics* 21:17–26.

Jones, O.R., A. Scheuerlein, R. Salguero-Gómez et al. 2014. Diversity of ageing across the tree of life. *Nature* 505:169–173.

Kim, H.J., M. Iwabuchi, Y. Sakakura and A. Hagiwara. 2017. Comparison of low temperature adaptation ability in three native and two hybrid strains of the rotifer *Brachionus plicatilis* species complex. *Fisheries Science* 83:65–72.

King, C.E. and M.R. Miracle. 1980. A perspective on aging in rotifers. *Hydrobiologia* 73:13–19.

Kotani, T. 2017. The current status of the morphological classification of rotifer strains used in aquaculture. In *Rotifers*, A. Hagiwara and T. Yoshinaga (Eds.), pp. 3–14. Singapore, Springer Nature.

Kubanek, J. and T.W. Snell. 2008. Quorum sensing in rotifers. In *Chemical Communication among Microbes*, S.C. Winans and B.L. Bassler (Eds.), pp. 453–461. Washington, DC, ASM Press.

Lee, J.S., R.O. Kim, J.S. Rhee et al. 2011. Sequence analysis of genomic DNA (680 Mb) by GS-FLX-Titanium sequencer in the monogonont rotifer, *Brachionus ibericus*. *Hydrobiologia* 662:65–75.

Lubzens, E., O. Zmora and Y. Barr. 2001. Biotechnology and aquaculture of rotifers. *Hydrobiologia* 446/447:337–353.

Mark Welch, D.B. 2018. The potential of comparative biology to reveal mechanisms of aging in rotifers. In *Conn's Handbook of Models for Human Aging*, J.L. Ram and P.M. Conn (Eds.), pp. 497–505. Amsterdam, the Netherlands, Elsevier.

McDaniel, M. and T.W. Snell. 1999. Probability distributions of toxicant sensitivity for freshwater rotifer species. *Environmental Toxicology* 14:361–366.

Michaloudi, E., S. Mills, S. Papakostas et al. 2017. Morphological and taxonomic demarcation of *Brachionus asplanchnoidis* Charin within the *Brachionus plicatilis* cryptic species complex (Rotifera, Monogononta). *Hydrobiologia* 796:19–37.

Michaloudi, E., S. Papakostas, G. Stamou, V. Nedela and E. Tihlarikova. 2018. Reverse taxonomy applied to the *Brachionus calyciflorus* cryptic species complex: Morphometric analysis confirms species delimitations revealed by molecular phylogenetic analysis and allows the (re)description of four species. *PLoS ONE* 13(9):e0203168.

Mills, S., J.A. Alcantara-Rodrıguez, J. Ciros-Perez et al. 2017. Fifteen species in one: Deciphering the *Brachionus plicatilis* species complex (Rotifera, Monogononta) through DNA taxonomy. *Hydrobiologia* 796:39–58.

Nussey, D.H., H. Froy, J.-F. Lemaitre, J.-M. Gaillard and S.N. Austad. 2012. Senescence in natural populations of animals: Widespread evidence and its implications for biogerontology. *Ageing Research Reviews* 12:214–225.

Ratcliff, M.J. 2000. Wonders, logic, and microscopy in the eighteenth century: A history of the rotifer. *Science in Context* 13:93–119.

Rico-Martinez, R., M.A. Arzate-Cardenas, J. Alvarado-Flores, I.A. Perez-Legaspi and G.E. Santos-Medrano. 2017. Rotifers as models for ecotoxicology and genotoxicology. In *Ecotoxicology and Genotoxicology: Non-Traditional Aquatic Models*, Issues in Toxicology no. 33, M.L. Larramendy (Ed.), pp. 48–69. London, UK, Royal Society of Chemistry.

Sarma, S.S.S., S. Nandini, J. Morales-Ventura, I. Delgado-Martinez and L. Gonzalez-Valverde. 2006. Effects of NaCl salinity on the population dynamics of freshwater zooplankton (rotifers and cladocerans). *Aquatic Ecology* 40:349. doi:10.1007/s10452-006-9039-1.

Segers, H. 2008. Global diversity of rotifers (Rotifera) in freshwater. *Hydrobiologia* 595:49–59.

Serra, M. and T.W. Snell. 2009. Sex loss in rotifers. In *Lost Sex: The Evolutionary Biology of Parthenogenesis*, K. Martens, I. Schon and P. van Dijk (Eds.), pp. 73–85. Berlin, Germany, Springer.

Snell, T.W. 2011. A review of the molecular mechanisms of monogonont rotifer reproduction. *Hydrobiologia* 662:89–97.

Snell, T.W. 2014. Rotifers as a model for the biology of aging. *International Review Hydrobiology* 99:84–95.

Snell, T.W. 2017. Analysis of proteins in conditioned medium that trigger monogonont rotifer mictic reproduction. *Hydrobiologia* 796:245–253.

Snell, T.W. and E.M. Boyer. 1988. Thresholds for mictic female production in the rotifer *Brachionus plicatilis* (Muller). *Journal of Experimental Marine Biology and Ecology* 124:73–85.

Snell, T.W. and R.K. Johnston. 2014. Glycerol extends lifespan of *Brachionus manjavacas* (Rotifera) and protects against stressors. *Experimental Gerontology* 57:47–56.

Snell, T.W. and H.S. Marcial. 2017. Using rotifers to diagnose the ecological impacts of toxicants. In *Rotifers*, A. Hagiwara and T. Yoshinaga (Eds.), pp. 129–147. Singapore: Springer Nature.

Snell, T.W., A.M. Fields and R.K. Johnston. 2012. Antioxidants can extend lifespan of *Brachionus manjavacas* (Rotifera), but only in a few combinations. *Biogerontology* 13:261–275.

Snell, T.W., R.K. Johnson, K.E. Gribble and D.B. Mark Welch. 2015. Rotifers as experimental tools for investigating aging. *Invertebrate Reproduction and Development* 59:5–10.

Snell, T.W., R.K. Johnston, A.B. Matthews, H. Zhou, M. Gao and J. Skolnick. 2018a. Repurposed FDA-approved drugs targeting genes influencing aging can extend lifespan and healthspan. *Biogerontology* 19:145–157.

Snell, T.W., R.K. Johnston and A.B. Matthews. 2018b. Utilizing *Brachionus* biodiversity in marine finfish larviculture. *Hydrobiologia*. doi:10.1007/s10750-018-3776-8.

Standard Methods for the Examination of Water and Wastewater. 1998. Estimating chronic toxicity using Rotifers. In *Standard Methods for the Examination of Water and Wastewater*, A.E. Greenberg, S.L. Clesceri and A.D. Eaton (Eds.), Washington, DC: American Public Health Association, American Water Works Association and Water Environment Federation.

Suga, K., D.B. Mark Welch, Y. Tanaka, Y. Sakakura and A. Hagiwara. 2008. Two circular chromosomes of unequal copy number make up the mitochondrial genome of the rotifer *Brachionus plicatilis*. *Molecular Biology and Evolution* 25:1129–1137.

Wallace, R.L. and T.W. Snell. 2010. Rotifera. In *Ecology and Systematics of North American Freshwater Invertebrates*, J.H. Thorp and A.P. Covich (Eds.), pp. 173–235. New York: Academic Press.

Wallace, R.L., T.W. Snell, C. Ricci and T. Nogrady. 2005. *Rotifera:* Volume 1—*Biology, Ecology and Systematics*. 2nd ed. SPB Academic Publishing: The Hague, the Netherlands.

Wells, P.G., K. Lee and C. Blaise. 1997. *Microscale Testing in Aquatic Toxicology*. Boca Raton, FL: CRC Press.

Won, E.J., J. Han, D.H. Kim, H.U. Dahms and J.S. Lee. 2017. Rotifers in ecotoxicology. In *Rotifers*, A. Hagiwara and T. Yoshinaga (Ed.), pp. 149–176. Singapore: Springer Nature.

5 The Caenorhabditis Genetics Center (CGC) and the *Caenorhabditis elegans* Natural Diversity Resource

Aric L. Daul, Erik C. Andersen and Ann E. Rougvie

CONTENTS

Abstract: Research using *C. elegans* has led to fundamental insights into basic biological mechanisms, including the genetic basis of programmed cell death and cell signaling, the discovery of microRNAs, and the identification and subsequent elucidation of the mechanism of RNA interference in animals, and has been used to increase our understanding of the mechanisms of cancer progression and other diseases including Alzheimer's and Parkinson's. *Caenorhabditis elegans* is an androdioecious (hermaphrodite-male) species with a short generation time. The worms develop externally and are transparent, allowing observation of developmental events throughout an animal's entire life history. The *Caenorhabditis* Genetics Center (CGC) was established in 1979 at the University of Missouri and subsequently moved to the University of Minnesota. During its first year in operation in MO, the center distributed a mere 15 strains. In 2013, 31,242 strains were shipped by the CGC. The CG Center's collection contains ~20,700 genetically distinct strains of *C. elegans* in addition to more than 40 species in the genus. The work of the CGC is complemented by the activities of Wormbase, WormAtlas, and the *Caenorhabditis elegans* Natural Diversity Resource.

INTRODUCTION

In 1963, Sydney Brenner turned to nematodes in his quest to "tame a small metazoan organism to study development directly" (Brenner 1988). Concerned that the "classical problems of molecular biology have either been solved or will be solved in the next decade," Brenner shrewdly predicted that the future of biology would lie in the molecular biology of development, and in particular, the nervous system. Coming from a background in microbiology, he believed the nematode offered advantages similar to microbial systems due to its small size, short life cycle, and ease of cultivation in the lab (Figure 5.1a and b; Brenner 1988). Although Brenner initially proposed *Caenorhabditis briggsae* as the model system in which to examine developmental processes, he later settled on the closely related species *Caenorhabditis elegans* because it was easier to cultivate in his laboratory (Félix 2008). *Caenorhabditis elegans* is a small, free-living nematode commonly found feeding on bacteria in decaying organic matter, such as compost piles or rotting fruits (Félix and Braendle 2010, Kiontki et al. 2011).

By the mid-1970s, Brenner's focus expanded beyond neuronal development into the developmental biology of the worm. In his seminal paper (Brenner 1974), he laid the foundation for the use of *C. elegans* as a model system for elucidating molecular regulation of development. In the years since that initial paper, researchers throughout the world have exploited the worm's essentially invariant cell lineage, simple anatomy, and transparent body to their advantage (Kimble and Hirsch 1979, Sulston and Horvitz 1977, Sulston et al. 1983).

ABOUT *C. ELEGANS*

Caenorhabditis elegans is an androdioecious (hermaphrodite-male) species with a short generation time, maturing from embryos to adults in about three days (Figure 5.1a). A single hermaphrodite typically produces ~300 progeny through self-fertilization and can quickly give rise to a hearty population (Wood 1988). Under

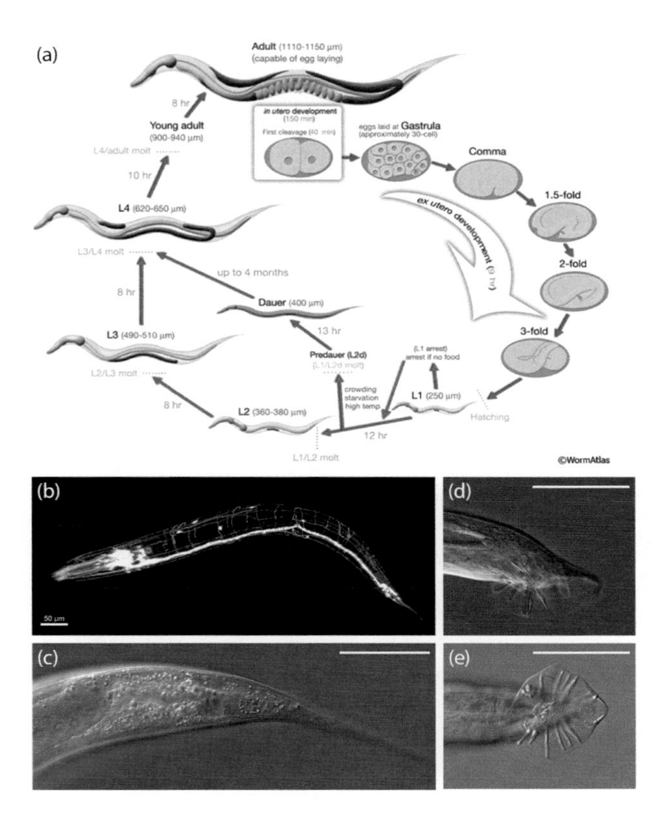

FIGURE 5.1 (See color insert.) *Caenorhabditis elegans* life cycle, nervous system, and sexually dimorphic tail. (a) Hermaphrodite life cycle. Under ideal growth conditions, *C. elegans* will develop from a fertilized egg to an adult in just over two days. The first few hours of development occur *in utero*; the egg is then laid and embryogenesis continues until the basic body plan is complete and the first stage (L1) larva hatches. Development continues, punctuated by molts, through three additional larval stages (L2 to L4) until the sexually mature adult stage is reached. When conditions are harsh, for example due to high temperature, starvation, and/or crowding, *C. elegans* L1 larvae can enter an alternate developmental stage known as a dauer larva, which confers resistance to environmental stress and promotes survival. If conditions become favorable, dauer larvae can resume development. Reprinted with permission from WormAtlas. (b) The simple *C. elegans* nervous system is visualized by pan-neuronal expression of the GFP transgene *evIs111* [*F25B3.3::GFP* + *dpy-20*(+)]. (c, d, and e) Micrographs showing tail morphology. Scale bars are 50 µm. A hermaphrodite's tail tapers gently to a point (c), whereas an adult male tail has a fan-like copulatory structure shown in lateral (d) and ventral (e) views.

typical laboratory culture conditions, a healthy hermaphrodite with ample food will have a lifespan of several weeks. However, stressful conditions such as crowding, starvation, or high temperature can induce *C. elegans* larvae to enter an alternate developmental state known as dauer diapause (Antebi et al. 1998, Cassada and Russell 1975, Golden and Riddle 1984). Dauer larvae form a specialized cuticle that protects them from environmental stresses, including desiccation, and allows them to survive for several months. When conditions become more favorable, dauer larvae will resume development and mature much as they would if conditions had always been tolerable.

Though primarily a hermaphroditic species, males can be found in most populations of *C. elegans* and can arise spontaneously from chromosomal nondisjunction of the X chromosome during meiosis. Sex in *C. elegans* is determined by the ratio of autosomes to X chromosomes: hermaphrodites carry two copies of the X chromosome (XX), whereas males carry a single copy (XO) (Hodgkin 1987). Males are significantly different from hermaphrodites in morphology and behavior; some anatomical differences can be observed in early development, though most become more apparent in later larval stages when the asymmetric gonad and distinctive tail fan (a specialized copulatory structure) are formed (Figure 5.1c–e).

Caenorhabditis elegans is ideally suited as a model for developmental studies. Working with *C. elegans* in a lab often requires little more than a basic dissection microscope with which to observe worms and a worm "pick" (an implement somewhat akin to a bacterial loop) used to manipulate worms (Figure 5.2). The worms are small enough (0.25 mm long at hatching to ~1 mm long as adults) to be grown on agar-filled petri dishes, yet large enough to be easily observed under a dissecting microscope. Most strains can be grown at a wide range of temperatures from roughly 15°C to 25°C, meaning room temperature is often sufficient for maintaining stocks. The rate of development is influenced by temperature (it is faster at warmer temperatures), providing some control over the timing of development by altering ambient temperature. Worms are usually grown on a standardized agar nutrient medium (nematode growth medium, or NGM) supplemented with a lawn of OP50, an auxotrophic strain of *E. coli* selected because it forms a thin, translucent lawn in which worms are still visible (Figure 5.2b). Alternatively, large populations of worms can be grown in liquid culture to facilitate biochemical work. Perhaps most critical from the standpoint of a stock center, *C. elegans* is amenable to cryogenic storage, which allows the collection of strains to be preserved indefinitely, reducing genetic drift that occurs while maintaining actively growing stocks over generations. The worms develop externally and are transparent, allowing the observation of developmental events throughout an animal's entire life history (Figure 5.1a). Researchers can identify developmental defects in mutants by following cell lineages to identify precisely when and where cell divisions go awry. Advances in microscopy, microfluidics, and computing technology can now allow 4D (3D time-lapse) imaging of cells during development, automating the process of tracking cell divisions, even during larval stages (Dutta et al. 2015, Keil et al. 2017, Zacharias and Murray 2016). Synchronized populations can be easily established through several means, including selection of embryos using hypochlorite (the eggshell is resistant to bleach), selection of dauer larvae with SDS (dauer cuticles are resistant to the detergent), or even automated sorting by size or fluorescent protein expression in flow-based large particle sorters (Pulak 2006, Stiernagle 2006).

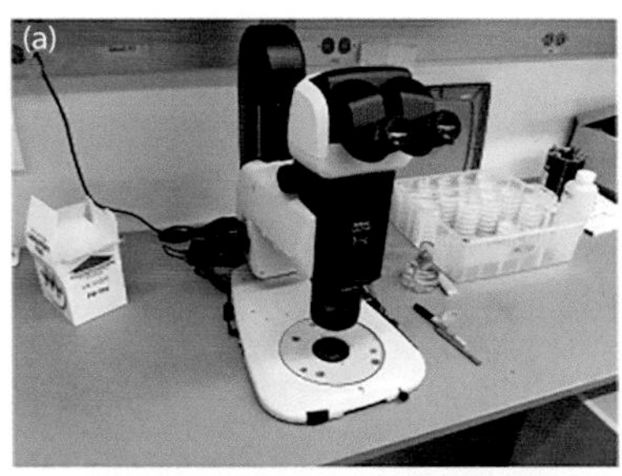

FIGURE 5.2 Tools for basic *C. elegans* manipulation and storage. (a) A typical stereomicroscope station. *Caenorhabditis elegans* are typically observed using a dissecting microscope with variable magnification and equipped with visible light transillumination. (b) Nematodes grown on petri dishes are manipulated using wire picks while viewing them through a dissecting microscope with ample working distance. Similar to an inoculation loop, a worm pick (arrow) is typically fashioned from a piece of platinum wire embedded into a handle of sorts, for example, with the wire melted into the tip of a glass Pasteur pipet. (c) A partial view of the active Caenorhabditis Genetics Center collection. Stocks of *C. elegans* can remain viable for several months if kept in a starved condition on sealed petri dishes stored at cool temperatures.

Caenorhabditis elegans was the first animal to have its DNA completely sequenced, revealing a relatively compact genome of ~100 Mb containing ~20,500 protein-coding genes (Hillier et al. 2005, The *C. elegans* Sequencing Consortium 1998), paving the way for molecular analysis, bioinformatics, and reverse genetics. The use of RNA interference (RNAi), a molecular phenomenon first identified in *C. elegans* (Fire et al. 1998), to knock down gene function in animals is trivial in *C. elegans*, facilitating high-throughput screening for genes regulating developmental processes of interest. RNAi can be administered in worms through injection of double-stranded RNA (dsRNA), feeding the worms bacteria expressing the dsRNA, or even soaking the worms in a solution of dsRNA (Tabara et al. 1998, Timmons and Fire 1998, Timmons et al. 2001). More recently, genome engineering using CRISPR/Cas9 to edit endogenous gene function has taken reverse genetics to the next level, allowing researchers to not only knock-out gene function, but to make specific modifications to endogenous genes, mimicking known disease models or dissecting loci to identify key regulatory elements or functional domains (reviewed in Dickinson and Goldstein 2016).

IMPACT OF *C. ELEGANS*

Research on *C. elegans* has led to fundamental insights into basic biological mechanisms, including the genetic basis of programmed cell death and cell signaling, the discovery of microRNAs, and the identification and subsequent elucidation of the mechanism of RNA interference in animals (Fire et al. 1998, Horvitz 1999, Lee et al. 1993). The nematode has also proved important for understanding mechanisms of cancer progression and other diseases including Alzheimer's and Parkinson's, as well as for revealing basic mechanisms underlying human development (Ewald and Li 2010, Johnson et al. 2010). In addition, *C. elegans* serves as a key model for broadening our understanding of parasitic nematodes and was used to identify the mode of action of the first new livestock anthelmintic brought to market in twenty-five years (Kaminsky et al. 2008, Waterman et al. 2010). Researchers in all of these areas, and a myriad of others, depend upon the availability of strains critical for their work.

HISTORY OF THE CAENORHABDITIS GENETICS CENTER (CGC)—MEETING A DEMAND

By the mid-1970s, leaders in the field had already recognized that worm strains were important reagents for research, and that the sharing of stocks among labs would greatly benefit the entire community (R. Herman, personal communication). In response, they created a central repository for key strains, thereby protecting them from loss and making them freely available to all members of the *C. elegans* community. With the goal to promote research on the nematode *C. elegans* by acquiring, maintaining, and distributing genetically characterized nematode stocks upon request, the CGC was born. It was believed that these services would enhance research progress by making key strains carrying mutations and transgenic lines readily and more widely available. Distribution of strains from a centralized location would remove the burden of strain maintenance and distribution then resting on individual research labs.

CGC operations began in 1979 at the University of Missouri under the direction of Dr. Donald Riddle. Strains were sent out in response to written requests from researchers, and thanks to the financial support from the National Institutes of Health (NIH), were initially distributed free of charge to academic researchers. A mere fifteen strains were distributed in that first year, but demand for strains grew slowly and steadily over the next decade as the field began to mature (Figure 5.3). In 1992, before Dr. Riddle transferred directorship of the CGC to Dr. Robert Herman, the CGC distributed nearly 1400 strains in its final year at the University of Missouri.

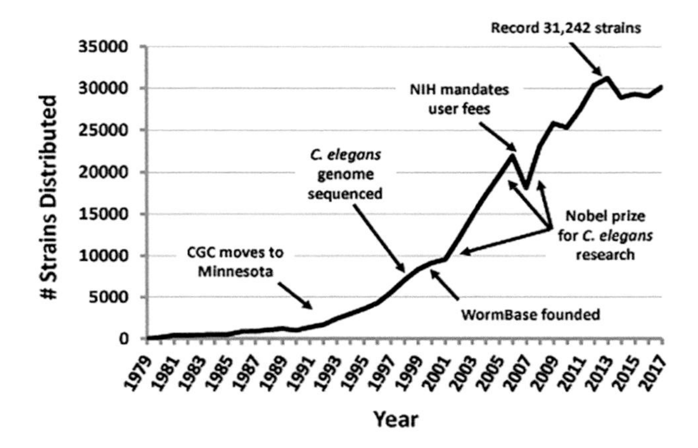

FIGURE 5.3 Caenorhabditis Genetics Center strain distribution. The number of strains shipped per year is graphed. Demand grew annually until 2007, when it dropped briefly in response to implementation of user fees and strain charges mandated by the NIH. Recent demand has remained steady around 30,000 strains shipped per year. Several key events within the worm community are indicated, including the sequencing of the *C. elegans* genome, the launch of WormBase, and the awarding of three Nobel Prizes for research on *C. elegans* (Sydney Brenner, Robert Horvitz, and John Sulston, Physiology or Medicine, 2002; Andrew Fire and Craig Mello, Physiology or Medicine, 2006; and Martin Chalfie, together with Roger Tsien and Osamu Shimomura, Chemistry, 2008).

To transfer the collection to Minnesota, Mark Edgley,the highly efficient curator of the CGC under Dr. Riddle, loaded 1600 strains frozen in liquid nitrogen (LN) into a rented moving van. He transported these strains more than 500 miles to personally deliver the strain collection to Dr. Herman and its new home at the University of Minnesota's St. Paul Campus. In 2003, the collection made a much shorter trip, moving to its current home on the University of Minnesota's Minneapolis (East Bank) campus. Dr. Herman oversaw CGC activities for the next fifteen years, a period of rapid growth in the CGC and worm community overall, until his retirement in 2007 (Figures 5.4 and 5.5). Demand for strains continued to climb steadily, each year setting a new record for the number of orders received and strains distributed by the CGC.

In June 2007, Dr. Ann Rougvie assumed directorship of the CGC, at which time the NIH mandated implementation of a cost recovery program as part of the CGC funding mechanism. This change in funding strategy meant that for the first time, academic users were required to pay an annual lab fee (essentially a one-year subscription for CGC services) in addition to a nominal fee per strain ordered. An audible gasp from the crowd in attendance was heard when the change in policy was announced at the International Worm Meeting that year, followed by the first ever drop in strain distributions (Figure 5.4). However, the community quickly adjusted to the new policy, and the demand for strains resumed its steady climb. An all-time high of 31,242 strains shipped by the CGC was achieved in 2013. Demand has remained steady since then, leveling off at about 30,000 strains distributed annually, with approximately 115 strains shipped each business day. The single day record was set on January 20, 2009, when a staggering 619 strains were shipped.

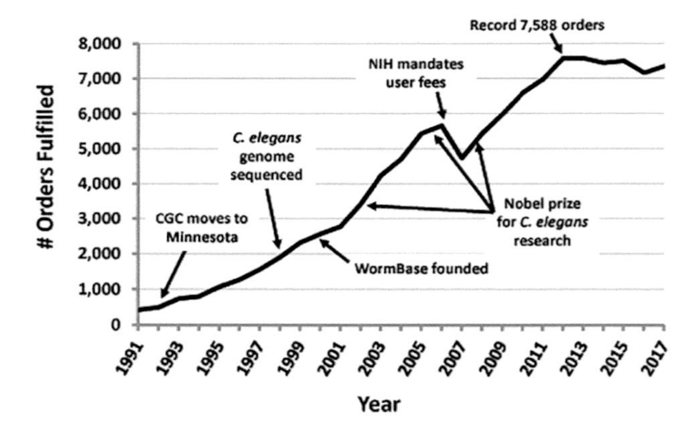

FIGURE 5.4 The number of orders processed by the Caenorhabditis Genetics Center per year. The number of orders grew steadily each year until the implementation of user fees in 2007. Recently, demand has remained steady at about 7500 orders annually, and for the last 25 years, the average number of strains requested per order has remained remarkably consistent at about four.

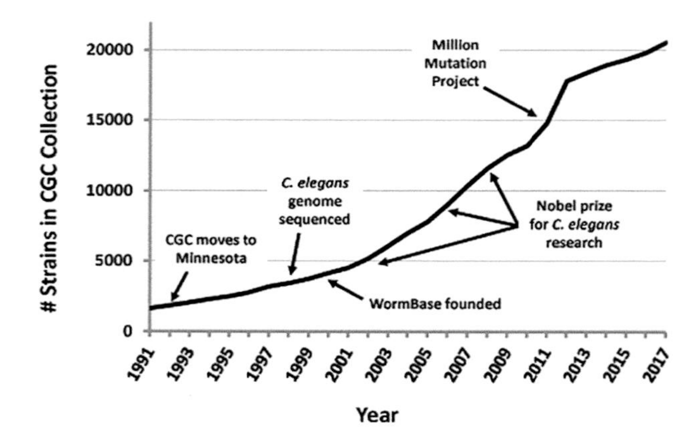

FIGURE 5.5 Expansion of the Caenorhabditis Genetics Center collection. Growth of the Caenorhabditis Genetics Center collection is graphed. The Caenorhabditis Genetics Center collection continues to grow steadily, with an average of about 700 new strains added each year. Large collections such as the Million Mutation Project (Thompson et al. 2013) strains have recently boosted the collection size.

The number of orders processed annually has followed a similar trend, averaging roughly 7500 requests in recent years (Figure 5.4). The historic average is around four strains requested per order, a figure that has remained fairly constant from year to year. The largest single order processed was for 3894 strains in 2009. Over 90% of the strains in the collection have been distributed at least once, a remarkable figure considering several large collections have been added to the catalog in recent years. More than three quarters of the strains in the collection have been requested within the last four years, showing continued demand for older strains as well.

WORMBASE AND WORMATLAS

In 2000, the establishment of WormBase (wormbase.org), a founding member of the Alliance of Genome Resources (alliancegenome.org), fueled the revolution in worm research that had begun with the collection and dissemination of systematic genetic, bibliographic, and nomenclatural data by the CGC. Composed of an international consortium of biologists and genomic specialists, WormBase provides the research community with an easy to use, centralized online compendium of freely available, up-to-date information covering all things worm. Information available includes the genetics, genomics, and the general biology of *C. elegans* and related nematodes; the academic history of past and present lab members; and a comprehensive bibliography of *C. elegans* publications and meeting abstracts (Lee et al. 2018). WormBase is nicely complemented by WormAtlas (http://www.wormatlas.org/), an online database featuring behavioral and structural anatomy of *C. elegans* that includes numerous detailed drawings and high-resolution micrographs.

CGC HOLDINGS AND RECENT ACQUISITIONS

Though *C. elegans* is by far the most widely studied species (more than 98% of the >20,700 genetically distinct strains available in the CGC catalog are *C. elegans*), the genus *Caenorhabditis* is composed of more than 50 species, and nearly 40 of these are represented in the CGC. *Caenorhabditis briggsae*, the second-most popular by strain count, is represented by only ~100 strains in the CGC collection, and most other species are represented by a few isolates. The CGC collection is composed of ~50 species (including members of related genera). Most species in the *Caenorhabditis* clade are gonochristic (female-male), but the two most commonly used species—*C. elegans* and *C. briggsae*—are androdioecious (Haag 2005). From the standpoint of the stock center, this provides a significant advantage as most *C. elegans* and *C. briggsae* stocks can be maintained as hermaphrodites, eliminating the need for intersex matings to propagate lines.

As other new molecular techniques are developed or adapted for use in *C. elegans*, they drive demand for genetic toolkits or certain sets of strains. For example, the CGC recently acquired sets of *C. elegans* and *C. briggsae* recombinant inbred lines (RILs), strains carrying fluorescent markers for cell lineage and expression studies, genetic mapping kits, and a series of strains with transposon insertion sites at defined positions throughout the genome (Bi et al. 2015, Frøkjær-Jensen et al. 2014, Murray et al. 2008, Sarov et al. 2006, Zhong et al. 2010). It is not feasible, of course, for the CGC collection to include all promoter constructs that might be created for use in these systems, but as additional technologies are adapted for use in *C. elegans*, the CGC will strive to make basic toolkits available by providing critical component strains of such systems with the aim of acquiring additional strains as demand might warrant.

The National BioResource Project (NBRP; headed by Dr. Shohei Mitani, Tokyo Women's Medical University School of Medicine) generously provided to the CGC for distribution an exceptional collection of chromosomal inversions that, in combination with preexisting balancers, covers nearly 90% of the genome (Dejima et al. 2018). Many of these new balancer chromosomes are enhanced by fluorescent markers so that users do not have to rely on phenotypic markers, which are sometimes not apparent until the later stages of development, or that can be difficult to score in some mutant backgrounds.

The CGC has long maintained a close relationship with the consortia of labs generating gene knock-out mutations in *C. elegans*. Strains generated by the Vancouver (and previously, Oklahoma) group as part of the *Caenorhabditis elegans* Gene Knockout Consortium (2012) are sent to the CGC for distribution. An additional sizable collection of deletion mutants isolated by NBRP are available directly from the Mitani Lab upon request, and once characterized, may be sent to the CGC with permission from the NBRP. In addition, a small consortium of labs is also working to develop Crispr-Cas9 pipelines to delete genes of specific interest to the human disease and parasitic nematode communities, and those deletions have begun to enter the CGC collection (Au et al. 2018, Wang et al. 2018).

The CGC catalog also includes a set of ~2000 strains comprising the Million Mutation Project (MMP: Thompson et al. 2013), a joint project by the Moerman and Waterston Labs to identify multiple mutations in virtually every *C. elegans* gene (Thompson 2013). Because many of these mutations are single nucleotide changes, it is important that researchers analyzing these strains have an appropriate control strain (and genome sequence) to use in their experimentation. While a single strain from Bristol, England, known as N2 (Sterken 2015) has served as the standard wild-type reference strain among the community for many years, the *C. elegans* genome was originally assembled piecemeal from sequence obtained from multiple but different N2 isolates, resulting in some haplotypes reported in the reference genome that are not found in the stock of N2 maintained by the CGC. Researchers using high-throughput, short-read DNA, or RNA sequencing have identified genetic differences between the reference genome and stocks of N2 maintained in their individual labs, suggestive of genetic drift in some labs' stocks. With the introduction of the MMP strains, the need for a true reference strain and corresponding genome became apparent. The CGC is collaborating with a consortium led by Dr. Shinichi Morishita, Dr. Andy Fire, and Dr.Erich Schwarz to establish a new fully sequenced wild-type reference strain derived from N2 (Yoshimura J, In Press. The new strain, PDLO74, was strategically chosen as the parental strain used in the Million Mutation Project. The CGC has many aliquots of the PDL074 reference strain, with protocols in place for maximizing the number of archived samples while minimizing the number of generations grown each time an original aliquot is thawed. These steps will ensure a large supply of worms frozen at essentially the same time to provide researchers with a long-term source of a standardized reference genome control strain that is common to all labs in the community. The CGC has already begun distributing PDI074 to the community, and the annotated sequence will soon be available on WormBase.

CGC OPERATING PRACTICES AND POLICIES

Acquisition and Handling of New Strains

The primary aim of the CGC is to obtain a reference allele of every identified gene and all useful chromosome rearrangements generated in the laboratory strain N2.

The CGC is actively seeking strains carrying mutations in genes not yet represented in the collection: null alleles of genes absent from the collection or represented only by hypomorphic alleles, unusual alleles of a gene (e.g., gain-of-function or conditional alleles, tagged-endogenous loci), difficult-to-make mutant combinations, and useful genetic or molecular tools (balancer chromosomes, deficiencies, inducible or tissue-specific transgenes, etc.). However, the CGC will consider strains not fitting these criteria if the user provides adequate justification of why a strain should be maintained by the CGC. Many new strains are obtained through donation requests from labs looking to deposit popular strains in the CGC to relieve the burden of distributing the strains themselves. These are often key reagents from recent papers that are in high demand from other research labs. A small number of new strains are acquired in response to requests from members of the community contacting the CGC looking for particular alleles or transgenes significantly different from those already available in the CGC catalog.

All donors must complete an online strain donation form available on the CGC homepage. The CGC curator vets all strain donation requests before a strain is accepted as part of the collection. Each donation request must include enough information for the curator to determine the value of adding that strain to the CGC catalog, including the complete genotype and a brief description of the phenotype, and provide enough information in the strain description that it will be meaningful to other users in the community. The description of a transgenic line must include all components of the transgenic array, including any transformation or co-injection markers used, as well as instruction on how to maintain the array in the population (picking individuals with appropriate markers or how to select against animals that have lost the transgene). For balanced strains, the description should include information about how to identify heterozygous animals to pick to maintain the stock and the expected classes segregating among their progeny. When considering donations of characterized wild isolates, required information includes where a strain was isolated, from what habitat it was isolated, when it was isolated, how it was isolated, and by whom it was isolated. The CGC advisory board is often consulted when vetting donations of non-*C. elegans* species and deciding whether strain submissions should be redirected to CeNDR.

Though the CGC aims to preserve unique strains and make them available to the community far into the future, it is not feasible to accept every strain that a lab has generated or published. With the rapid expansion of the collection in recent years (Figure 5.5), the CGC has been forced to become more selective about additions to the collection. This selectivity has become a greater challenge in recent years as the first wave of PIs dedicated to worm research reach retirement age and seek a long-term home for their lab's collections of strains, many of which are not preserved anywhere else outside their own laboratories. The CGC has also seen an increased number of requests from labs closing or changing research direction with no former post-docs or other lab members interested in or able to assume responsibility for the lab's frozen stocks. In such cases, the CGC will work with the PI or lab manager to narrow down the list of potential donations to strains that fill the needs of the CGC. The selected strains will be those that enhance the current CGC catalog while avoiding redundancy in the collection by excluding multiple lines of similar transgenes and collections of genotypically or phenotypically similar alleles.

The CGC is dedicated to the long-term preservation and maintenance of strains and discourages contributions of stocks not suitable for long-term stability, such as transgenes with especially poor transmission rates or that are prone to silencing, mutants with unstable genomes or mortal germ lines, unbalanced heterozygous mutants, etc. Although important strains may be excluded by these criteria, it is not practical for the CGC staff to dedicate the time and resources necessary to maintain these difficult stocks. Additionally, it is misleading to users to advertise the availability of strains that are unlikely to be successfully recovered from frozen stocks.

New strains are normally sent to the CGC on small agar petri plates (Figure 5.2b). Upon arrival, plates are inspected for any sign the stock might be compromised, such as crushed or broken plates that might have allowed different populations to mingle. A visual inspection is performed, examining each plate for mites, which can prey on nematodes and carry worms or embryos between plates, before unsealing the plate. Worms are transferred to new plates to establish a stable population before attempting to decontaminate the stock, and gross mutant phenotypes are observed. Nearly all incoming stocks are treated with hypochlorite to eliminate unwanted bacteria or other micro-organisms present on the original plate received (Stiernagle 2006). Once the stock is clean, the strain is evaluated to determine if the observed phenotypes match the description of strain provided by the donor. In some cases, published references are checked for additional information about phenotypes and expression patterns. If inconsistencies are noted between observed phenotypes and submitted strain descriptions or published descriptions, the lab submitting the strain is contacted and asked for clarification.

Frozen stocks of each strain are prepared as soon as possible after receipt. At least seven frozen stocks of each strain are made: one is thawed to check the quality of the frozen stocks, and the other frozen aliquots are distributed across four locations for long-term storage in both –80°C and LN freezers on site in the CGC lab space, in LN freezers in a neighboring building, and in LN freezers at the USDA-National Animal Germplasm Program (NAGP) in Fort Collins, CO. Samples tested at the CGC have remained viable when thawed after more than 30 years of frozen storage. Utilizing multiple storage facilities ensures the long-term viability of the collection to make it a permanent community resource. Multiple locations at the University of Minnesota protect the collection against localized loss due to fire or equipment failure, while maintaining a complete mirror copy of the collection at the NAGP provides insurance in event of catastrophic loss at the primary location in Minnesota.

When a potential problem with a strain is reported, attempts are made in-house to verify if the stock is correct. When needed, the originating lab is consulted for further information or asked to validate the stock. If deemed necessary, a replacement is requested from the originating lab. CGC staff will always work with users to troubleshoot potential problems with strains obtained from the CGC.

Large collections or sets of strains (genetic toolkits) may be donated to the CGC but require special consideration before acceptance. This practice was initially reserved for collaboration with large community resource projects, such as the *C. elegans* Knock-out consortium (The *C. elegans* Deletion Mutant Consortium 2012) and the *C. elegans* Gene Expression Project (Hunt-Newbury et al. 2007), each of which submitted more than 2000 strains to the CGC. However, individual research labs may also now submit collections of strains in this manner as they develop tool kits, such as genetic mapping strains, transposon-insertion strains, strains for tissue-specific expression (or knock-down of expression), etc. In these cases, strain information for large collections may be submitted as a spreadsheet or CSV file using a template provided by the CGC, allowing the data to be uploaded by CGC curators *en masse*. Each strain being donated must meet the same requirements for inclusion as if it were being donated individually. When submitting large collections of strains, the CGC request that the donors send them as frozen aliquots rather than on NGM petri plates. The CGC can provide a cryoshipper to streamline the process. Once the frozen stocks are received, the strains will appear in our online catalog and be available to order. Frozen stocks can be simply stored in a freezer until the strain is ordered, allowing CGC staff to process the new stocks as needed, rather than processing dozens (or hundreds) of strains simultaneously upon receipt. The first time a strain from a large collection is ordered, the original aliquot is thawed, the population is expanded, and the strain is cleaned and processed as any other newly acquired strain would be before the stock is sent to the requesting lab and refrozen in aliquots for CGC freezer stocks.

CGC DISTRIBUTION PRACTICES AND POLICIES

Preparation

All strains prepared for orders are transferred to new plates of food (bacteria), decontaminated (if necessary), and assessed for proper phenotypes before shipment. Homozygous strains usually require little maintenance beyond examining the plate to confirm phenotypes. Heterozygous strains and strains carrying chromosomal duplications or extrachromosomal transgenic arrays are checked for correct segregation of progeny. Mutants with temperature-sensitive alleles are periodically shifted to restrictive temperatures to confirm phenotypes. Strains carrying fluorescent reporters are checked for expression when practical, using dissecting microscopes equipped for epifluorescence. Animals are not normally mounted for examination at high magnification unless a strain is being verified in response to suspected problems such as silencing of transgene expression or inappropriate expression patterns.

Shipping

Strains are normally shipped via first class US post at no additional cost to user; shipping costs are built into the annual operating budget. The CGC will ship strains with a private courier such as FedEx or DHL when requested by the user, but the user must provide an account number or prepaid label to use when setting up the shipment. The agar-filled petri dishes on which the strains are raised and shipped are sealed with paraffin film to protect against desiccation and contamination, and the small petri dishes (35 mm diameter) are packed in padded envelopes, allowing strains to be shipped at relatively low cost (most packages weigh only a few ounces). Strains are normally sent in a starved condition because starved animals are much more tolerant of environmental stress than fed animals. Starvation makes the worms more likely to survive swings in temperature and prolonged shipping times, requiring no special shipping conditions for most orders—a key advantage for a stock center sending samples around the globe. Still, if local weather conditions are forecast to be warmer than 85°F or colder than 0°F, the CGC will hold shipments until weather is more favorable to prevent death of strains during shipping (though our comfortably cool winter temperatures are embraced by most Minnesotans, our small nematode friends typically do not survive prolonged exposure below zero).

Recovery after Shipping

Upon receiving their requested worm strains, users are encouraged to recover strains by transferring individual animals (or a chunk of agar containing one or more animals) to a new plate using a flame-sterilized scalpel or spatula (Stiernagle 2006). Even if no motile animals are seen on the surface of the agar plate, there are often starved animals that have burrowed into the agar that will crawl out when moved to a plate with bacterial food. Unfortunately, conditions encountered during shipping are sometimes too extreme for the worms to survive. In the rare event that the user is unable to recover a strain within a few days after transferring to a fresh plate, the CGC will typically send a replacement at no cost if notified in a timely manner. Similarly, if a user finds a problem with a strain received from the CGC, the CGC requests notification as soon as possible to help ensure the validity of strains for all users.

Order Processing

When strains were distributed free of charge, the CGC shipped strains in response to written or e-mailed requests from researchers. Once user fees were imposed (in 2007), the CGC developed an online ordering system to accommodate credit card payment and simplify invoicing. In 2017, the CGC moved to a custom database software package that merged key large databases (strains information, freezer locations, shipping history, and lab information) and integrated them with the ordering system, greatly increasing the efficiency of daily operations.

New Orders Are Processed Daily

Requested strains are thawed from frozen stocks or prepared from stocks already out in the lab for other orders. At any given time, roughly 15%–20% of the strains in the collection are actively growing in the lab as they are being prepared for orders, or stored on starved plates at 11°C (Figure 5.2c). A strain that is actively growing in the lab might require as little as 1–2 business days to prepare for shipping. Requested strains not already actively growing in the lab take longer because they must be thawed from frozen stocks. Aliquots stored in –80°C freezers contain a small amount of agar in the freezing medium, allowing a portion of the frozen sample to be removed with a sterile microspatula and placed on an NGM plate. Thawing survival rates of mutant strains and non-*C. elegans* species vary somewhat, but the small size of the worms makes it possible to preserve hundreds or thousands of animals in each frozen aliquot. Thawed strains are allowed to recover for 1–3 days before animals are transferred to new plates for verification and further processing.

Intellectual Property Rights

Implicit in any strain donation to the CGC is the agreement that the strain may be distributed to academic labs without constraints. The CGC does not use material transfer agreements for any strains it receives or distributes but will respect requests from labs specifying that their strains not be provided to commercial groups or used in human research as required by their funding sources or institutions. Any such restrictions are clearly noted in a strain's description in our online catalog. The CGC does not retain any intellectual property (IP) rights or otherwise regulate how strains are used by recipients after distribution. Any potential problems or conflicts over IP rights should be resolved between the user and lab in which the strain originated.

ACCESSING THE CGC COLLECTION

All strain orders must be submitted through the CGC online ordering system, and credit card payments are strongly encouraged. In cases of financial hardship, users who do not have funds available may email the director or curator to request a waiver of the fees. Waivers are considered for specific strains requested on a case-by-case basis. If the request is approved and a waiver granted, the user is provided with a one-time code to remove the fees when submitting their order through our website. Waivers are capped at a maximum of 25 strains per lab within a year.

The online searchable catalog of the CGC strain collection is available through the CGC website and ordering system (cgc.umn.edu), which are designed to enhance the user's experience while simplifying the ordering process. Changes to the strain collection are now updated in real time. These changes might include modifications to strain descriptions based on newly acquired information such as the identification of molecular breakpoints or linked background mutations, additional aspects of phenotypes, reported problems currently being examined, etc. Newly acquired strains are automatically added to special pages. The site provides users with advanced search functions in the on-line catalog and the ability for lab members to set up their own individual sub-accounts to view their personal order history, including order status and shipping dates. PIs or lab managers have administrative regulatory privileges over any user sub-accounts linked to their lab.

A NOTE ON *C. ELEGANS* STRAIN NOMENCLATURE

Caenorhabditis elegans has had a strict set of rules regulating its nomenclature since its introduction as a model system (Horvitz et al. 1979, Tuli et al. 2018). Drawing from annotations used in the bacterial, yeast, and fly communities, the founders of the worm community established a system of standardized nomenclature ensuring that worm strains and alleles are named in a manner that makes the unique identity and source of each strain unambiguous. A strain is defined as a set of individuals of a particular genotype with the capacity to produce more individuals of the same genotype. Each laboratory dedicated to long-term use of *C. elegans* is assigned a set of unique alphabetic laboratory codes by WormBase (genenames@wormbase.org) and is registered with the CGC for the naming new strains and alleles generated in that lab. A lab code is assigned to the principal investigator (PI) of each lab because they are ultimately responsible for maintaining the records of named strains and alleles in their lab and ensuring that their strains are accurately described in publications, on WormBase, and when distributed to the scientific community, CGC, or CeNDR.

A PI's lab code also serves as a unique identifier for the CGC to use for tracking the lab's account, including the lab's location and billing information, their history of strains received from the stock center, strains they have contributed to the CGC collection, and so forth. Once assigned, a code is permanently affiliated with that researcher. If a lab relocates to a different institution, the lab code moves with the PI to preserve the historical records of strains and alleles generated, as well as their order history at the CGC. Codes are not reassigned to other groups if a lab ceases to work on *C. elegans*, for example, due to retirement or a change of direction in their research. Post-docs will receive their own code when establishing their lab or research group (for many new PIs, being assigned their lab code is a rite of passage).

Labs maintaining bacterial strains used in conjunction with the worm may also assign names to these strains using their strain designation followed by "b" to distinguish bacterial stocks from nematode strains. A small collection of such bacterial strains is maintained and distributed by the CGC. In *C. elegans*, genes are given names consisting of three or four italicized letters, a hyphen, and an Arabic number, as in the example *hbl-1*. Assignment of new gene names is tightly regulated to prevent different groups from using the same name to describe more than one gene or using different names to describe the same gene, and requires approval from a senior member of the worm community who acts as the registrar of gene names. Requests for new gene names or species prefixes should be directed to genenames@wormbase.org.

Dr. Robert Herman regulated genetic nomenclature for a brief period in the late 1970s until the founding of the CGC in 1979. For the next thirteen years, Don Riddle and Mark Edgley oversaw genetic nomenclature as well as the strain curation while the CGC was housed at the University of Missouri. When CGC operations moved to the University of Minnesota in 1992, administration of gene names and nomenclature was transferred to Dr. Jonathan Hodgkin (Oxford University), who retained the role through 2013, an era spanning multiple landmark events that reshaped the worm community, notably the complete sequencing of the *C. elegans* genome in 1999, and the establishment of WormBase in 2000. In 2013, Dr. Hodgkin handed responsibilities to Dr. Tim Schedl (Washington University), who currently gives final approval of new gene names.

The CGC cooperates extensively with WormBase on matters of genetic nomenclature and information about strains in our collection. The CGC website interfaces with WormBase regularly to keep databases up to date, track changes in gene names, curate molecular data and descriptions of phenotypes associated with alleles and transgenes, and keep strain lists synchronized. Changes in reported genotypes are tracked and updated to ensure strain information is accurate and consistent between the two sites. This cross-talk between sites also provides users with alternate means of locating strains needed for their work, allowing WormBase users to quickly see which strains of interest are available to order from the CGC (Figure 5.6a), and CGC users to more efficiently search our catalog by having our search engine use synonymous gene names (such as the sequence ID or alternative gene names). Furthermore, each strain entry in the CGC catalog (Figure 5.6b) provides a link to that strain and

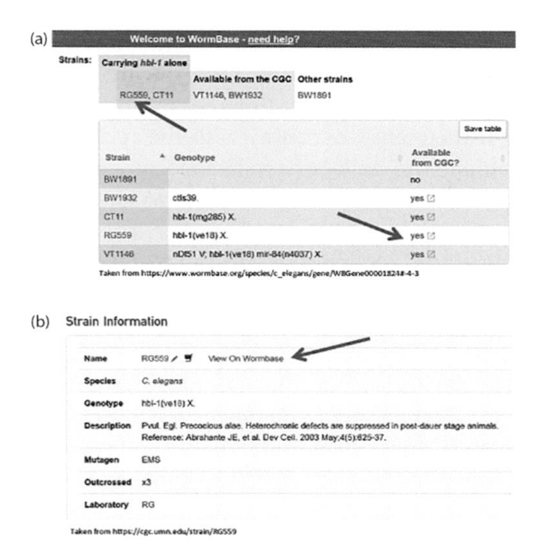

FIGURE 5.6 Crosstalk between the Caenorhabditis Genetics Center website and WormBase. The Caenorhabditis Genetics Center works closely with WormBase to ensure that strain information is kept current and synchronized between the databases. (a) Selecting the Genetics tab from the *hbl-1* gene page reveals, in part, this typical WormBase page view listing curated strains containing mutant alleles of *hbl-1*. At a glance, users learn which strains are available through the Caenorhabditis Genetics Center and can identify which strains are *hbl-1* single mutants. The table below links directly to Caenorhabditis Genetics Center strain pages, as shown in (b) for *hbl-1(ve18)*, and the user can quickly add the strain to their order. The Caenorhabditis Genetics Center strain information pages provide essential information about strains and also link to WormBase, allowing users to access additional information.

the alleles it carries on WormBase, allowing users to quickly access whatever additional information might be available beyond what is provided in the CGC description, including additional publication references.

SERVING THE WORLD COMMUNITY

Approximately 6000 labs are registered users of the CGC. The overwhelming majority of these groups are academic research labs, though *C. elegans* is growing in popularity for instructional purposes in teaching colleges and high schools. Only a small percentage of registered users are commercial (for-profit) research groups. More than half of all the labs registered with the CGC are located within the US. However, there is steady growth in the number of new labs ordering from the CGC both from within the US and abroad. Historically, most of the strains shipped to labs outside of the US were sent to members of the worm community in the United Kingdom and Europe, but the CGC has seen a substantial increase in the number of requests from academic research labs in Asia, especially within China, in recent years.

The CGC has sent strains to researchers in 74 countries, some of which present special challenges due to extreme climate conditions or import regulations. Recipients residing in extremely hot or cold climates will sometimes make special shipping accommodations to ensure the requested strains arrive in good condition. The CGC will accommodate such requests when feasible, for example, arranging temperature-controlled shipping (at recipient's expense) with the recipient's preferred courier, or having the order shipped to collaborator located in a more temperate region of the destination country. Additionally, some users who have experienced difficulties in clearing their shipments through customs prefer to arrange shipment through particular freight-forwarding agencies or customs brokers contracted with their university to aid in the process, but the user is responsible for making any such arrangements and dealing with any problems that might arise during transit.

International delivery is generally reliable, but can be slow, with customs clearance often being the largest hurdle. Despite the non-hazardous and non-infectious nature of *C. elegans*, many countries have specific regulations for the import of live specimens. Carrier rules and import regulations are prone to changes and can vary between destination countries (and sometimes even at different ports of entry within the same country), so in some situations a private courier will typically be more successful than US post when shipping to some countries and vice versa. Through experience, CGC staff members have learned to avoid pitfalls in the shipping process that are likely to trigger red flags at certain destinations and give each shipment its best chance of arriving in a timely manner. Regardless of which shipping method is used, the CGC will declare the full dollar amount of order and contents of the shipment are accurately described as required by each courier. If additional permits, special shipping forms, or declaration letters are required for importing strains into the destination country, the recipient is responsible for providing any such documentation.

ENHANCING CGC OPERATIONS AND AVAILABLE RESOURCES

Though the primary focus of the CGC is the collection and distribution of strains, the CGC funding was switched from a government contract to a grant in 2012, adding a small research component and aligning the CGC with other NIH-supported biological resource centers. As part of our research component, CGC staff members are actively developing new strains and molecular tools that will benefit researchers throughout the community. For example, until recently, the CGC collection of available balancer chromosomes lacked complete coverage. CGC staff have used CRISPR/Cas9 engineering to insert fluorescent markers into some of the unlabeled inversions from the NBRP, as well as many of the preexisting classical balancers. This provides users with new genetic tools to quickly and easily distinguish homozygous mutants from heterozygous animals in a population, even in relatively early stages of development.

THE *CAENORHABDITIS ELEGANS* NATURAL DIVERSITY RESOURCE (CeNDR)

Technological breakthroughs in whole-genome sequencing have led to newfound appreciation for understanding how genetic variation can affect biological processes. The identification, preservation, and distribution of wild isolates (isolated from nature) are critical for research in comparative genomics, evolutionary genetics, statistical genetics, and other fields utilizing whole-genome analyses. Although a few wild isolates are maintained in the CGC collection, detailed curation of the stocks and their genomic information is beyond the scope of CGC operations. To meet the needs of a growing number of investigators interested in incorporating studies of natural variation across wild populations into their work, Dr. Erik Andersen (Northwestern University) spearheaded the establishment of the *Caenorhabditis elegans* Natural Diversity Resource. The CeNDR (Cook et al. 2017) is a resource center that complements the role of the CGC by specializing in the curation, maintenance, and distribution of wild isolates of *C. elegans* and their whole-genome sequences. By focusing on natural isolates, CeNDR is able to maintain these stocks in ways that avoid genetic drift or behavioral changes in response to laboratory conditions, thereby easing the burden on the CGC and ensuring that these key resources are available.

CeNDR provides a specialized platform for collecting, distributing, and maintaining strains isolated from nature. Currently, CeNDR contains 766 wild strains that make up 330 distinct isotypes (genome-wide haplotypes that are shared across strains from the same isolation location). *C. elegans* researchers or citizen scientists can donate wild strains to CeNDR through a web form. Following the receipt of new strains, a single hermaphrodite animal is propagated to ensure that the genotype is genetically distinct from a potentially heterogeneous wild population.

Goals

The overwhelming success of *C. elegans* as a model system results from the use of strain N2 as the standard wild-type strain throughout the community. This standardized genetic background has enabled fine-scale analyses of experimental perturbations without concerns for the occurrence of divergent genetic backgrounds between labs. However, because differences between populations are caused by allelic variation, the *C. elegans* community has many more discoveries yet to be made by studying the natural diversity present within this species (Braendle and Felix 2009, Felix and Braendle 2010, Frezal and Felix 2015, Nigon and Felix 2017, Sterken et al. 2015). To address this gap in the community's experimental toolkit, a large population of wild strains has been collected from nature by *C. elegans* researchers and citizen scientists worldwide (Figure 5.7) (Andersen et al. 2012, Cook et al. 2016). These strains provide a reservoir of natural genetic variation that can be used to understand the genetic drivers of evolutionary processes and identify the underlying molecular mechanisms for traits relevant to genetic differences among individuals in a population. CeNDR was created to maintain a comprehensive database on this collection of strains and provide a set of tools for examining natural variation in *C. elegans* wild strains to perform genome-wide association (GWA) mappings (Cook et al. 2017, Zdraljevic et al. 2017). Since the fall of 2016, CeNDR (www.elegansvariation.org) has had over 45,000 unique visitors investigating one of the three main areas of focus. This overwhelmingly positive response and growth of CeNDR has moved studies of *C. elegans* natural variation to many laboratories across the community.

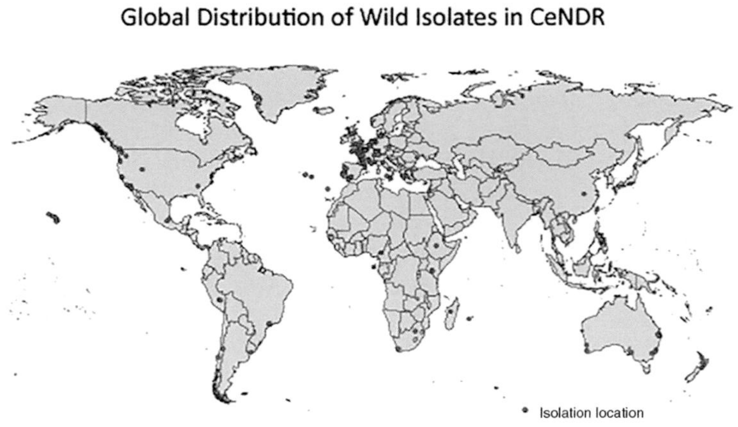

Global Distribution of Wild Isolates in CeNDR

• Isolation location

FIGURE 5.7 Global origins of wild *C. elegans* isolates available in the *Caenorhabditis elegans* Natural Diversity Resource. Partial world map indicates the origins of some of the 766 wild isolates of *C. elegans* available through *Caenorhabditis elegans* Natural Diversity Resource. The clickable map (available through the CeNDR homepage or at https://www.elegansvariation.org/strain/global-strain-map/) allows users to quickly access more information about specific strains isolated from regions of geographical interest.

Future Acquisitions

Over the next few years, CeNDR expects to incorporate over 1000 wild *C. elegans* strains and associated whole-genome sequence data. These strains come from many different locations across the globe, but collections from the Pacific Rim are encouraged, as this region harbors the highest genetic diversity in the species (Andersen et al. 2012, Cook et al. 2016, Zdraljevic et al. 2017). GWA mappings will be enhanced by the addition of two new features. CeNDR will update the genetic markers used for GWA mappings to better define genomic regions and make better predictions of haplotype structures. Most importantly, CeNDR hopes to expand to wild strains from two related androdioecious *Caenorhabditis* species, *C. briggsae* and *C. tropicalis*. Analyses of these two species will enable powerful comparative approaches of natural variation across the genus.

Distribution

CeNDR provides strains to the community in three different ways. First, individual strains are sent on agar NGM plates when individual strains are requested. Second, a collection of 12 divergent wild strains are sent for users to optimize quantitative trait assays. These strains are sent cryopreserved in strips of 12-well tubes. Third, six unique collections of 48 wild strains are sent for users to perform species-wide quantitative trait assays. Again, these strains are sent cryopreserved but in 48-well formats. Over 2000 wild strains have been sent to the community since CeNDR was announced in the fall of 2016.

In Support of Genetics and Genomics

CeNDR offers whole-genome sequence and variant data for all isotype reference strains, along with metadata on gene conservation and functional studies. Because most reproduction in *C. elegans* occurs through self-fertilization by hermaphrodites, most natural substrates are colonized by genetically clonal individuals. Sometimes, genetically distinct strains are found in the same isolation location. Therefore, the concordance of genetic variation among strains is examined and whole-genome sequence data for identical or nearly identical strains is combined into isotypes, which represent genetically distinct genome-wide haplotypes from the same isolation location. The strain set for future experiments comprises a single representative strain (called the isotype reference strain) from each isotype set. By combining sequence coverage of all strains within an isotype, CeNDR obtains high-coverage sequence data that are aligned and used to perform variant calling. CeNDR sequences all wild strains to a minimum of 30-fold depth of coverage, a level that facilitates reliable calling of single-nucleotide variants (SNVs) along with other classes of genomic variants. All variant data are available through an API or can be downloaded in tab-delimited format or Variant Call Format (VCF) files, and aligned sequence data is available in binary alignment map (BAM) format. For users familiar with genomic applications, these file types and data enable individual studies beyond just looking at single-nucleotide variation. Additionally, CeNDR has developed a genome browser for querying and visualizing genetic variation across the *C. elegans* species, allowing users to toggle different tracks that detail genomic information including genes, SNVs identified in individual strains, variant effects predicted with SnpEff (Cingolani et al. 2012), raw sequence reads from the BAM files, and conservation scores across nematode species (e.g., phyloP and phastCons). This functionality gives users the ability to query their favorite gene and discover whether natural variation might affect its function, suggesting future phenotypic studies if these alleles have interesting mutational profiles (gain-of-function or reduction-of-function). Importantly, these studies set the context for what *C. elegans* does in nature. The laboratory environment is drastically different than the nature niche, and the roles of genes and pathways might be very different than the standard laboratory-adapted N2 genetic background.

CeNDR's GWA mapping process is optimized for *C. elegans* and combines whole-genome genotype data with measurements of quantitative traits to perform association mappings, an approach that has been used successfully in many studies (Cook et al. 2016, Zdraljevic et al. 2017). So far, approximately 3100 unique GWA mappings have been performed on CeNDR. The GWA mapping portal is designed for non-experts and has several user-defined options along with drag-and-drop capabilities. Multiple traits can be submitted simultaneously and organized within an easy-to-read report. Within these reports, users are presented with figures, tables, interactive elements, and data in a tab-delimited format. Once a quantitative trait locus from GWA mapping is identified, CeNDR provides tools to browse the genes and potential functional connections underlying that genomic region, including integrated functional studies based on RNA interference (RNAi) screens and biochemical pathway predictions. Lastly, system features enable CeNDR to interact with other services (Wormbase and the CGC) and allow access to the underlying databases through an API, which can be used to query, among other things, genetic variants, strain information, mapping report data, and *C. elegans* genes and homologs.

Documentation

The CeNDR collects information on each strain, including its isolation location, date of collection, the substrate where the nematodes were found, elevation, etc. These data are integrated into the CeNDR database and can be browsed via a geographic interface on the website. This dataset is also available for download or accessible through an application programming interface (API). After isolation and propagation of the strains, the population is split to cryopreserve animals for long-term storage and to isolate DNA for whole-genome sequencing. This step ensures that the genotype information obtained from whole-genome sequencing can be connected directly back to a specific strain. Sample mix-ups and strain contamination are possible when managing many strains and samples, especially when most wild strains look phenotypically identical. However, CeNDR's ability to retain frozen stocks allows verification of the genetic identity of strains should the need arise. CeNDR checks all frozen samples by low-coverage whole-genome sequencing to ensure that genotypes are as reported. If strain fidelity issues are discovered, as has been observed with other wild strains in the past (Andersen et al. 2012, McGrath et al. 2009), stains are recovered from long-term cryopreserved (LN) stocks. Data checks and cryopreservation improve data fidelity for downstream genome-wide association (GWA) mappings. Passaging of strains drives evolution and adaptation to laboratory conditions, so CeNDR strains are never passaged more than three times in a given year to reduce these effects.

A LOOK AHEAD

The future of the worm community is bright. *C. elegans* continues to grow in popularity as a model organism, and with users registering an average of six new CGC lab accounts in each week over the last year, shows no signs of slowing down. Many new users are entering the field without coming from experience in an established worm lab and are instead using *C. elegans* as a complement to the other organisms that are the primary system in their laboratory. The use of *C. elegans* in high school and college biology labs is introducing new generations of scientists to the worm. The worm's versatility will only continue to grow as new technologies are adapted for use in this efficient, easy to use, and cost-effective model system. The simplicity, speed, and relative ease of working with worms continue to provide advantages as new technologies are applied to worm models. The compact genome and worldwide distribution of worms in the wild makes Caenorhabditis ideally suited for comparative genomics studies, which will only gain traction as community resources such as CeNDR grow.

The collaborative spirit that has invariably been a hallmark of the worm community endures in new generations of researchers openly sharing exciting new reagents with their colleagues around the globe, and it is just as important today, as it was 40 years ago, that these shared reagents continue to be made available to researchers worldwide. Although technology is making the world a smaller place by making it easy for scientists in different countries to communicate and share information, customs and international shipping regulations are continually becoming more restrictive, threating the exchange of reagents between these same researchers. Resource centers such as the CGC and CeNDR provide vital services for the community, not only by maintaining stocks and strain information, but also by using their experience and knowledge to ensure key strains will be successfully delivered to the researchers who rely upon them for their research.

CONCLUSIONS

After 40 years, the CGC continues to play a unique role in promoting research on the nematode *C. elegans* by collecting strains from research labs and distributing them within the worldwide community of Caenorhabditis researchers. The CGC and CeNDR continuously work to meet the rising demands of the community while maintaining high standards with small but efficient staffs.

ACKNOWLEDGMENTS

The Caenorhabditis Genetics Center (CGC) is supported, in part, by a grant from the National Institutes of Health—Office of Research Infrastructure Programs (P40 OD010440). CeNDR is funded, in part, by a Weinberg College of Arts and Sciences Innovation Grant. Thanks to Dr. Robert Herman and Mark Edgley for providing personal insight into the history of the CGC. Fluorescent confocal microscopy for Figure 5.1b was performed in the University of Minnesota—University Imaging Centers (http://uic.umn.edu) with the assistance of Dr. Guillermo Marques.

REFERENCES

Andersen, E.C., J.P. Gerke, J.A. Shapiro et al. 2012. Chromosome-scale selective sweeps shape *Caenorhabditis elegans* genomic diversity. *Nature Genetics* 44:285–290.

Antebi, A., J.G. Culotti and E.M. Hedgecock. 1998. daf-12 regulates developmental age and the dauer alternative in *Caenorhabditis elegans*. *Development* 125:1191–1205.

Au, V., E. Li-Leger, G. Raymant et al. 2018. Optimizing guide RNA selection and CRISPR/Cas9 methodology for efficient generation of deletions in *C. elegans*. *BioRxiv*. doi:10.1101/359588.

Bi, Y., X. Ren, C. Yan, J. Shao, D. Xie and Z. Zhao. 2015. A genome-wide hybrid incompatibility landscape between *Caenorhabditis briggsae* and *C. nigoni*. *PLoS Genetics* 11:e1004993.

Braendle, C. and M.A. Felix. 2009. The other side of phenotypic plasticity: A developmental system that generates an invariant phenotype despite environmental variation. *Journal of Biosciences* 34:543–551.

Brenner, S. 1974. The genetics of *Caenorhabditis elegans*. *Genetics* 77:71–94.

Brenner, S. 1988. Forward. In *The Nematode Caenorhabditis Elegans*, B.W. Wood (Ed.). Cold Spring Harbor, New York: Cold Spring Harbor Laboratory.

Cassada, R.C. and R.L. Russell. 1975. The dauerlarva, a post-embryonic developmental variant of the nematode *Caenorhabditis elegans*. *Developmental Biology* 46:326–342.

Cingolani, P., A. Platts, L. Wang et al. 2012. A program for annotating and predicting the effects of single nucleotide polymorphisms, SnpEff: SNPs in the genome of *Drosophila melanogaster* strain w1118; iso-2; iso-3. *Fly* (Austin) 6:80–92.

Cook, D.E., S. Zdraljevic, R.E. Tanny et al. 2016. The genetic basis of natural variation in *Caenorhabditis elegans* telomere length. *Genetics* 204:371–383.

Cook, D.E., S. Zdraljevic, J.P. Roberts and E.C. Andersen. 2017. CeNDR, the *Caenorhabditis elegans* natural diversity resource. *Nucleic Acids Research* 45:D650–D657.

Dejima, K., S. Hori, S. Iwata et al. 2018. An aneuploidy-free and structurally defined balancer chromosome toolkit for *Caenorhabditis elegans*. *Cell Reports* 22:232–241.

Dickinson, D.J. and B. Goldstein 2016. CRISPR-based methods for *Caenorhabditis elegans* genome engineering. *Genetics* 202:885–901.

Dutta, P., C. Lehmann, D. Odedra, S. Singh and C. Pohl. 2015. Tracking and quantifying developmental processes in *C. elegans* using open-source tools. *Journal of Visualized Experiments* 106:e53469. doi:10.3791/53469.

Ewald, C.Y. and C. Li. 2010. Understanding the molecular basis of Alzheimer's disease using a *Caenorhabditis elegans* model system. *Brain Structure and Function* 214:263–283.

Félix, M.A. 2008. RNA interference in nematodes and the chance that favored Sydney Brenner. *Journal of Biology* 7:34. doi:10.1186/jbiol97.

Félix, M.A. and C. Braendle. 2010. The natural history of *Caenorhabditis elegans*. *Current Biology* 20:R965–R969.

Fire, A., S. Xu, M.K. Montgomery, S.A. Kostas, S.E. Driver and C.C. Mello. 1998. Potent and specific genetic interference by double-stranded RNA in *Caenorhabditis elegans*. *Nature* 391:806–811.

Frézal, L. and M.A. Félix. 2015. The natural history of model organisms: *C. elegans* outside. *Elife* 4. doi:10.7554/eLife.05849.

Frøkjær-Jensen, C., M.W. Davis, M. Sarov. et al. 2014. Random and targeted transgene insertion in *Caenorhabditis elegans* using a modified Mos1 transposon. *Nature Methods* 11:529–534.

Golden, J.W. and D.L. Riddle. 1984. A pheromone-induced developmental switch in *Caenorhabditis elegans*: Temperature-sensitive mutants reveal a wild-type temperature-dependent process. *Proceedings of the National Academy of Sciences* (USA) 81:819–823.

Haag, E.S. 2005. The evolution of nematode sex determination: *C. elegans* as a reference point for comparative biology. In *WormBook*, B.J. Meyer (Ed.), pp. 2–9. The *C. elegans* Research Community, doi:10.1895/wormbook.1.120.1.

Hillier, L.W., A. Coulson, J.I. Murray, Z. Bao, J.E. Sulston and R.H. Waterston. 2005. Genomics in *C. elegans*: So many genes, such a little worm. *Genome Research* 15:1651–1660.

Hodgkin, J. 1987. Primary sex determination in the nematode *C. elegans*. *Development* 101 (Supplement):5–16.

Horvitz, H.R., S. Brenner, J. Hodgkin and R.K. Herman. 1979. A uniform genetic nomenclature for the nematode *Caenorhabditis elegans*. *Molecular and General Genetics* 175:129–133.

Horvitz, H.R. 1999. Genetic control of programmed cell death in the nematode *Caenorhabditis elegans*. *Cancer Research* 59:1701s–1706s.

Hunt-Newbury, R., R. Viveiros, R. Johnsen et al. 2007. High-throughput in vivo analysis of gene expression in *Caenorhabditis elegans*. *PLoS Biology* 5:e237. doi:10.1371/journal.pbio.0050237.

Johnson, J.R., R.C. Jenn, J.W. Barclay, R.D. Burgoyne and A. Morgan. 2010. *Caenorhabditis elegans*: A useful tool to decipher neurodegenerative pathways. *Biochemical Society Transactions* 38:559–563.

Kaminsky, R., P. Ducray, M. Jung et al. 2008. A new class of anthelmintics effective against drug-resistant nematodes. *Nature* 452:176–180.

Keil W, L.M. Kutscher, S. Shaham and E.D. Siggia. 2017. Long-term high-resolution imaging of developing *C. elegans* larvae with microfluidics. *Developmental Cell* 40:202–214.

Kimble, J. and D. Hirsh. 1979. The postembryonic cell lineages of the hermaphrodite and male gonads in *Caenorhabditis elegans*. *Developmental Biology* 70:396–417.

Kiontki, K, M.A. Félix, M. Ailion et al. 2011. A phylogeny and molecular barcodes for *Caenorhabditis*, with numerous new species from rotting fruits. *BMC Evolutionary Biology*. 11:339. doi:10.1186/1471-2148-11-339.

Lee, R.C., R.L. Feinbaum and V. Ambros. 1993. The *C. elegans* heterochronic gene lin-4 encodes small RNAs with antisense complementarity to lin-14. *Cell* 75:843–854.

Lee, R.Y.N., K.L. Howe, T.W. Harris et al. 2018. WormBase 2017: Molting into a new stage. *Nucleic Acids Research* 46:D869–D874.

McGrath, P.T., M.V. Rockman, M. Zimmer et al. 2009. Quantitative mapping of a digenic behavioral trait implicates globin variation in *C. elegans* sensory behaviors. *Neuron* 61:692–699.

Murray, J.I, Z. Bao, T.J. Boyle et al. 2008. Automated analysis of embryonic gene expression with cellular resolution in *C. elegans*. *Nature Methods* 5:703–709.

Nigon, V.M. and M.A. Félix. 2017. History of research on *C. elegans* and other free-living nematodes as model organisms. In *WormBook*, The *C. elegans* Research Community (Ed.), WormBook, doi:10.1895/wormbook.1.181.1.

Pulak, R. 2006. Techniques for analysis, sorting, and dispensing of *C. elegans* on the COPAS flow-sorting system. *Methods in Molecular Biology* 351:275–286.

Sarov, M. S. Schneider, A. Pozniakovski et al. 2006. A recombineering pipeline for functional genomics applied to *Caenorhabditis elegans*. *Nature Methods* 10:839–844.

Sterken, M.G., L.B. Snoek, J.E. Kammenga and E.C. Andersen. 2015. The laboratory domestication of *Caenorhabditis elegans*. *Trends in Genetics* 31:224–231.

Stiernagle, T. 2006. Maintenance of *C. elegans*. In *WormBook*, The *C. elegans* Research Community (Ed.), WormBook, doi:10.1895/wormbook.1.101.1.

Sulston, J.E. and H.R. Horvitz. 1977. Post-embryonic cell lineages of the nematode, *Caenorhabditis elegans*. *Developmental Biology* 56:110–156.

Sulston, J.E., E. Schierenberg, J.G. White and J.N. Thomson. 1983. The embryonic cell lineage of the nematode *Caenorhabditis elegans*. *Developmental Biology* 100:64–119.

Tabara, H., A. Grishok and C.C. Mello. 1998. RNAi in *C. elegans*: Soaking in the genome sequence. *Science* 282:430–431.

The *C. elegans* Sequencing Consortium. 1998. Genome sequence of the nematode *C. elegans*: A platform for investigating biology. *Science* 282:2012–2018.

The *C. elegans* Deletion Mutant Consortium. 2012. Large-scale screening for targeted knockouts in the *Caenorhabditis elegans* genome. *G3: Genes, Genomes, Genetics* (Bethesda). 2:1415–1425.

Thompson, O., M. Edgley, P. Strasbourger et al. 2013. The million mutation project: A new approach to genetics in *Caenorhabditis elegans*. *Genome Research* 23:1749–1762.

Timmons, L. and A. Fire. 1998. Specific interference by ingested dsRNA. *Nature* 395:854. doi:10.1038/27579.

Timmons, L., D.L. Court and A. Fire. 2001. Ingestion of bacterially expressed dsRNAs can produce specific and potent genetic interference in *Caenorhabditis elegans*. *Gene* 263:103–112.

Tuli, M.A., A.L. Daul and T. Schedl. 2018. *Caenorhabditis* nomenclature. In *WormBook*, The *C. elegans* Research Community (Ed.). WormBook Early Online doi:10.1895/wormbook.1.183.1.

Wang H., J. Liu, S. Gharib et al. 2017. cGAL, a temperature-robust GAL4-UAS system for *Caenorhabditis elegans*. *Nature Methods* 14:145–148.

Wang, H., H. Park, J. Liu and P.W. Sternberg. 2018. An efficient genome editing strategy to generate putative null mutants in *Caenorhabditis elegans* using CRISPR/Cas9. *G3: Genes, Genomes, Genetics* (Bethesda).

Waterman, C., R.A. Smith, L. Pontiggia and A. DerMarderosian. 2010. Anthelmintic screening of Sub-Saharan African plants used in traditional medicine. *Journal of Ethnopharmacology* 127:755–759.

Wood, W. 1988. *The Nematode* Caenorhabditis elegans. Cold Spring Harbor, New York: Cold Spring Harbor Laboratory.

Yoshimura, J., K. Ichikawa, M. Shoura et al. 2019. Recompleting the *C. elegans* genome.

Zacharias, A.L. and J.I. Murray. 2016. Combinatorial decoding of the invariant *C. elegans* embryonic lineage in space and time. *Genesis* 54:182–197.

Zdraljevic, S., C. Strand, H.S. Seidel, D.E. Cook, J.G. Doench and E.C. Andersen. 2017. Natural variation in a single amino acid substitution underlies physiological responses to topoisomerase II poisons. *PLoS Genetics* 13:e1006891. doi:10.1371/journal.pgen.1006891.

Zhang, L., J.D. Ward, Z. Cheng and A.F. Dernburg. 2015. The auxin-inducible degradation (AID) system enables versatile conditional protein depletion in *C. elegans*. *Development* 142:4374–4384.

Zhong, M., W. Niu, Z.J. Lu et al. 2010. Genome-wide identification of binding sites defines distinct functions for *Caenorhabditis elegans* PHA-4/FOXA in development and environmental response. *PLoS Genetics* 6(2):e1000848. doi:10.1371/journal.pgen.1000848.

6 The *Chlamydomonas* Resource Center

Paul A. Lefebvre, Matthew Laudon and Carolyn Silflow

CONTENTS

Abstract: The unicellular green alga *Chlamydomonas reinhardtii* serves as a unique model system for studies of many aspects of cell biology. To facilitate the use of *Chlamydomonas*, the Chlamydomonas Resource Center (CRC) was founded in 1979 to provide wild-type and mutant strains, plasmids, and related items and services to the user community. The core collection currently holds >4000 mutant strains in addition to wild-type field isolates. The CRC also holds numerous other collections of *Chlamydomonas* provided by research programs utilizing this organism. The CRC offers educational kits, complete with reagents and instructions, that enable students to grow and perform experiments with *Chlamydomonas* in the classroom and at home. This chapter describes the origin of the CRC, its current holdings, available resources, and its operational policies and procedures.

INTRODUCTION

The unicellular green alga *Chlamydomonas reinhardtii* serves as a unique model system for studies of photosynthesis, flagellar biogenesis and function, signal transduction, and other aspects of cell biology. *Chlamydomonas* allows researchers to apply the powerful genetic techniques available in a haploid eukaryote, such as tetrad analysis, to an organism with chloroplasts and flagella. Unique among algal systems, *Chlamydomonas* has a large collection of mutants available and a complete genetic map linked to a molecular map. To facilitate research using this powerful model system, and to assist students and educators in incorporating *Chlamydomonas* into laboratory experiments, the Chlamydomonas Resource Center (CRC), originally known as the Chlamydomonas Genetics Center (CGC) was founded in 1979 to provide wild-type and mutant strains, molecular reagents for research, teaching kits, and sources of information for *Chlamydomonas* researchers around the world.

WHY *CHLAMYDOMONAS*?

With the advent of rapid genomic sequencing methods, and more recently with the ability to manipulate genes using clustered regularly interspaced short palindromic repeats (CRISPR) and other gene targeting approaches, many organisms can make some claim to being "genetic systems." However, it is still clear that when many researchers focus on a few key model systems, discoveries can be made more rapidly. Membership in the elite group of "model systems" is fluid and sometimes controversial. *Chlamydomonas* stakes its claim to membership in this group based on two unique features of its lifestyle.

First, *Chlamydomonas reinhardtii* is a photoheterotroph, meaning that it can grow using either photosynthesis or using an exogeneous carbon source (in this case acetate) for growth. As a result, *Chlamydomonas* mutants defective in any aspect of photosynthesis are auxotrophic mutants rather than being lethal. The largest number of named mutants in the *Chlamydomonas* database carry the name "ac" for acetate-requiring. Although *Chlamydomonas* in the context of this chapter (and the CRC) refers to *Chlamydomonas reinhardtii*, some of the earliest mutants in the genus *Chlamydomonas* were isolated from the distantly related species *Chlamydomonas moewusii*. However, *C. moewusii* is an obligate phototroph, having never acquired the ability to grow on an exogenous carbon source. As a result, after a promising start, *C. moewusii* fell out of favor as a model system for genetic studies.

Second, *Chlamydomonas* assembles a pair of eukaryotic flagella. The two flagella of *Chlamydomonas* reside in a peninsular position on the cell, and they can be isolated in large quantities of very high purity without damaging the cell. Cells readily regrow their flagella after amputation. Cell growth can be readily synchronized by carefully controlling illumination conditions, supporting studies of circadian rhythms and the cell cycle (Harris et al. 2009). Flagella are essential for swimming. The ability to swim is required in order to accomplish the two key functions of mating and the phototactic response to light. Cells of *Chlamydomonas* grow and divide well with missing or defective flagella (a term used interchangeably with "cilia"). Hence, it is easy to screen for such mutants by allowing those cells to sink in a tube of liquid medium. This has allowed for dozens of mutants to be isolated and different genes required for flagellar function to be identified (Tam and Lefebvre 1995). An example of the variety of flagellar mutant phenotypes that have been important in elucidating flagellar function is shown in Figure 6.1.

Wild-type cells have flagella that are of equal length, approximately 12 micrometers, while many different mutants have been described that have long flagella, short flagella, and flagella of unequal length. Eukaryotic flagella are remarkably conserved in structure and function and, as a result, studies of *Chlamydomonas* motility have made major contributions to our understanding of human diseases associated with ciliary defects, or "ciliopathies" (Adams 2008, Hildebrandt et al. 2011, Reiter and Leroux 2017). For example, an unexpected finding of sequence similarity between the gene required for normal kidney function in mammals and a gene required for flagellar assembly in *Chlamydomonas* led to uncovering the key role cilia play in kidney development (Pazour et al. 2000). Temperature-sensitive assembly mutants of *Chlamydomonas* have also been isolated (Adams et al. 1982).

In addition to studies of photosynthesis and flagellar function, *Chlamydomonas* has made important contributions to other research areas including cell cycle and circadian rhythm, cell-to-cell interactions, cell signaling, metal ion transport and

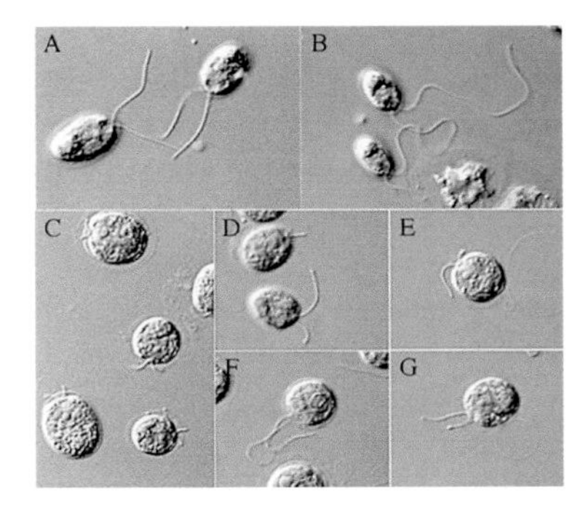

FIGURE 6.1 Nomarski differential interference microscopy of wild-type (A) and mutant (B through G) *Chlamydomonas reinhardtii* cells. The mutants in panels B through G have flagella in which the control of the flagellar assembly is defective.

utilization, cell metabolism, and others. Among the most important contributions of this haploid green alga to metazoan biology was the discovery and characterization of light-sensitive ion-selective channels known as channelrhodopsins (e.g., Nagel et al. 2005). It was established that the channelrhodopsins *ChlR1* and *ChlR2* could be used to depolarize many different types of cells, including neurons, in response to light. As a result, these genes have been used in hundreds of studies on the regulation of neuronal function in vivo in mammalian systems and have given rise to an entirely new field of neurobiology that has come to be known as optogenetics (reviewed in Wietek and Prigge 2016).

Technical Advantages

Chlamydomonas research is facilitated by the toolkit of powerful genetic, molecular, biochemical, and cell biological techniques available for use with this organism. For example, although tetrad analysis has certainly seen its most elegant applications in fungal systems, it has been used to enormous advantage in *Chlamydomonas* genetic studies. The ability to physically isolate and cultivate all four progeny of a single meiotic event enables rapid mapping of genetic markers and centromeres in ways not readily achieved in other systems. *Chlamydomonas* is a haploid organism and so phenotypes of individual recessive mutants are readily scored without having to use crosses to generate homozygosity.

Numerous attempts at developing gene disruption methodologies using homologous recombination after transformation have met with limited success. However, transforming DNA readily integrates into the *Chlamydomonas* genome by non-homologous integration. This has allowed the isolation of genes of a desired phenotype by insertional mutagenesis (Tam and Lefebvre 1993). In a large-scale mutagenesis and sequencing project funded by the NSF, Martin Jonikas and colleagues (Princeton University) have tagged a majority of the genes in the genome (Li et al. 2016) and have made these mutants available through the CRC.

Progress in *Chlamydomonas* Genomics

Since the completion of the *Chlamydomonas* nuclear genome sequence by the Joint Genome Institute (JGI) of the Department of Energy (DOE; Blaby et al. 2014, Merchant et al. 2007), the *Chlamydomonas* research community has continued to add scientific value to the sequence by refining gene model prediction and annotating genes with functional information. This information is made available to the community through the JGI *Chlamydomonas* site, version 5.5 (https://phytozome.jgi.doe.gov/pz/portal.html #!info?alias=Org_Creinhardtii), as well as through a newly developed *Chlamydomonas* genome site "ChlamyCyc" (https://www.plantcyc.org/databases/chlamycyc/8.0) supported by the NSF and the DOE as part of the Plant Metabolic Network at the Department of Plant Biology, Carnegie Institution for Science.

The completion of the genome sequence *Chlamydomonas reinhardtii* has led to new avenues for exploring the function of *Chlamydomonas* genes. For example, in silico subtraction experiments have been used to identify genes involved in the assembly of flagella and basal bodies. With the completion of the genome sequence it has become possible to identify and compare genes that are present in organisms that have cilia and flagella (*Chlamydomonas*, mice) and those that don't (*Arabidopsis*). Using this approach, Dutcher and colleagues were able to identify 688 genes that were present in the genomes of organisms that have basal bodies and flagella but missing from the *Arabidopsis* genome (Li et al. 2004). Among these genes were most of the known flagellar proteins, suggesting that this approach may facilitate the rapid identification of genes involved in the assembly and function of basal bodies and cilia.

Chlamydomonas for Bioenergy and Protein Production

The burgeoning interest in algae as potential sources for bioenergy through the production of biodiesel and hydrogen has led to an explosion of interest in *Chlamydomonas* as the premier algal genetic system. The CRC provides wild-type and mutant *Chlamydomonas* strains to industrial and academic users to facilitate bioenergy research (Table 6.1). Of potential interest to applied researchers are the various wild-type strains in the collection originally obtained as field isolates.

Chlamydomonas has been shown to produce substantial quantities of hydrogen when placed in anoxic conditions, and protocols for the continuous production of hydrogen for extended periods have been described (Melis et al. 2007). A complex network of fermentation pathways has been revealed in studies with wild-type cells and a hydrogenase-deficient mutant strain (Dubini et al. 2009). Of interest among the useful strains in the CRC collection are mutants with reduced chlorophyll antenna size. The reduced antenna size allows light to penetrate dense cultures to a significantly

TABLE 6.1
Source of Requests, Number of Orders Received/Items Shipped per Year (2013–2017) by the Chlamydomonas Resource Center

	Year				
Source	**2013**	**2014**	**2015**	**2016**	**2017**
US Academic[a]	331/900	267/674	276/640	361/1002	336/1027
Foreign Academic[a]	170/568	168/722	171/535	279/1310	232/931
US research[b]	20/53	13/21	24/55	25/81	17/31
Foreign research[b]	65/250	52/204	47/209	57/325	46/229
US government	14/28	10/42	9/17	14/42	9/63
Foreign government	12/54	12/71	23/96	28/122	30/125
US industry	11/11	11/22	14/29	4/5	8/10
Foreign industry	13/19	9/36	13/33	10/17	3/4
Miscellaneous[c]	41/45	23/25	34/20	27/23	26/10

Note: Includes plasmids and cultures only.

[a] Universities and undergraduate colleges.

[b] Public and private research institutes.

[c] Includes requests from high schools, for personal use, and requests from other *Chlamydomonas* collections.

greater depth and increases the efficiency of light harvesting (Polleet al. 2003). It is interesting to note that the most popular of the CRC's kits for teaching is the hydrogen production kit. This kit allows high school and college students to grow *Chlamydomonas* and produce hydrogen in response to anaerobic conditions without requiring sophisticated laboratory equipment or environmental conditions.

An exciting new development in applied research using *Chlamydomonas* is the development of techniques for the production of large quantities of foreign proteins in chloroplasts. Recent technical advances have made it possible to express proteins of interest to levels of 5% of total protein or more (Fletcher et al. 2007). Proteins, such as monoclonal antibodies, have been produced in the chloroplast and assembled into active form (Tran et al. 2009). The CRC collection also contains mutants with altered starch accumulation (e.g., Wattebled et al. 2003).

HISTORY OF THE CRC

Research in *Chlamydomonas* has greatly benefited from the care that researchers in the early years of the genetic analysis of this organism (1950s and 1960s) took to maintain and share interesting mutants with other researchers. Of particular note are the roles of Ruth Sager (Rockefeller University), Ralph Lewin (Scripps Institution of Oceanography), and Paul Levine (Harvard University) in establishing collections of mutants with defects in motility and photosynthesis, and their sharing of these with other laboratories. In 1978, Nick Gillham and John Boynton (Duke University) obtained large collections of mutants from Levine and Ebersold (University of California at Los Angeles). These formed the basis for the culture collection. The following year, with Elizabeth Harris as Director, the CGC was founded using a 4.5-year grant from the National Science Foundation's (NSF) Genetic Biology Program. The funding was renewed for three more five-year periods receiving support from the Living Stock Collections Program at NSF. With the retirement of Drs. Gillham and Boynton in 2000, the CGC, now the CRC to reflect its expanded offerings of reagents and teaching kits beyond the original genetic strains, continued under the sole direction of Dr. Elizabeth Harris. In 2004, the CRC moved to the University of Minnesota's Department of Plant Biology in the laboratories of Pete Lefebvre and Carolyn Silflow. In 2005, Matt Laudon was hired as curator of the project. Dr. Harris continued to serve as co-director of the CRC and co-principal investigator on the support grants. She also continued to provide information to the user community and maintain the CRC website (chlamycollection.org) until her retirement in 2012. All of the CRC functions are now housed at the University of Minnesota.

In addition to her key role in establishing and supervising the CGC and the CRC, Dr. Harris made valuable contribution to *Chlamydomonas* research and published a comprehensive monograph *The Chlamydomonas Sourcebook*. *The Chlamydomonas Sourcebook* (Harris 2008) and its updated three volume second edition (Harris et al. 2009) provide a useful introduction to all of the information new and experienced users need in order to utilize this organism in their research. The book brings together basic information about the biology of this organism with a comprehensive assembly of techniques and tools used for *Chlamydomonas* research.

CRC AVAILABLE BIORESOURCES

The Core Collection

The *Chlamydomonas* core collection contains 4000 + wild-type and mutant strains. More than 30 different field isolates are contained in the collection. These strains have served as a valuable resource for researchers interested in quantitative trait loci, such as those affecting lipid content. Among the mutant strains are various auxotrophic mutants, drug-resistant mutants, large numbers of acetate-requiring mutants, mutants with missing or defective flagella, and many others with mutant phenotypes. Nuclear, cytoplasmic, and mitochondrial mutants are available. The most frequently requested strains from the CRC core collection are shown in Table 6.2. Not surprisingly, a number of different wild-type strains top the list. Cell wall-less mutants are included among the most popular strains requested because such mutants are easier to transform and to isolate cellular components from for biochemical analysis. Strain CC-503 (cw92 mt+) was used by the JGI for genome sequencing.

Two identical sets of core collection strains are maintained as slants in constant temperature rooms (20°C) on different floors of the Bioscience Center at the University of Minnesota. Slants offer relatively large amounts of cells for transfer and so it is often possible to fill orders within a day of receipt. All domestic shipments are sent using overnight delivery. The handling of international orders is dependent on the customs regulations of each country.

TABLE 6.2
Plasmids and Strains Most Frequently Requested from the Chlamydomonas Resource Center

Category/ID	Description/Use	Number of Times Requested[a]
Plasmid		
pSRSapI	Chloroplast expression plasmid	36
pASapI	Chloroplast expression plasmid	27
pHyg3	Confers hygromycin resistance	22
pHS-SaCas9 (pPH187)	Expression plasmid for cas9 enzyme	16
pBR9 GFP	GFP[b] expression plasmid	16
pLM006	Generating mCherry 3x His fusion proteins	13
Strain		
CC-125	Wild type mt+ [137c]	206
CC-124	Wild type mt− [137c]	93
CC-1690	Wild type mt+ [Sager 21 gr]	91
CC-503	cw92 mt+	69
CC-400	cw15 mt+	62
CC-5168	cw15 ΔpsbH [strain TN72]	38

[a] 2015–2018.

[b] Green fluorescent protein.

Additional Strains Available

In keeping with the history of the CRC, which began when mutant collections were obtained as a result of the retirements of Professors Levine and Ebersold in 1978, the CRC accepts and manages mutant collections received from individual research programs.

The CRC distributes strains of the Insertional Mutant Collection generated through the Chlamydomonas Library Project (CLiP). Strains from this collection of mapped insertional mutants of *Chlamydomonas* generated by Martin Jonikas and collaborators are available through the CRC. More than 83% of the genes in the *Chlamydomonas* genome are represented by at least one insertional allele (Li et al. 2016) in this collection. The sites of insertion in CLiP mutants have been determined by deep sequencing of pools of mutants and the results made available to users on a searchable website maintained in the Jonikas lab (https://www.chlamylibrary.org). The CRC maintains multiple copies of the CLiP strain collection (62,289 mutants) on agar plates and as frozen (backup) stocks. When users identify mutants with insertions in genes of interest, they order them on the website and the orders are filled from the CRC. More than 2100 orders for individual CLiP mutants have been filled. No single CLiP mutant strain has been ordered more than 7 times, indicating that users are finding hundreds of mutants of interest in the collection.

A collection of more than 300 mutants was recently acquired from the Spreitzer laboratory (University of Nebraska) where research was focused on nuclear and chloroplast mutants affecting ribulose-1,5-bisphosphate carboxylase/oxygenase (Satagopan and Spreitzer 2004). This enzyme is the most abundant protein on the planet and is critical for CO_2 fixation. More recently, the Chlamydomonas Spatial Interactome (CSI) collection of ~200 strains was generated by Luke Mackinder (University of York) and Martin Jonikas to allow investigators to use fluorescence from yellow (YFP) or green (GFP) fluorescent protein tags to localize proteins involved in the carbon concentrating mechanism of *Chlamydomonas*. Tulin and Cross (2014) deposited a collection of more than 450 mainly temperature-sensitive mutants for studies of the cell cycle.

Investigators using *Chlamydomonas* are urged to deposit all published mutant strains of *Chlamydomonas* in the CRC collection upon manuscript acceptance. The database entry for each strain includes information about the source (provider), genetic background and method of mutagenesis, growth and selection conditions, and a citation for a reference describing the strain. Single mutants, as well as strains in which genetic crosses have brought numerous mutants together, are welcome. Plasmids are also welcome for deposit in the CRC collection. The contribution of plasmids to the CRC ensures that the resource will be available to other researchers, and that the laboratory that generated the plasmid is spared the task and expense of distributing the plasmids to other laboratories.

MOLECULAR TOOLS AVAILABLE FROM THE CRC

cDNA and Genomic Libraries

Once the putative function of a given gene has been suggested by mutation, either in a traditional mutant or a CLiP mutant, it is important to confirm the assignment of the phenotype of the mutant to the candidate gene by rescuing the phenotype. This is done by transformation of the mutant with the cloned wild-type gene. To this end, a bacterial artificial chromosome (BAC) library of more than 15,000 clones from *Chlamydomonas* was prepared and arrayed in 40 plates of 384 wells/plate (Kathir et al. 2003). Each end of every BAC was sequenced by the JGI, and the sequence reads were aligned with the *Chlamydomonas reinhardtii* genome sequence (https://phytozome.jgi.doe.gov/pz/portal.html#!info?alias=Org_Creinhardtii). Users can identify BAC clones flanked by individual end sequences. A tool on the CRC web page can then be used to identify the specific BAC clone. In addition, for each of the more than 300 molecular markers placed on the *Chlamydomonas* genomic map, a series of overlapping BACs forming a contig have been identified. Users can use these to find BACs of interest and order them from the CRC. In addition, the CRC provides a limited number of cDNA libraries prepared from mRNA isolated under a diverse set of environmental and physiological conditions (e.g. post-deflagellation, during sexual differentiation, stress conditions, etc.) for screening.

Cloned Genes and Organelle Genome Fragments

The first molecular reagents provided to the CRC by users were plasmid clones containing overlapping DNA fragments that, taken together, covered the chloroplast genome. Later, this was extended to include the mitochondrial genome. Other genes of interest including ribosomal RNA encoding genes, tubulins, and many other nuclear genes, are also available from the CRC.

An interesting feature of *Chlamydomonas* biology is that unlike the nuclear genome in which homologous gene replacement has proven difficult, chloroplast genome gene replacement is routinely accomplished (Boynton et al. 1988). This technical advantage has been used to great effect in the study of chloroplast gene expression and function. A large number of mutants in which specific amino acid changes in chloroplast genes of interest (such as the large subunit of Ribulose-1,5-bisphosphate carboxylase/oxygenase) (Satagopan and Spreitzer 2004) are available from the CRC.

Plasmids for Protein Visualization and Genome Alteration

The mission of the CRC is not to just maintain and distribute mutants and molecular tools produced previously, but to also encourage and facilitate the use of the newest techniques, as they become available. A key role of the CRC is to bring together all of the most recent plasmid tools needed for genetic and cell biology experiments (Table 6.2). Some of the most important plasmid groups that the CRC provides include the following:

- *Selectable Markers*: These include plasmids that can confer resistance to drugs such as paromomycin, hygromycin, and xeocin, as well as *Chlamydomonas* genes that can rescue the auxotrophic defects in genes such as nitrate reductase (*NIT1*) and argininosuccinate lyase (*ARG2/7*).
- *Fluorescent Proteins*: Fluorescent proteins such as YFP and GFP are being increasingly used for tracking proteins and their movements in vivo. The CRC provides a large number of different fluorescent proteins, often conjugated

to proteins of biological interest such as components of the intraflagellar transport machinery involved in flagellar assembly and function (Table 6.2).

- *Chloroplast Gene Expression and Targeting*: Many plasmids using different selectable markers and fluorescent protein tags are available for visualizing and manipulating chloroplast gene expression. These are also used for targeting exogenous proteins to the nucleus, chloroplast, endoplasmic reticulum, and the mitochondria (Rasala et al. 2014).

Oligonucleotide Primers for Mapping

With the completion of a detailed molecular map with more than 300 PCR-based markers identifying multiple regions of each of the 17 *Chlamydomonas* linkage groups (Kathir et al. 2003, Vysotskaia et al. 2001), it is possible to genetically map newly identified mutants with phenotypes of interest by crossing them with a highly polymorphic field isolate (S1-D2, CC-1952) and recovering recombinant progeny. Co-segregation of the phenotype of interest and the appropriate polymorphic alleles generated by PCR can place the mutant on the genetic map, and the map can thus be refined in the region of interest.

QUALITY CONTROL/SECURITY

One advantage to using *Chlamydomonas* as a model system is that cell color is an easily scored measure of a culture's health. Each slant in the core collection is examined once per week. All strains in the core collection that are fading from a dark green to a paler green color are identified. These are then replaced by streaking a new slant from the old tube. The frequency of transfer required is determined by the strain genotype. Some strains need to be renewed monthly, whereas others remain healthy for as long as a year. The most difficult to maintain strains in the core collection, the so-called dims and darks mutants that are hyper-sensitive to illumination, are frozen in cryovials in liquid nitrogen (LN). The entire core collection is in the process of being stored in cryovials.

Each of the 251 plates in the CLiP collection is replicated using a Singer RoToR robot every 6 weeks. Fungal contamination is an ever-present threat when dealing with plates of agar medium. Plates are wrapped in Parafilm immediately after replication and are visually inspected weekly for contamination. Three copies of the CLiP library plates are stored frozen in two LN storage tanks for deep security. These are not routinely accessed, except for periodic test thaws to monitor continued viability. If CLiP mutants are lost from the working collection, they are replaced using the frozen stocks.

EDUCATIONAL RESOURCES AVAILABLE FROM THE CRC

The CRC offer various kits in support of teaching objectives. These are described below.

Hydrogen Project Kit

This kit demonstrates how *Chlamydomonas* cells produce hydrogen when exposed to anaerobic conditions during sulfur starvation. The kit contains two *Chlamydomonas reinhardii* wild-type cultures. Also included are stock solutions for the preparation of complete and sulfur-deficient media, plastic tubing, connectors, two 1-Liter plastic soda bottles, and detailed instructions for production of hydrogen after sulfur starvation. Students are encouraged to grow the cells at home using filtered water and the supplied media components.

Phototaxis Mutant Kit

Chlamydomonas cells rapidly swim toward light and this movement can be observed on a macro scale in a matter of minutes. Cultures of wild-type and mutant strains of *Chlamydomonas* in liquid media are provided for observation. No preparation of media or subculturing is required. For larger classes, the CRC ships the cultures on agar and provides stock solutions for making the liquid medium. Background material and references are also included.

Circadian Rhythm Kit

Some strains of *Chlamydomonas* show a strong circadian rhythm of phototaxis. Wild-type strain CC-124 is included in this kit sent for the demonstration of circadian rhythm. It also includes stock solutions for culture medium, and instructions.

Motility Kit

This kit contains a selection of motility mutants and reference materials. With access to simple phase contrast microscopes, students can learn to distinguish various types of motility mutants (e.g., flagella-less, short flagella, long flagella, paralyzed flagella) by observing living cells.

Chlamydomas Mating and Dikaryon Rescue Kit

Chlamydomonas cells are heterothallic, expressing one of two allelic mating types (mt+ and mt-). Gametogenesis is induced by nitrogen starvation. Within minutes of mixing gametes, large clumps of cells, vigorously interacting via their flagella, can be observed. Cell fusion produces dikaryon cells with four flagella as a precursor to zygote formation. If two different mutants with paralyzed flagella are crossed, a remarkable rescue of both mutant phenotypes can occur within minutes by complementation (Dutcher 2014). This kit includes gamete cultures in liquid media or on agar, complementing non-motile mutants and wild-type controls for the dikaryon rescue experiment, and instructions.

Uniparental Inheritance Kit

The *Chlamydomonas reinhardtii* chloroplast genome is inherited exclusively from the mt+ parent. Mutants in the chloroplast genome are inherited uniparentally upon mating while mutations (often with identical phenotypes) in the nuclear genome are inherited in a traditional Mendelian (2:2) ratio. The kit contains cultures of antibiotic resistant mutants, antibiotics, stock solutions for culture media, and instructions. This kit is intended for college level laboratories equipped for microscopic dissection of tetrads and the preparation of sterile media.

OPERATIONAL PROCEDURES/OVERSIGHT

The CRC is supervised by co-directors, Carolyn Silflow and Paul Lefebvre, professors in the Department of Plant and Microbial Biology at the University of Minnesota. The operation of the CRC is under the direction of the Curator (Matt Laudon) with the assistance of Matthew LaVoie. The Curator is responsible for maintaining the CRC website, passing the CLiP collection, accepting and filling orders, consulting with users, and supervising laboratory personnel.

A great advantage to housing a program like the CRC on the campus of a large research university is the availability of a sizeable pool of enthusiastic undergraduate assistants for handling routine duties. The College of Biological Sciences at the University of Minnesota has a large pool of talented undergraduate students eager for an introduction to laboratory research. The CRC hires undergraduate students, usually in their first or second years, and provides them with the opportunity to work in a research laboratory. Students with a demonstrated interest in, and talent for, research are encouraged to take on increasing levels of responsibility. This often leads to independent research projects.

A Center Oversight Committee (COC) of five research biologists consults with the Co-directors and the Curator of the CRC on pricing and policies. Four members of the COC are selected from among the major users of the collection and are *Chlamydomonas* researchers. One member is chosen from among the directors of collections of other organisms. The COC selects its own members through meetings held at the biannual International Meetings on the Cell and Molecular Biology of *Chlamydomonas*. Dr. Anne Rougvie, Professor of Genetics, Department of Cell and Developmental Biology at the University of Minnesota and Director of the *C. elegans* Genetic Center is the current "outside" member of the COC. The COC members, as representatives of the larger *Chlamydomonas* research community, work to insure the long-term continuation of the CRC.

IMPACT OF THE CRC: LEVEL OF USE

One of the most documentable measures of the importance of the CRC to the international *Chlamydomonas* research community is the number of published papers that cite resources obtained from the CRC (or variations on that name, including "stock center" and "genetics center"). Since 2017, 258 published manuscripts have acknowledged the CRC for providing essential research materials. The demand for resources from the CRC remains strong (Table 6.3). During a recent 5 years (2013 through 2017), the CRC received orders from 1947 requestors for plasmids and cultures resulting in >10,000 items shipped over this time period.

The current website launched in 2015, and detailed analytics then became available. From 2015 and through 2017, 2123 orders were filled by the CRC. These resulted in ~7000 unique items requested and shipped (Table 6.1).

TABLE 6.3
Data on Resources Utilization and Fees Collected at the Chlamydomonas Resource Center Including the Number of Requestors, Items Shipped (Cultures and Plasmids) and Total Orders Received per Year from 2013 to 2017

	Year				
Category	**2013**	**2014**	**2015**	**2016**	**2017**
Requestors	403	357	364	427	396
Cultures	1577	1584	1314	2624	2182
Plasmids	351	233	332	300	268
Total orders	677	565	618	805	710
Fees collected	$46,760	$53,145	$68,603	$18,735	$144,943

The single most popular item was the most commonly used wild-type strain of *Chlamydomonas reinhardtii*, 137c mt+ (CC-125) that was provided to the collection by the Levine lab in the early 1960s. The 137c strains together (plus and minus mating types) were ordered 299 times. Reflecting a historic split that occurred during the evolution of *Chlamydomonas* research, the second most ordered wild-type strain (21gr+) originally came from the Sager laboratory. An interesting historic note is that most researchers working on flagellar motility use the Sager strain while most laboratories working on metabolism and photosynthesis use the Levine strain.

Among the top ten orders of all types from the CRC are those for Hutners trace metals and the hydrogen evolution teaching kits (Table 6.4). The top mutant strain in

TABLE 6.4
Teaching Kits and Miscellaneous Items Shipped by the Chlamydomonas Resource Center per Year from 2013 to 2017

	Year				
Item	**2013**	**2014**	**2015**	**2016**	**2017**
Teaching kits					
Circadian rhythm	9	10	15	5	9
Hydrogen project	35	22	33	27	32
Mating/dikaryon	18	6	1	6	2
Motility	22	11	17	12	12
Photoaxis	9	9	15	6	9
Uniparental inheritance	2	0	2	1	3
Miscellaneous					
Mapping kit	4	2	1	0	3
Trace metals	4675 mL	4990 mL	5445 mL	5470 mL	7410 mL

terms of number of orders was cw92 mt+ (CC-503), which was requested 69 times. This is a reflection of the fact that cw92 mt+ was the strain sequenced by the JGI (Merchant et al. 2007).

FINANCIAL SUPPORT AND FEES

Grants from the NSF were critical to the establishment of the CRC, and they continue to provide approximately 50% of the budget. The remainder of the budget is provided by user fees. Regular users with continuing grant support can take advantage of a subscription plan in which, for an annual fee, they can receive up to 5 orders of up to 50 cultures or plasmids, with free shipping. A premier subscription entitles users to 15 orders of up to 150 cultures/plasmids. Subscriptions are only available to academic or other non-commercial users, and do not include CLiP mutant strains.

In order to generate increased revenue and provide users with the option of saving research time and effort, the CRC offers a continually expanding line of products for *Chlamydomonas* research. These include concentrated stocks of different formulations, Hutners trace metals solution, and various growth media in liquid or agar tube form.

ACCESSING RESOURCES AVAILABLE FROM THE CRC

Resources available through the CRC can be accessed using the CRC website (chlamycollection.org). The website allows users to search for strains and other resources offered through the CRC, and to place orders using a shopping cart. Payment is accomplished by credit card. The website regularly posts general information such as job openings and meeting announcements and serves as a bulletin board for the international *Chlamydomonas* community. Pull-down menus offer access to technical information such as genetic and molecular genome maps, software tools, commonly used laboratory methods for *Chlamydomonas* research, and laboratory exercises for teaching.

FUTURE DIRECTION OF THE CHLAMYDOMONAS RESEARCH CENTER

The CRC will continue to do its utmost to serve both experienced users and others whose interests lead them to utilize *Chlamydomonas* for addressing their research interests. Given that the prospects for enhanced US government support for living collections seems unlikely in the near future (McCluskey et al. 2017), the CRC will continue to look for ways to expand its revenue base through new offerings and cost savings. The freezing of cultures in LN is one attempt to reduce maintenance costs. Cultures that have been ordered infrequently will no longer be maintained on agar slants after test thaws have shown that the culture responds well to freezing.

Depending on the level of interest in the user community, the CRC hopes to offer services that users might not undertake on their own. For example, research on genome manipulation/editing in *Chlamydomonas* using a CRISPR/cas9 approach has demonstrated that this system of electroporating the cas9 enzyme with the appropriate small RNA is very effective in inducing targeted gene disruption. With appropriate templates, it can be used for site-directed mutagenesis in vivo (e.g., Greiner et al. 2017). Another new service to be evaluated is the "retrofitting" of BAC clones with selectable markers, for phenotypic rescue experiments. Currently, if a user wants to determine whether a BAC clone contains a wild-type gene that rescues a mutant of interest, it is necessary to co-transform it with a selectable marker to introduce the BAC DNA into the cell. Co-transformation with a large BAC is very inefficient with, at most, a few percent of cells expressing the selectable marker. Using so-called recombineering techniques (Ting and Feng 2014) to introduce a small selectable marker gene, such as *AphVIII* conferring paromomycin resistance, into BACs would mean that the user would have to only assay a handful of drug-resistant transformants to rescue the mutant phenotype.

CONCLUSIONS

Like many other organismal resource centers, the CRC faces mounting financial pressure as costs rise and as access to research grants becomes increasingly competitive. The need for the CRC, however, continues to grow. The power of the model green eukaryote *Chlamydomonas* for studies of processes such as ciliary motility, photosynthesis, lipid production and many others will lead to more and more researchers choosing to pursue research using this genetic system. However, the enormous number of mutants and molecular clones that have been produced in *Chlamydomonas* laboratories over the span of decades prevents any individual laboratory from being able to maintain even a small fraction of the available resources. With continued support from the *Chlamydomonas* community in the form of user fees, and from the federal government through research grants, the CRC intends to continue to make critical resources available to researchers around the world.

REFERENCES

Adams, G.M., B. Huang and D.J. Luck. 1982. Temperature-sensitive, assembly-defective flagella mutants of *Chlamydomonas reinhardtii*. *Genetics* 100:579–586.

Adams, M., U.M. Smith, C.V. Logan and C.A. Johnson. 2008. Recent advances in the molecular pathology, cell biology and genetics of ciliopathies. *Journal of Medical Genetics* 45:257–267.

Blaby, I.K., C.E. Blaby-Haas, N. Tourasse et al. 2014. The *Chlamydomonas* genome project: A decade on. *Trends in Plant Science* 19:672–680.

Boynton, J.E., N.W. Gillham, E.H. Harris et al. 1988. Chloroplast transformation in *Chlamydomonas* with high velocity microprojectiles. *Science* 240:1534–1538.

Dubini, A., F. Mus, M. Seibert, A.R. Grossman and M.C. Posewitz. 2009. Flexibility in anaerobic metabolism as revealed in a mutant of *Chlamydomonas reinhardtii* lacking hydrogenase activity. *Journal of Biological Chemistry* 284:7201–7213.

Dutcher, S.K. 2014. The awesome power of dikaryons for studying flagella and basal bodies in *Chlamydomonas reinhardtii*. *Cytoskeleton* 71:79–94.

Fletcher, S.P., M. Muto and S.P. Mayfield. 2007. Optimization of recombinant protein expression in the chloroplasts of green algae. *Advances in Experimental Medicine and Biology* 616:90–98.

Greiner, A., S. Kelterborn, H. Evers, G. Kreimer, I. Sizova and P. Hegemann. 2017. Targeting of photoreceptor genes in *Chlamydomonas reinhardtii* via zinc-finger nucleases and CRISPR/Cas9. *Plant Cell* 29:2498–2518.

Harris, E.H. 2008. *The Chlamydomonas Sourcebook, Vol. 1: Introduction to Chlamydomonas and Its Laboratory Use*. New York: Academic Press.

Harris, E., D. Stern and G. Witman. 2009. *The Chlamydomonas Sourcebook*. 2nd ed. New York: Academic Press.

Hildebrandt, F., T. Benzing and N. Katsanis. 2011. Ciliopathies. *New England Journal of Medicine* 364:1533–1543.

Kathir, P., M. LaVoie, W.J. Brazelton, N.A. Haas, P.A. Lefebvre and C.D. Silflow. 2003. Molecular map of the *Chlamydomonas reinhardtii* nuclear genome. *Eukaryotic Cell* 2:362–379.

Li, J.B., J.M. Gerdes, C.J. Haycraft et al. 2004. Comparative genomics identifies a flagellar and basal body proteome that includes the BBS5 human disease gene. *Cell* 117:541–552.

Li, X., R. Zhang, W. Patena et al. 2016. An indexed, mapped mutant library enables reverse genetics studies of biological processes in *Chlamydomonas reinhardtii.Plant Cell* 28:367–387.

McCluskey, K., K. Boundy-Mills, G. Dye et al. 2017. The challenges faced by living stock collections in the USA. *Elife* 13:6.

Melis, A., M. Seibert and M.L. Ghirardi. 2007. Hydrogen fuel production by transgenic microalgae. *Advances in Experimental Medicine and Biology* 616:110–121.

Merchant, S.S., S.E. Prochnik, O. Vallon et al. 2007. The *Chlamydomonas* genome reveals the evolution of key animal and plant functions. *Science* 318:245–250.

Nagel, G., T. Szellas, S. Kateriya, N. Adeishvili, P. Hegemann and E. Bamberg. 2005. Channelrhodopsins: Directly light-gated cation channels. *Biochemical Society Transactions* 33:863–866.

Pazour, G.J., B.L. Dickert, Y. Vucica et al. 2000. *Chlamydomonas* IFT88 and its mouse homologue, polycystic kidney disease gene tg737, are required for assembly of cilia and flagella. *Journal of Cell Biology* 151:709–718.

Polle, J.E.W., S.-D. Kanakagiri and A. Melis. 2003. *tla1*, a DNA insertional transformant of the green alga *Chlamydomonas reinhardtii* with a truncated light-harvesting chlorophyll antenna size. *Planta* 217:49–59.

Rasala, B.A., S.S. Chao, M. Pier, D.J. Barrera and S.P. Mayfield. 2014. Enhanced genetic tools for engineering multigene traits into green algae. *PLoS One* 9:e94028.

Reiter, J.F. and M.R. Leroux. 2017. Genes and molecular pathways underpinning ciliopathies. *Nature Reviews Molecular and Cell Biology* 18:533–547.

Satagopan, S. and R.J. Spreitzer. 2004. Substitutions at the Asp-473 latch residue of *Chlamydomonas* ribulosebisphosphate carboxylase/oxygenase cause decreases in carboxylation efficiency and CO_2/O_2 specificity. *Journal of Biological Chemistry* 279:14240–14244.

Tam, L.W. and P.A. Lefebvre. 1993. Cloning of flagellar genes in *Chlamydomonas reinhardtii* by DNA insertional mutagenesis. *Genetics* 135:375–384.

Tam, L.W. and P.A. Lefebvre. 1995. Insertional mutagenesis and isolation of tagged genes in *Chlamydomonas*. *Methods in Cell Biology* 47:519–523.

Ting, J.T. and G. Feng. 2014. Recombineering strategies for developing next generation BAC transgenic tools for optogenetics and beyond. *Frontiers in Behavioral Neuroscience* 8:111.

Tran, M., B. Zhou, P.L. Pettersson, M.J. Gonzalez and S.P. Mayfield. 2009. Synthesis and assembly of a full-length human monoclonal antibody in algal chloroplasts. *Biotechnology and Bioengineering* 104:663–673.

Tulin, F. and F.R. Cross. 2014. A microbial avenue to cell cycle control in the plant superkingdom. *Plant Cell* 26:4019–4038.

Vysotskaia, V.S., D.E. Curtis, A.V. Voinov, P. Kathir, C.D. Silflow and P.A. Lefebvre. 2001. Development and characterization of genome-wide single nucleotide polymorphism markers in the green alga *Chlamydomonasreinhardtii*. *Plant Physiology* 127:386–389.

Wattebled, F., J.P. Ral, D. Dauvillée et al. 2003. STA11, a *Chlamydomonas reinhardtii* locus required for normal starch granule biogenesis, encodes disproportionating enzyme. Further evidence for a function of alpha-1,4 glucanotransferases during starch granule biosynthesis in green algae. *Plant Physiology* 132:137–145.

Wietek, J. and M. Prigge. 2016. Enhancing channelrhodopsins: An overview. *Methods in Molecular Biology* 1408:141–165.

7 The Zebrafish International Resource Center

April Freeman, Ron Holland, Jen-Jen Hwang-Shum, David Lains, Jennifer Matthews, Katrina Murray, Andrzej Nasiadka, Erin Quinn, Zoltan M. Varga and Monte Westerfield

CONTENTS

Abstract: The genus *Danio* (zebrafish) includes up to 84 small freshwater species that originate from the Indian subcontinent and neighboring countries. In the past decades, zebrafish have become one of the most frequently used organisms in many aspects of biomedical research including chemical screens for drug development and as a model organism for research on a variety of human diseases. The Zebrafish International Resource Center (ZIRC) has expanded from a few lines (in 1998) to more than 11,000 lines (40,300 alleles) as of 2018. The collection contains point mutations induced by ethyl nitrosourea, viral insertion mutants, y-ray induced chromosomal rearrangements, CRISPER/cas9 edited strains, transgenic lines, enhancer traps, and wild-type lines. In addition to live and frozen stocks of wild-type, mutant, and transgenic lines, ZIRC provides antibodies, gene probes, live feed paramecia cultures, and many protocols for genotyping and husbandry. Resources are shipped to hundreds of laboratories in more than 40 countries each year. Active research at ZIRC has identified and characterized the most common diseases that affect laboratory stocks of zebrafish, and in addition to research publications, ZIRC maintains an up-to-date online manual for the prevention, diagnosis, and treatment of diseases affecting zebrafish. ZIRC also provides pathology and consultation services, including diagnostic services and health status testing for zebrafish laboratories around the world.

ZEBRAFISH (*DANIO RERIO*) – THE ORGANISM

The genus *Danio* includes up to 84 small freshwater species by scientific name and at least 26 by International Code of Zoological Nomenclature-based taxonomy. These species originate from the Indian subcontinent and neighboring countries (http://www.fishbase.org/search.php; ver. 06/2018). The genus belongs to the family of Cyprinidae and the Subfamily Danioninae. The name *Danio* is derived from the Hindi word dhan = paddy (an irrigated or flooded field where rice is grown), and dhani in the related Bengalese vernacular implies "of the rice paddy." (http://dict.hinkhoj.com/paddy-meaning-in-hindi.words; personal communication, Sunit Dutta.)

Between 1807 and 1814, Francis Buchanan (later called Hamilton, 1762–1829) conducted an extensive survey of Bengal for the British East India Company. Ten *Danio* species, among them *Cyprinus rerio* "of the Danio kind," were first described in his *An Account of the Fishes Found in the River Ganges and Its Branches* (1822). Hamilton (Buchanan) described and characterized small cyprinid fish species in Eastern India (*Cyprinis Danio*) as inhabiting small, stagnant to slow moving bodies of water. By his account, *Danio rerio* were described as "with several blue and silver stripes on each side; with the body much compressed, and with four tendrils, of which two are a little longer than the head" (Hamilton 1822, pp. 323/324).

As more species were identified and added, the genus was divided into the subgenus *Brachydanio* (Weber and de Beaufort 1916), which included *B. rerio*, for species with seven branched dorsal fin rays and incomplete or absent lateral lines. The remaining species of the genus *Danio* were characterized as having 12–16 branched dorsal fin rays and complete lateral lines. However, because these criteria turned out to be too variable and therefore insufficient to distinguish a precise subgenus delimitation, Barman (1991) proposed to again synonymize all species under the original genus name *Danio*.

Recent field studies show that zebrafish are benthopelagic, inhabit a wide range of habitats, and are extremely adaptable to different climatic and rapidly changing water conditions (Engeszer et al. 2007, Parichy 2015). Zebrafish have been described from the Krishna River system in south-central India up into the Bengal region of north-east India, and farther north into Bangladesh (reviewed in Jayaram 1981). They have also been identified in Pakistan, Sri Lanka, Burma, and Nepal (reviewed in Barman 1991). An exhaustive review of the geographical range, habitats, and behaviors of zebrafish has been recently provided by Engeszer et al. (2007) and Parichy (2015). Due to their hardy nature and adaptability, zebrafish have become a favorite of pet hobbyists and can be propagated with ease in captivity.

WHY ZEBRAFISH?

Zebrafish were selected for genetic experimentation in the 1970s by George Streisinger (University of Oregon) because they offered key experimental advantages such as:

- Large number of progeny (up to several hundred) per clutch,
- Optical transparency during embryonic stages allowing for microscopic observation of cells, tissue layers, and developing organs, and
- Rapid development from embryo to hatching larvae (48–72 hours) in a relatively short generation time of circa 3 months.

Additionally, it soon became apparent that zebrafish were amenable to powerful genetic (Golin et al. 1982, Grunwald and Streisinger 1992, Streisinger et al. 1981) and cellular manipulations (Ho and Kane 1990), and easy observation of developmental phenotypes induced by mutation (Driever et al. 1996, Eisen 1996, Haffter et al. 1996). In the past decades, zebrafish have become one of the most frequently used organisms in virtually all aspects of biomedical research (Kinth et al. 2013, Patton et al. 2014), including chemical screens for drug development and as model organisms for a variety of human diseases (Ali et al. 2011).

THE ORIGIN OF THE ZEBRAFISH INTERNATIONAL RESOURCES CENTER (ZIRC)

The need for a centralized zebrafish resource stock center was recognized at the first open international zebrafish meeting held at Cold Spring Harbor in 1994. Large- and small-scale mutant screens were ongoing in many laboratories resulting in a rapid increase in the number of genetic lines. Individual laboratories were taxed with the

FIGURE 7.1 The "temporary," post–WWII Quonset hut in August 2018 that housed Streisinger's first zebrafish tanks. The ZIRC building's northeast corner and front entrance is in the background.

burden of maintaining these lines. Although cryopreservation was utilized by a few researchers, the majority could not afford the space or the cost of maintaining lines they were no longer actively utilizing. A centralized resource center soon became a top priority within the zebrafish community in order to ensure the highest possible level of quality, uniformity, health, and long-term security of genetic stocks.

Initially, negotiations were conducted to identify an appropriate location for the center. While several locations were considered, the University of Oregon was ultimately selected and it agreed to take on the task of establishing the program. The National Institutes of Health (NIH), National Center for Research Resources (NCRR), which later became the Office of Research Infrastructure Programs (ORIP), agreed to consider an application for funding. The initial grant for the Zebrafish International Resource Center (ZIRC) was awarded in 1998. Various buildings in which to house ZIRC on or near the University of Oregon campus in Eugene were considered. However, it was agreed that a new building would be constructed next to the site of George Streisinger's original fish facility—a World War II temporary Quonset hut that still stands today (Figure 7.1). The bulk of the funds for construction were provided by bonds issued by the State of Oregon. An NIH NCRR infrastructure construction grant provided additional funds. The facility was designed with help from local architectural and engineering firms, based on the faculty's collective fish husbandry experience, and visits to other large zebrafish and Medaka (*Oryzias*) laboratories. Construction began in 1999, and the official opening of ZIRC took place in 2000.

The Zebrafish Information Network (ZFIN; http://zfin.org) was established in 1994, after the Cold Spring Harbor meeting, to provide a model organism database for zebrafish. It was initially funded by NSF. The initial NIH grant to support ZIRC also included funding for ZFIN, ensuring that ZIRC's initial database development was well integrated into ZFIN's information database about zebrafish genes, mutants, and the genome. ZFIN received independent funding in 2002, and ZIRC has supported its own database since then while maintaining close integration with ZFIN.

THE ZIRC COLLECTION

ZIRC is funded by the NIH for the purpose of acquiring and preserving genetic zebrafish strains and to function as a repository of mutant, transgenic, and wild-type lines, making these lines readily available to the research community. In the United States, laboratories that develop zebrafish strains using federal funding are required to make these genetic lines publicly available by providing them to ZIRC for future re-distribution and research. In addition to live and frozen stocks of wild-type, mutant, and transgenic lines, ZIRC provides antibodies, gene probes, live feed paramecia cultures, and many protocols for genotyping and husbandry. Resources are shipped to hundreds of laboratories in more than 40 countries each year. Active research at ZIRC has identified and characterized the most common diseases that affect laboratory stocks of zebrafish, and in addition to research publications, ZIRC maintains an up-to-date online manual for the prevention, diagnosis, and treatment of diseases affecting zebrafish. ZIRC also provides pathology and consultation services, including diagnostic services and health status testing for zebrafish laboratories around the world.

ZIRC has expanded from a few lines provided by the local laboratory to more than 11,500 lines contributed by numerous research programs (Figure 7.2). The number of zebrafish lines maintained at ZIRC has increased steadily over the past three decades. As of 2018, the ZIRC's collection is approaching 40,300 alleles. These include point mutations induced by ethyl nitrosourea, viral insertion mutants, γ-ray

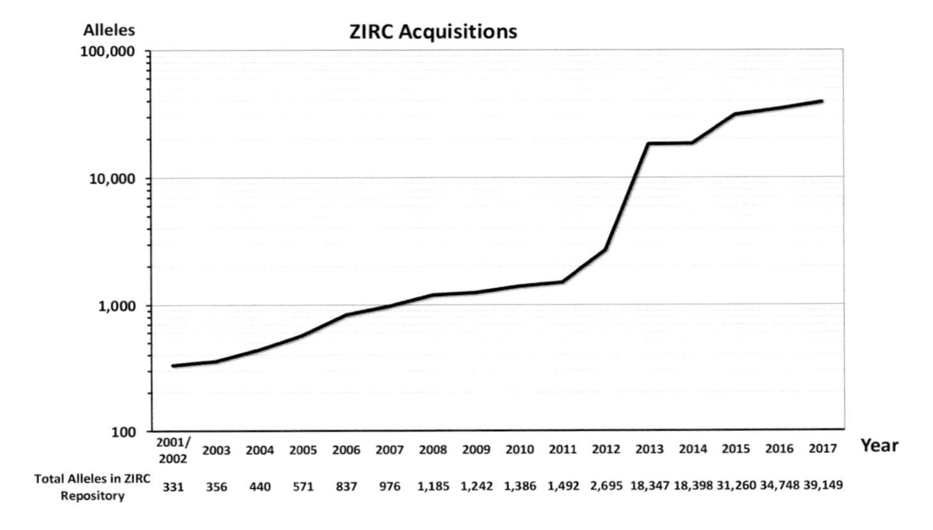

FIGURE 7.2 Acquisition of fish lines (alleles) by the ZIRC between 2001 and 2017. Between 2001 and 2011 and after 2015 again, a fish line typically contained a single allele. Between 2011 and 2015, ZIRC imports also included multi-allelic *sa*- and *la* lines which contained dozens of alleles per line. *sa* lines were generated through ENU mutagenesis by the Zebrafish Mutation Project at the Sanger Institute (Hinxton, UK). *la* lines were generated, by a collaboration of the laboratories of Shuo Lin at UCLA, Los Angeles and Shawn Burgess at the NHGRI (NIH). See also the section "Mutant Lines."

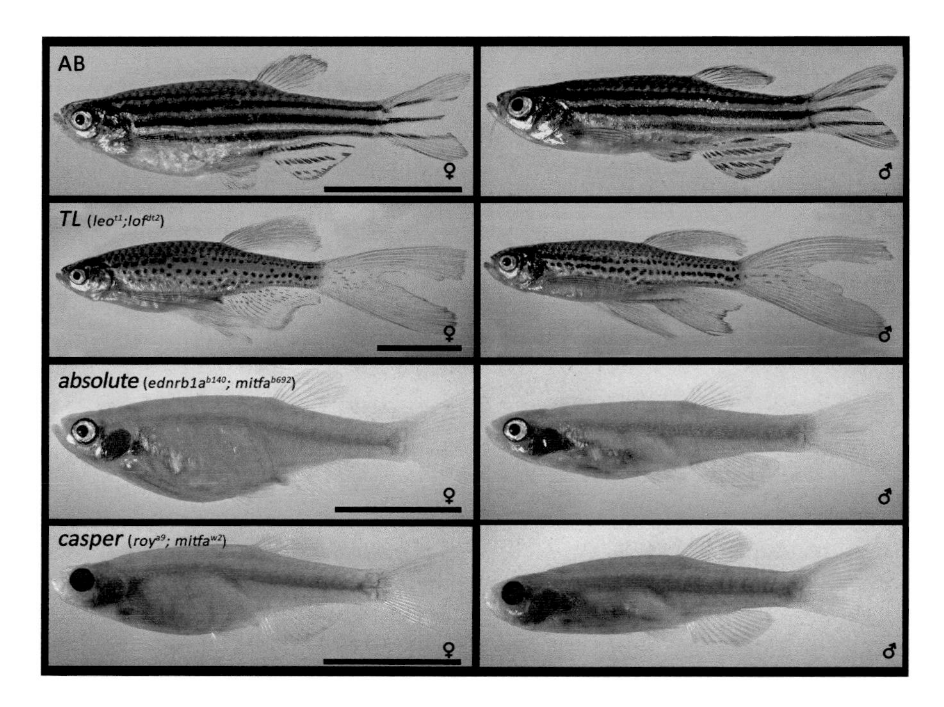

FIGURE 7.3 **(See color insert.)** Adult wild-type and pigment double-mutant strains maintained at ZIRC. Top row, AB wild-type strain. Characteristic dimorphisms of the sexes include white protruding belly, yellow dorsal fin, and urogenital papilla (females) and usually slender males, with a yellowish (sometimes reddish) belly and anal fins, clear dorsal fin. Second row from top: Tüpfel-longfin (*leo*^{t1};*lof*^{dt2}) adults have dotted (dots = Tüpfel in German) pigmentation instead of stripes due to the *leopard* (*leo*) mutation. *lof*^{dt2} is a dominant homozygous viable mutation causing long fins. Third row from top: *absolute* (*ednrb1a*^{b140};*mitfa*^{b692}) double mutant fish lack melanophore, xantophore, and most iridophore cells. Iridophores are present in the iris. Bottom row: *casper* (*roy*^{a9};*mitfa*^{w2}) lack melanophores, most iridophore, and xantophore cells. Melanocytes are present in the eyes, which however, lack iridophores. Scale bars: 1 cm; each bar corresponds to female (left) and male (right) in the same row.

induced chromosomal rearrangements, CRISPER/cas9 edited strains, transgenic lines, enhancer traps, and wild-type lines (Figure 7.3).

The ZIRC staff includes 18 members who are organized into 9 workgroups. Because several staff members occupy roles in multiple workgroups, and because communication and workflow among workgroups is dynamic and extensive, a circular arrangement was adopted (Figure 7.4; wide, gray circle). Several workgroups interact directly with the research community, e.g., when distributing fish lines or providing fish health diagnostic services (double-headed arrows). The public website and the internal fish line inventory database (IT) are extensively crosslinked with ZFIN to provide seamless transition between research data and acquisition of relevant genetic lines.

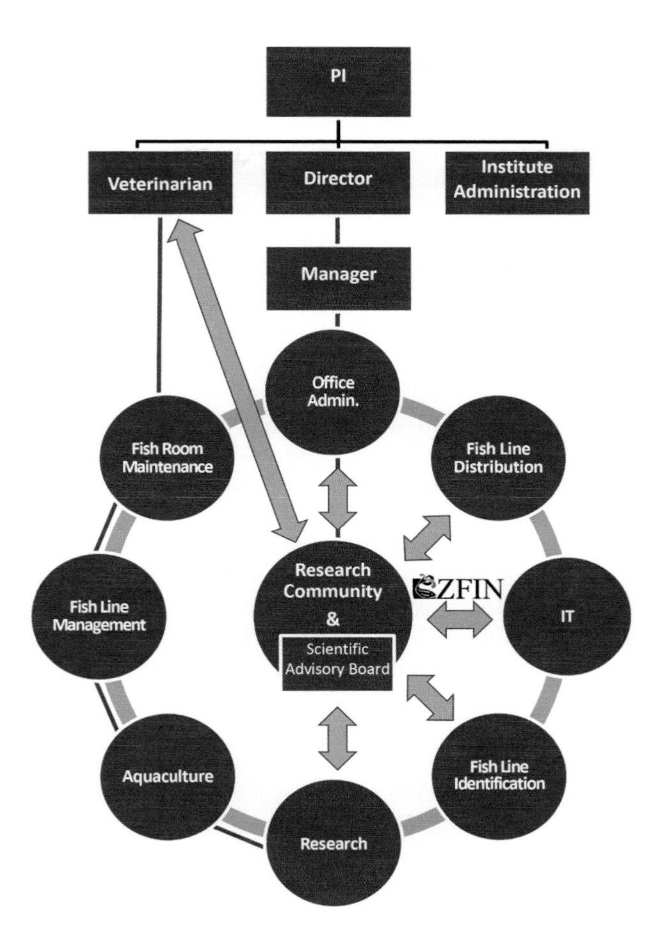

FIGURE 7.4 ZIRC's Organizational Chart. The Resource Center is organized into several workgroups (grey circles and veterinarian) that are focused on its key operational processes. The workgroups fulfill the following key functions (and more): Veterinarian: biosafety, in-house fish health monitoring, and diagnostic services; Manager: coordination and import of fish lines, personnel, and building management; Office Administration: customer care, travel arrangements, purchasing, and shipping support; Fish Line Distribution: order processing, fish packaging and shipping; IT: development and maintenance of the public website (zebrafish.org), the internal inventory database, and freezer and life support systems monitoring; Fish Line Identification: review of acquisition data, genotyping, phenotyping and documentation of line identities at times of import and during routine line propagation or amplification of frozen stocks; Aquaculture: source water conditioning, system water chemistry, water filtration, and monitoring of life support systems; Fish Line Management: propagation of live fish lines, cryopreservation of rarely requested lines, thawing, in-vitro fertilization, resource amplification, and weekly production of adult and embryo shipments; Fish Room Maintenance: student worker supervision, fish nursery care, fish feeding, and tank and facility cleaning. Research: fish health management strategies, identification of novel pathogens, fish reproduction and conditioning, and optimization and improvement of cryopreservation methods.

ZIRC POLICIES/PROCEDURES REGARDING THE ACQUISITION OF NEW LINES

All new importations are subject to an evaluation process. The overall goal of the evaluation process is to acquire the most useful lines for the research community and to ensure that ZIRC obtains all the information necessary to maintain and propagate the acquired lines. Researchers, who wish to make a zebrafish research line available should contact ZIRC via the website (zebrafish.org).

To minimize redundancies and manage acquisitions efficiently, criteria for acceptance have been established. These include determining relevance for current or future research needs and the potential scientific significance of each allele (if multiple alleles per locus are submitted or if some alleles already exist in the inventory). Additional information required for assessment and possible acceptance include the availability of information for the genotypic and/or phenotypic identification of the strain, husbandry requirements, and the submitter's ability to maintain the strain until it is fully established at ZIRC.

Depending on the number of accepted lines and the overall efficiency of importation, ZIRC staff determine whether the line should be imported as live fish or cryopreserved sperm samples. This decision is usually based on the number of lines acquired per submission. Submissions range from a single line to thousands of lines per laboratory or research project. Fifty or fewer lines are typically imported as live fish, whereas submission of more than 50 lines typically requires importation as cryopreserved sperm samples. Other factors such as budgetary considerations and experience with cryopreservation methods may contribute to the import strategy. Ultimately, the submitter is provided with an acceptance letter and the importation details are communicated and established between both parties.

Many lines in the repository have been imported as frozen sperm samples. When sperm is frozen offsite, the frozen samples are sent to ZIRC in a LN shipping vessel and transferred to the repository for storage. In recent years, ZIRC has imported a large number of so-called multi-allelic lines from large-scale mutation screens (Dooley et al. 2013, Varshney et al. 2013) as cryopreserved sperm samples. Because these lines are often imported with only a few samples per line, a single order can trigger the need for line rederivation and sample amplification.

Lines are also imported as live fish for sperm cryopreservation. Imports of live fish (typically 10 males per line) are held in a quarantine room for 2–4 weeks prior to sperm freezing, depending on the age and size of the fish. At the time of cryopreservation, all fish are euthanized and pooled sperm is collected by stripping and testis dissection. A fin clip is also collected from each male for genotype confirmation. The fish are then fixed and processed for histopathology.

Occasionally, sperm have already been cryopreserved at the submitting laboratory before ZIRC is notified of an intended submission. In such instances, the laboratories might have used a less recent version of the cryopreservation protocol, or their own protocol, for cryopreservation. The lack of a standardized cryopreservation protocol has sometimes created challenges in sample storage and recovery as considerable differences exist in post-thaw fertilization rates of samples. This can impact both the efficiency of line recovery at ZIRC and the response time for shipping lines after a frozen line has been requested.

ZIRC provides training in cryopreservation of sperm. In these cases, cryopreservation is conducted by members of the submitting laboratory, as their schedule permits, and the samples are collected over a period of time before being shipped to ZIRC. Ideally, the submitting laboratory uses ZIRC's optimized cryopreservation protocol (Matthews et al. 2018). Budget permitting, ZIRC staff members travel to the submitting laboratory, or someone from the submitting laboratory visits ZIRC, for hands-on cryopreservation training. This training develops proficiency with the most current methods, including quality control assessment of pre- and post-thaw samples, fin-clipping for genotyping, and bio-safety training.

The scientific information for each line is collected in a spreadsheet file with worksheets for the submitted line type (e.g., mutant, transgenic, wild-type), and data fields for line name and allele designation, relevant publications, phenotypic and genotypic information, and husbandry details. The data are then passed to curators at ZFIN to ensure that the nomenclature conforms to up-to-date nomenclature conventions and that genotype records have been created in the database.

Fish must be confirmed carriers of the specified genotype, healthy, and of breeding age. For wild-type strains, both male and female fish are required for embryo production. Transport of the fish to ZIRC is coordinated in advance with the submitting laboratory, and ZIRC staff provide shipping materials and instructions to ensure that live fish are packaged in compliance with the International Air Transport Association (IATA) and local animal use protocols (Institutional Animal Care and Use Committee).

In some cases, ZIRC will accept lines that can be visually identified by phenotype or gene expression. However, if visual or experimental identification is too subtle or if ZIRC is not equipped to unequivocally identify the genetic modification, then a line may not be accepted for import at the time of submission data review. Examples of such cases are mutations affecting blood plasma protein levels, circadian rhythm, or behavioral changes, or subtle visual phenotypes that cannot be detected by conventional bright field and/or epifluorescence microscopy.

Following acceptance into the collection, health and licensing information are provided prior to arranging the import shipment. The submitted information is provided to the University of Oregon Innovation Partnership Services team, and they communicate with the corresponding technology transfer office at the submitting institution to determine whether a licensing agreement is required. At the same time, the ZIRC veterinarian consults with the submitting laboratory and/or the institution's attending veterinarian to obtain health information that will be used to determine the necessary biosafety procedures for the import of live fish or cryopreserved sperm. The objective of this consultation is to identify any biosecurity risks at the time of import and to minimize the potential for importing pathogens present in external zebrafish research colonies.

MAINTENANCE OF ZEBRAFISH BIORESOURCES

Cryopreservation of Sperm

Cryopreservation is the most efficient and cost-effective method for large-scale, long-term storage of important genetic materials. At ZIRC, cryopreservation plays a critical role in maintenance and distribution of an expanding catalog of genetic lines. ZIRC maintains alleles as cryopreserved sperm (>77,000 samples) in an onsite liquid nitrogen (LN) repository. Liquid nitrogen freezers are operated in vapor-phase to provide biosecure storage. The freezers are monitored continuously using a Supervisory Control and Data Acquisition (SCADA) system and are filled automatically every 3 days from an external bulk LN tank. The bulk tank utilizes radio telemetry monitored by the commercial LN supplier. When the bulk tank reaches 30% of its capacity, the supplier automatically schedules a refill. The National Center for Genetic Resources Preservation (NCGRP, U.S. Department of Agriculture) in Fort Collins, CO, acts as an additional emergency backup repository for ZIRC frozen stocks. Two samples from all lines are sent to the NCGRP.

Upper and lower limits are set on the number of samples to be stored in the in-house repository. The lower limit prompts line regeneration and resource amplification, such as when an order is placed and the low inventory threshold is reached. The goal for the number of samples to obtain when freezing a line depends on a number of factors including popularity of the line, (genetic) stability of the transgene(s), line identification requirements, the number of alleles carried in the strain, and the number of available males. Hence, the number of samples frozen from a line can range from 15 to 50.

When line regeneration is required due to low sample numbers, a sample is thawed and used for in vitro fertilization (IVF) of wild-type (AB) eggs. If an order is pending, some embryos are shipped to the customer (if inventory permits) and a portion are raised at ZIRC for sample amplification. A number of different freezing strategies are used for line regeneration and sperm sample amplification as determined by the number of alleles present in a line and the number of available males.

- If a line has one to three alleles, the alleles will be identified in carrier-males prior to freezing.
- If there are four to nine alleles present in a line, the line is frozen as individual males (with 2–4 samples/male) and a fin-clip is taken at the time of freezing. The alleles present in each sample are then determined after freezing by PCR-based genotyping of genomic DNA from the fin tissue.
- For lines generated from thaws of sperm samples that are one generation removed from the mutagenesis and that have 10 or more alleles, the line is frozen without genotyping, as single samples from individual males. These samples are used later for line rederivation. After the males have rested for a few weeks, sperm samples are collected again from the same 16–20 males and are pooled in extender solution (typically 20 samples) and used for the distribution of the line. Because pooling of samples guarantees uniformity of cell-density and motility, as well as later genotyping, quality assessments will be representative for the entire pool.

Conditioning males prior to sperm collection can increase the amount of sperm collected. Male conditioning is also used to more efficiently move lines through the cryopreservation process. In the conditioning process, males are separated from females when they are 4–5 months of age (as soon as their sex can easily be determined), held at a lower tank density (1.5–3.5 males/L) and provided additional dry food daily. Males are conditioned from 2 to 6 weeks prior to sperm collection and are separated from females for at least one day to prevent spawning on the morning of sperm collection. Extra feeding has been found to be an important factor in the conditioning process.

Lines must pass a thaw test before they are considered to be successfully cryopreserved at ZIRC (Figure 7.5). A representative frozen sample of a line is thawed and used for IVF of wild-type (AB) eggs. A sperm sample passes quality requirements if at least 10% of the eggs are fertilized. ZIRC uses computer-assisted sperm analysis to analyze sperm motility to assess the quality of the test thaw. This allows samples to be assessed before or after cryopreservation without the need for anesthetizing and squeezing females for eggs, further reducing the number of animals used and improving animal welfare. Our research on cryopreservation, thawing, and IVF methods has resulted in significant protocol developments and a progressive improvement in the average fertilization rates from test thaws (Matthews et al. 2018).

Live Fish Maintenance

Tank space at ZIRC limits the maintenance of live fish to approximately 400 families. Genetic lines can be provided from live or cryopreserved stocks within a few days. Hence, the pressure to maintain live lines is relatively low. Live fish populations at the ZIRC have been reduced by almost 50% over the past 8 years (from an annual census of 65,000 fish to 38,000, currently). To further ensure increases in the efficiency of operations, live and cryopreserved stocks are turned over at a high rate, and only lines that fulfill specific criteria are maintained live. Among the criteria considered are the number of requests received, or the anticipated number of requests based on the research or publication that accompanies a submission. Live lines are evaluated every 6 months to determine whether they should be

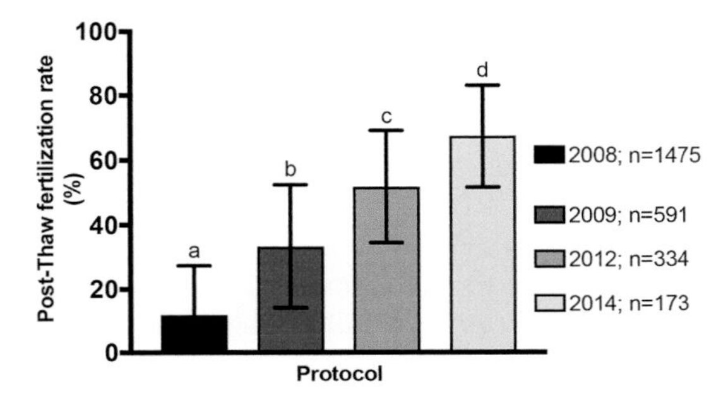

FIGURE 7.5 Test thaws between 2007 and 2016 indicate continual improvement of post-thaw fertilization rates with the implementation of protocol changes. Letters indicate that averages are statistically significantly different (p = 0.001).

retired (and be made available only as embryos from frozen sperm) or maintained as live stocks (available as embryos or adults). Currently, lines are maintained as embryos or adults if they are requested ten times or more within 6 months. Lines with fewer requests than this are cryopreserved and a new generation is not generated. However, the adult fish are maintained until they are a year old in case they need to be bred or shipped. Lines are always regenerated when a cryogenic stock has reached its low sample number threshold. At that time, the resource is re-amplified.

The husbandry practices and the environment of the live colony have been described previously (Varga 2016). Briefly, the ZIRC maintains its lines based on line-specific husbandry needs and their genetic characteristics. Thus, wild-type lines are typically maintained in a way that maximizes the number of contributing pairs, thereby reducing the rate of inbreeding. Mutant lines are maintained typically as heterozygotes, and transgenic lines and pigment mutations are maintained as heterozygotes, or as homozygotes if the genetic modification permits.

Wild-Type Lines

ZIRC has acquired and maintains several wild-type strains, including AB (Figure 7.3), Tübingen (TU), and WIK. Tüpfel longfin (TL) is a double mutant line with a recessive pigment (*leot1*) and a dominant long-fin (*lofdt2*) mutation (Figure 7.3). It is maintained as if it were a wild-type line because historically it has been used in that manner. In addition, two AB-TU hybrid lines, SAT and NHGRI-1, are also maintained. The SAT line was provided to ZIRC by the Sanger Institute, Hinxton, UK (Howe et al. 2013), and NHGRI-1 was obtained from the NIH's National Human Genome Research Institute (HGRI) in Bethesda, MD (LaFave et al. 2014). In contrast to the AB line, which has lost the sex-linked loci on chromosome 4 (Wilson et al. 2014), the ZIRC also maintains the natural wild-type lines Cooch Behar and Nadia from the Institute of Neuroscience in Eugene, OR, as these lines still carry the sex-determining loci (Z/W females and Z/Z Males).

ZIRC distinguishes "in-house" wild-type stocks that are not distributed to the research community from "dedicated shipping stocks" created from in-house stocks specifically for the purpose of distribution. Dedicated shipping stocks are propagated at the same time and in the same manner as new in-house stock generations. Orders for wild-type embryos are filled using medium-to-large group crosses of 15–50 in-house fish.

To maintain the polymorphic character of wild-type lines, the propagation of TU, TL, WIK, SAT, and NHGRI-1 in-house stocks is based on combining the same number of embryos (e.g., 40) from each of a minimum of 25 small-group incrosses consisting of 3 females and 2 males. For SAT and AB, two lineages of in-house stocks are maintained in parallel. These lineages are maintained at regularly spaced intervals to produce new generations. Two lineages, AB-1 and AB-4, are currently maintained at ZIRC. Lineage 1 is the original ZIRC AB line acquired from the Institute of Neuroscience, Eugene OR, whereas lineage 4 was reimported from the Institute of Neuroscience in 2017. In the case of AB lines, any surplus fish are used for shipping, outcrossing, and egg production for sperm thaws. Should any of the lineages underperform, that lineage can be reconstituted from the remaining one.

Previously, natural breeding and IVF were alternated for each generation to maintain a line's capacity to spawn naturally and to provide gametes when stripped (Figure 7.3). The latter procedure involved pooling sperm from multiple males then rotating through multiple sperm pools to fertilize individual clutches of eggs. However, because this round-robin approach has the potential to disseminate recessive background mutations unchecked as compared with the natural breeding of pairs, the round-robin IVF approach has been discontinued for in-house stocks.

To preserve the quality of gamete stripping, IVF has been integrated into the process for AB line maintenance. AB populations are now designated specifically for the collection of stripped eggs for IVF with thawed sperm samples, and these populations are also generated primarily via IVF. Males that do not produce sperm upon first stripping are euthanized because these males typically do not become effective sperm producers over time. Females that release eggs during stripping are isolated from those that do not. When a new in-house stock is generated, the small-group crosses that had been stripped for eggs successfully and unsuccessfully in the database are tracked. Several criteria are considered when clutches are selected that will contribute to the next in-house generation. These include clutch size, fertility, and embryonic morphology. However, naturally spawned clutches from females that produced eggs when stripped are prioritized to contribute to the next generation.

Each single-pair derived batch of embryos of a new generation is screened for embryonic lethal mutations between days 1 and 5, and for swim bladder development at day 5. Applying the selection criteria established at the Institute of Neuroscience and maintaining the line's genetic diversity to the extent possible reduces inbreeding in the AB line. To better monitor the quality of the AB-1 population, an added level of screening was incorporated in 2017. In addition to selecting for normal embryo development, ZIRC also tracks the progeny of single-pair crosses until they reach approximately 4 months of age. This was initially implemented to address pigment mutations that began appearing in AB stocks but had not been detected in 5–6 days post-fertilization larvae typically screened only for swim bladder development. This extended young adult screen yielded additional benefits. ZIRC currently observes female bias (62% females on average) in the AB-1 population, which is not ideal due to the heavy reliance on AB for sperm cryopreservation. The sex of offspring can be determined at the 4 months evaluation, and we now choose families with balanced sex ratios to generate future stocks.

ZIRC has developed protocols to decrease disease susceptibility and increase the fertile period in the AB population. Previously, 5-month-old AB siblings were bred every 5 months, and new generations were established sequentially from the preceding ones. ZIRC is currently evaluating a new breeding plan that backcrosses the 5-month-old offspring of an AB stock to their 10-month-old parents to produce a new generation. In doing so, ZIRC will select for sustained breeding at 10 months of age. Furthermore, because health monitoring typically occurs around 8 months, and infected stocks are eliminated upon diagnosis, healthy fish with better disease resistance will be selected for.

Mutant Lines

ZIRC maintains a large number of lines carrying a variety of mutations, including point mutations, chromosomal deletions, translocations and complex chromosomal rearrangements, and virus insertional mutations.

Recessive zygotic lethal mutations maintained at ZIRC were predominantly generated in the original Tübingen, Boston, and Eugene screens as well as in the retrovirus-mediated insertional mutagenesis carried out in the Nancy Hopkins laboratory at the Massachusetts Institute of Technology (Amsterdam et al. 1999). The large-scale mutagenesis screens performed by the Zebrafish Mutation Project at the Sanger Institute (Hinxton, UK; Dooley et al. 2013) and the Burgess and Shuo Lin laboratories in the US (Varshney et al. 2013) have produced a significant number of lines with recessive viable mutations. The Resource Center also maintains lines with maternal-effect mutations received from the Mullins lab (Pelegri and Mullins 2016) in addition to mutations from many other laboratories.

Carriers of mutations with obvious visual defects are identified by their phenotype. Individuals carrying dominant visual traits can be distinguished easily from their wild-type siblings. Similarly, fish homozygous for viable recessive mutations with visible phenotypes can be readily distinguished and separated from their heterozygous and wild-type siblings. Identification of carriers of zygotic recessive lethal mutations is more complex because it involves the analysis of putative carrier progeny derived from single-pair incrosses. A genetic incross protocol is more complicated for identification of carriers of maternal recessive lethal mutations because it is carried out through two generations. Thus, phenotypic analysis of both zygotic and maternal recessive lethal mutations requires a larger number of tanks to accommodate single-pair crosses and to maintain incrossed pairs individually during the time required to phenotype their progeny. In addition, phenotypic analysis of these lines is labor intensive and depends on single-pair breeding, which can be a limiting factor for stocks that do not breed well or have a skewed sex ratio. For these reasons, many of the recessive lethal mutations whose lesions have already been characterized are identified by molecular genotyping.

Genotyping is independent of breeding and allows identification of a large number of mutation carriers in a relatively short period of time. However, some genotyping procedures rely on expensive reagents and, for some lines, the overall cost of genotyping significantly exceeds the cost of mutant identification via phenotyping. In addition, lines propagated solely on the basis of genotyping may accumulate background mutations that could eventually interfere with or obscure the original phenotype of the line. For this reason, some mutant lines that are propagated strictly by genotyping are also periodically analyzed visually to confirm their phenotype.

Transgenic Lines

The majority of the transgenic lines propagated and maintained at ZIRC can be classified into the following groups:

- Transgenic lines that contain exogenous promoters driving the expression of a reporter gene

- Lines generated for ectopic gene expression
- Enhancer traps
- Gene traps
- Chromosomal position markers

Homozygous and heterozygous transgenic fish expressing fluorescent reporter proteins are identified visually. Transgene carriers that express reporter genes during early development are typically identified as embryos or larvae derived from large group incrosses. In the case of lines with maternal transgene expression, transgene carriers are identified during the later stages of development when the maternal component is no longer detected. Otherwise, embryos lacking the transgene, but containing fluorescent reporter products deposited by the mothers, could be falsely identified as transgene carriers. Early embryonic stages can be used for transgene identification in lines with the maternal transgene expression only if the screened embryos are derived from an outcross in which transgenic males, but not females, are used. Transgenic lines that do not express visually detectable products are identified based on functional assays. For example, progeny of lines with *hsp70* promoter-dependent transgene expression are heat-shocked and then the effects of ectopic transgene expression on morphology are analyzed.

Transgene carriers for a number of lines are also identified by genotyping. Genotyping assays work well for transgenic lines for which the genomic location of the functional transgene has been characterized. This allows for the design of protocols in which transgene- and flanking region-specific primers can be used. Unfortunately, for a majority of transgenic lines maintained at ZIRC, the genomic sites of transgene integration are not known, and genotyping is based only on transgene-specific primers. This may lead to false positives if the line contains fragments or nonfunctional copies of the transgene. For this reason, transgene carriers identified by transgene-specific primers also need to be confirmed, if possible, by visual identification of transgene function.

Propagation of Mutant and Transgenic Lines

Only unambiguously identified carriers of mutations and transgenes are used in group crosses to propagate lines. Based on ZIRC's experience, crosses are more productive if the number of females exceeds the number of males. Different sizes of tanks are used for breeding based on the number of identified and available carriers, as well as on the demand for the line:

- 3–5-gallon tanks that hold up to 50 fish
- 1.5-gallon tanks for up to 15 females and 10 males
- Small group crosses in 0.5-gallon crossing cages with 3 females and 2 males

To maximize the number of mutation and transgene carriers in each new generation, incrosses rather than outcrosses are preferred for line propagation. Incrosses also ensure that homozygous individuals are generated. The demand is very high for homozygous fish, such as those carrying viable recessive mutations, or transgenes with reporter gene expression under the control of UAS sequence.

Line propagation solely by incrossing may lead to inbreeding, which significantly reduces the vigor and fecundity of a line. Thus, after three generations of incrosses, an outcross with AB wild types is used to generate the next offspring. Some lines, such as those carrying maternal dominant lethal mutations, can be propagated only by male outcrosses. Outcrossing is also frequently used for lines carrying alleles generated in large-scale screens in an attempt to "clean up" these lines by crossing out background mutations. Based on the finding that crossing-over events are suppressed in zebrafish males as compared to females (Singer et al. 2002), only females are used for outcrossing these lines. This practice assures a more effective elimination of background mutations residing on the same chromosome with the induced mutation.

CHARACTERIZATION OF FISH LINES

Information for line identification is collected during line acquisition from available publications and from the data provided by line submitters. This includes sequence and phenotype information for mutant lines, and transgene sequence and reporter gene expression for transgenic lines. The data are assembled into line-specific protocols that can be used to identify mutation and/or transgene carriers.

Visual Characterization

The observation of phenotypes is an important tool for accurate identification of genetic lines at ZIRC for which molecular information is not available. In addition, visual characterization of genetic modifications frequently complements the molecular characterization of mutant and transgenic lines. For example, mutant lines may have phenotypes that will be of interest to researchers but lack some or all information about the underlying molecular modifications of the genome at the time ZIRC reviews the submitted information.

Molecular analysis does not inform as to whether a particular transgene is functional, concatemerized, or whether it is expressed in a temporal and/or regional pattern consistent with endogenous genes. Visual characterization by bright field or epifluorescence microscopy can help distinguish between the expressivity and variability of phenotypes, as well as the patterning, functionality, and strength of expression of transgenes. The ZIRC uses a variety of stereomicroscopes and compound microscopes for visual characterization of lines. Curated images that complement data from publications or ZFIN records are added to the ZIRC fish lines database to aid future identification of the line. Embryos and larvae are typically inspected at 24, 48, 72, and 96 hours post-fertilization (hpf) and sometimes later as required by the nature of the transgene or mutation. It is sometimes not possible to visualize all of the characteristics exactly as illustrated in a publications (or as presented in ZFIN), so any complementary visual documentation provided by submitters or generated at the ZIRC is helpful for verification and future line identification.

Molecular Characterization

The genotypes of mutant and transgenic lines that are imported, distributed, or propagated in-house are identified by molecular methods. Assays used for line identification can be divided into two general groups. The first group includes assays whereby mutation and/or transgene carriers are identified visually based on their morphological phenotypes or transgenic products such as green fluorescent protein (GFP). The second group contains assays in which molecular techniques are utilized such as PCR, in situ hybridization, and immunocytochemistry. There are advantages and disadvantages for both types of analyses. For example, mutant lines analyzed solely by molecular techniques can accumulate background mutations that may obscure original phenotypes. Visual identification alone, on the other hand, may not always be suitable for distinguishing between different alleles affecting the function of the same gene. For this reason, whenever possible, a combination of both approaches is used for line identification.

Genotyping by PCR is the most common molecular approach for identification of mutation and transgene carriers, and different assays are used to genotype various mutant lines. Point mutations have typically been analyzed with Restriction Fragment Length Polymorphism (RFLP) assays (Botstein et al. 1980), derived Cleaved Amplified Polymorphic Sequence (dCAPS) assays (Neff et al. 1998), Allele-Specific Amplification (ASA) assays (Kwok et al. 1990, Newton et al. 1989), Kompetitive Allele Specific PCR (KASP) assays (LGC Group Inc.), and sequencing assays.

Mutations caused by deletions whose breakpoints have been characterized at the molecular level are typically genotyped by deletion-flanking primers. Mutations caused by deletions with uncharacterized breakpoints are genotyped with genomic markers covered by the deletion. Individual samples obtained from single haploid embryos are typically analyzed in these assays. The absence of marker-specific PCR products demonstrates that the sample contains the deletion. An additional primer set is used as an internal, positive control in these assays.

Insertional mutations are genotyped by a primer set designed so that one primer anneals to the insert DNA that caused the mutation, and the other primer hybridizes specifically to the genomic region flanking the mutation. Insertional mutations and transgenic lines for which genomic integration sites are unknown are genotyped with transgene-specific primers. It is possible that multiple copies of the transgene might have integrated into the genome during transgenesis and that some of these integrations are non-functional. Samples that contain only a non-functional transgene or a fragment of it will identify as false positives. For this reason, whenever possible, functional assays are used to verify individuals identified as positive in the genotyping assay. These assays include visual (e.g., GFP, morphology), molecular (e.g., in situ hybridization, immuno-cytochemistry), or biochemical (e.g., lacZ activity detection) identification procedures.

Mutations that have not yet been cloned are genotyped with Simple Sequence Length Polymorphism (SSLP or SSR) markers. Two closely linked SSLP markers that flank the mutation are selected for genotyping. These markers display inter-strain polymorphisms for the genetic background on which the mutation was induced and the background to which the mutation is outcrossed. An individual fish is considered to carry the mutation if no recombination is detected for both of these markers, and the PCR products are specific for the genetic background on which the mutation was originally induced. Whenever possible, fish identified by SSLP genotyping are verified by phenotype analysis or other procedures to confirm that they are indeed carriers of the mutation.

Designing PCR-Based Assays

Most genotyping assays are designed and tested in-house. ZIRC also tests and utilizes protocols provided by line submitters. To design PCR primers, MacVector (Accelrys) and Oligo 4.0-s (MBI) are used. MacVector software recognizes NCBI accession numbers and imports annotated sequences directly from GenBank. The accession numbers and the information about sequences are typically obtained from original publications, ZFIN, NCBI, or Ensemble websites. Primers for genomic DNA amplification are preferably designed for regions that are most likely to be conserved in various genetic backgrounds (e.g., exons). This is especially important for heterozygous samples containing mixed genetic backgrounds. For such samples, mismatches to primer sequences in one background often lead to preferential mono-allelic amplification and, as a result, the PCR product derives predominantly from the template containing no mismatches to primers. In addition to primer sequences, MacVector also provides calculations and predictions for the melting temperatures of the primers and the optimal annealing temperature for each primer set. Primer sets are designed with primer melting temperatures within 2°C of each other. MacVector and Oligo 4.0-s programs are also used to review and, if necessary, optimize existing protocols acquired from the provider of the line.

Line identification strategies used for line rederivation and cryopreservation. The majority of alleles (~95%) currently deposited at ZIRC have been imported as multiallelic mutant lines. Two general strategies are used for rederivation and cryopreservation of multiallelic lines. They are referred to as ID-Freeze and Freeze-ID.

The ID-Freeze strategy is chosen when the number of alleles to be genotyped in a stock is three or fewer. In this strategy, a stock is first genotyped and mutation carriers identified. Then, carriers with specific genotypes are grouped together and used for line cryopreservation.

The Freeze-ID approach is used when four or more alleles need to be genotyped in a single line. Using this strategy, a stock is first cryopreserved using unidentified males and then the line is genotyped using genomic DNA from fin-clips collected during cryopreservation. Once genotyping results for individual alleles are obtained, the resulting genotypes are assigned to each of the previously cryopreserved sperm samples. The Freeze-ID approach was introduced to handle the high number of genotypes present in stocks established for multiallelic lines. For example, a stock established for a line heterozygous for 5 alleles may contain individuals representing up to 32 different genotypes (i.e., allele combinations) which resulted from random segregation of each allele. If fish were genotyped first, each of these genotypes would have to be followed individually and a large number of separate stocks would need to be established for each multiallelic line.

PROPAGATION AND DISTRIBUTION OF THE RESOURCES

Line Propagation and Distribution Without Line Identification

There are over 40,300 alleles in the ZIRC inventory. The demand for lines is much higher than can be processed by ZIRC's genotyping capacity. To prevent delays in line distribution, some lines are propagated through one generation without genotyping.

In this procedure, sperm is cryopreserved from unidentified males. This sperm is then used for line rederivation and distribution. Two types of sperm samples are cryopreserved. The first are samples collected from individual males. When a line needs to be rederived, several samples of this group are thawed simultaneously, and multiple stocks are established per line. These stocks are then genotyped, and carriers of specific mutations and/or transgenes identified. Samples of the second type consist of sperm that are collected and pooled from 10 to 16 unidentified males. Depending on sperm quality and density, these pools are subsequently divided into several samples that are used for line distribution. Thaws from these samples generate progeny with approximately 25% of carriers, if the stock used for sperm cryopreservation originated from heterozygous individuals. Line distribution and propagation through one generation without genotyping not only supports timely distribution of resources but also makes line identification dependent on actual demand.

Assigning Line Identification Status to Sperm Samples

Line distribution and propagation without genotyping was introduced to accommodate large-scale acquisitions of multiallelic lines (Kettleborough et al. 2013, Varshney et al. 2013). These lines are referred to as F_1 lines because they were imported as sperm samples collected from F_1 males of the mutagenic screen (Kettleborough et al. 2013). Propagation of these lines with and without genotyping produced frozen samples with different states of line identification, which necessitated the definition of different sperm categories (C1–C5) in the inventory database.

- C1: samples derived directly from F_1 males
- C2: samples contain sperm pooled from several unidentified F_2 males
- C3: samples derived from individual, unidentified F_2 males
- C4: samples pooled from males that were genotyped at submitting institutions
- C5: samples derived from males genotyped at ZIRC

Assigning line identification status to sperm samples has streamlined the sample inventory and significantly increased the efficiency of line distribution and propagation.

ZIRC shares line identification resources with the zebrafish research community. Upon request, descriptions and images of phenotypes and expression patterns submitted to ZIRC are provided. ZIRC also shares sequence information, including transgene sequences, coordinates of transgene integration sites in the genome (when available), and the molecular nature of distributed mutations. Additionally, ZIRC posts genotyping protocols on the ZIRC website as PDF files. These protocols contain information about primer sequences, PCR program settings, PCR products, and, whenever applicable, information on restriction enzyme digestion. In addition to line-specific protocols, ZIRC has prepared a number of generic protocols describing the procedures, materials, and reagents used at ZIRC to genotype lines. ZIRC also conducts individual e-mail and phone consultations with regard to line identification. Finally, ZIRC works closely with ZFIN to assure that information shared with ZIRC is also available on ZFIN.

COLONY HEALTH AND BIOSECURITY

Biosafety and health monitoring, which are overseen by the ZIRC veterinarian, require the interaction of several workgroups (Figure 7.4, dark line, left side). The Diagnostic Health Service also interacts with members of the research community when providing diagnoses and consultations for improving the colony health at client facilities.

The health monitoring and biosecurity program at ZIRC encompasses three areas: (1) assessing colony health, (2) minimizing known pathogens, and (3) preventing entry of new pathogens. The success of this program is based on education and inclusivity. Every staff member is expected to be able to identify physical and behavioral signs of morbidity, have a basic understanding of how pathogens in the facility are transmitted, and have an appreciation for ZIRC's biosecurity objectives. We strive for a program in which all employees see themselves as members of a team with a goal of optimizing fish health. Results of our in-house health monitoring are made available quarterly in the Animal Health Report, which can be accessed through the webpage: https://zebrafish.org/wiki/health/health_reports/start.

Assessing Colony Health

Fish colony morbidity and mortality are monitored daily. Dead fish are removed from tanks immediately and recorded in the database. Moribund fish are removed within 72 hours of being identified. The number of fish removed from a tank and clinical signs are recorded in the database. The majority of moribund fish are fixed and processed for histopathology. If sufficient information about the health status of a tank is already known, additional moribund fish may be euthanized without further diagnostic testing (e.g., masses in fish with a mutation in the p53 gene or skinny fish in a tank diagnosed with microsporidiosis).

Tanks of sentinel fish are located in pre- and post-filtration locations on recirculating water systems that supply the main fish room. Sentinel fish are AB wild types that are the genetic background of the majority of fish in the facility. Fish are 3–4 months of age when they are placed in sentinel tanks and selected from a source tank that has been prescreened for the microsporidian parasite *Pseudoloma neurophilia*. Sentinel fish are sampled quarterly for histopathology after they have been exposed to system parameters for 6 months and 1 year. By comparing tissue changes and pathogen prevalence in fish exposed to pre- and post-filtration water, the program is able to evaluate the efficacy of our filtration and ultraviolet sterilizers at preventing pathogen recirculation.

The largest stocks maintained are of various wild-type lines. These fish are reared in 20-gallon (76-liter) tanks, as opposed to the majority of mutant and transgenic lines where small populations are reared in 1-gallon (3.8-liter) tanks. All 20-gallon tanks of wild-type fish are screened for *P. neurophilia* infection at 8 months of age. Random samples of fish are removed from the tanks and fixed for histopathology with Luna stain or frozen for PCR. Sample sizes are calculated such that diagnostic results provide 95% confidence of detecting a pathogen with a minimum prevalence of 25% or 50% (Murray et al. 2016).

Minimizing Known Pathogens

Diagnostic testing of moribund, sentinel, and randomly sampled wild-type fish has identified two pathogens in the main fish room. *Mycobacterium chelonae* is a low virulence, environmental bacterium that exists in fish tissue and biofilms. Most infections are subclinical and diagnosed by observation of acid-fast bacilli in histological sections of fish with no gross or behavioral indicators of morbidity (Murray et al. 2011). *M. chelonae* infections are minimized by maintaining optimized water quality parameters, instituting husbandry protocols to minimize fish stress, and limiting exposure to sources of mycobacteria. Dead fish are removed from tanks to prevent cannibalism of potentially infected tissue, and tanks are cleaned regularly to decrease exposure to biofilms (Whipps et al. 2012).

The second pathogen diagnosed in the colony is the microsporidian parasite *Pseudoloma neurophilia*. This is an obligate intracellular parasite that is transmitted horizontally by ingestion of the spore, infected fish tissue, or detritus, and vertically by intraovum infection (Kent and Bishop-Stewart 2003, Sanders et al. 2013). Treatment is not available for this infection. Thus, our control efforts are focused on the identification and removal of infected tanks and the prevention of transmission of infections between tanks (Murray et al. 2011). Infected tanks are identified by diagnostic testing of moribund fish and screening of wild-type populations. *P. neurophilia* spores can also be shed with gametes at spawning. Therefore, wild types that are used to outcross mutant and transgenic lines are not reused.

Various measures are in place to prevent transmission of disease between tanks. Emphasis is placed on detection of *M. chelonae* and *P. neurophilia*. However, these practices increase overall biosecurity and minimize the spread of unintentionally introduced new pathogens. Newly cleaned equipment is used on every tank during fish handling and tank cleaning. Tanks of wild-type stocks are arranged in vertical clusters to minimize cross-contamination by dripping water. Ultraviolet sterilizers deliver a minimum dose of 132,000 μWsec/cm2 to neutralize pathogens in recirculated water. Work areas are routinely cleaned with 70% ethanol or a dilute bleach solution. Tanks are washed with acid, alkaline, and hydrogen peroxide at 170°F. Nets and scrubbers are autoclaved between uses. The efficacy of disinfection methods is evaluated regularly by culture on RODAC plates and ATP monitoring (Hygiena ATP monitoring system).

All embryos are surface-sanitized by soaking for 10 minutes in a 30 ppm bleach solution. Although this will not neutralize *P. neurophilia* nor eliminate mycobacteria (Chang et al. 2015, Ferguson et al. 2007), it decreases overall bacterial and fungal organisms to which larval fish are exposed.

Preventing the Entry of New Pathogens

The greatest risk for the entry of pathogens occurs when new fish are introduced to the facility. New lines are introduced as live adult fish and cryopreserved sperm. The majority of new lines are shipped to ZIRC as sperm samples that were cryopreserved at the exporting facility. Studies have shown that most zebrafish pathogens can survive cryopreservation and remain viable when thawed (Norris et al. 2018). Prior to the arrival of the sperm samples, information about previous diagnostic testing and disease history is gathered from the exporting facility. Fixed specimens of moribund or pre-filtration sentinel fish are also requested from the facility and processed for histopathological examination. Diagnostic results from fixed specimens and health history information is used to evaluate the risk of importing pathogens with sperm samples and to assign a health status to them. Once received by ZIRC, sperm samples are stored in designated quarantine freezers. When the line is regenerated by IVF and needs to be reared to reamplify frozen sperm samples, the health status determines whether the first generation will be reared in the quarantine room or main fish room (Murray et al. 2016).

New lines are also imported as live fish into the quarantine room. For mutant and transgenic lines, only male fish are imported. After acclimation, sperm samples are collected and cryopreserved. Although pathogens can also be cryopreserved, the number of pathogens surviving and contaminating thawed sperm samples is likely less than that present during natural spawning. When sperm are collected, the male fish are then fixed and processed for histopathology examination and evaluated before the line is regenerated. Diagnostic results are used to determine whether the first generation will be reared in the quarantine room or the main fish room.

Wild-type lines cannot be cryopreserved. Therefore, both males and females must be imported and spawned. Wild-type lines are imported when there are no other fish in the quarantine room. After spawning, embryos are surface-sanitized and held in quarantine. Successfully spawned adults are euthanized and fixed for histopathology. If no or only low virulence (i.e., *P. neurophilia* or *M. chelonae*) pathogens are diagnosed in the adults, embryos are reared in the main fish room. If high risk pathogens (e.g., *M. marinum* or *M. haemophilum*) are diagnosed, the embryos are reared in the quarantine room and evaluated by diagnostic testing before their progeny are moved to the main fish room.

Within the facility, ZIRC recognizes that staff can be a vector for pathogen introduction into the main fish room, and protocols are in place to minimize that risk. Duties in the quarantine room are limited to a few full-time staff members. Designated facility shoes are worn by staff members, and shoe covers are used by guests. Sticky mats at all doorways prevent tracking detritus within the facility. Gloves are worn whenever working with fish or fish water. Personnel and guests are not permitted into the fish facility if they have been in another fish facility the same day.

Live feeds can also be a vector for pathogen introduction. To feed larval fish, ZIRC's in-house paramecia culture is occasionally supplemented with rotifers. *Artemia* spp. (brine shrimp) are fed beginning at 25 days post-fertilization. The brine shrimp are decapsulated in-house with a bleach solution to eliminate external contaminants and make them palatable. Samples of paramecia cultures are tested quarterly with a *Mycobacterium* spp. PCR panel assay. Rotifer cultures are also tested by PCR for *Mycobacterium* spp. contamination before they are fed to larval fish.

ZIRC DIAGNOSTIC SERVICES

ZIRC operates a diagnostic pathology service in collaboration with the Oregon Veterinary Diagnostic Laboratory. We provide histopathology, bacteriology, virology, parasitology, necropsy, and molecular assays for the diagnosis of moribund and sentinel zebrafish. Clients contact ZIRC through the ZIRC website. A summary of annual submissions and diagnostic findings is available for review from a link at (https://zebrafish.org/wiki/health/submission/report). This summary provides an overview of the prevalence of particular pathogens and pathologies in zebrafish facilities.

CURRENT AND FUTURE RESEARCH EFFORTS

Zebrafish pathogens have negative direct and indirect effects on research. Potential impacts include altered behavior (Spagnoli et al. 2015), repressed fecundity (Ramsay et al. 2009), massive mortalities (Hawke et al. 2013), and human infection by a zoonotic pathogen (Mason et al. 2016). Implementation of systematic and reliable pathogen detection and health monitoring in zebrafish facilities is crucial for effectiveness and efficiency in achieving scientific goals. Yet, platforms offering health screening are sometimes lacking, and the majority of facilities do not always perform systematic pathogen screening of the zebrafish populations (Lawrence et al. 2012). Zebrafish molecular diagnostic kits are available. However, detailed information about the detection assays, including their sensitivity and validation, are often proprietary. This can confound the interpretation of results and an assessment of the relative risks of false negatives and positives. If negative results are incorrectly reported to the user, this can provide a false sense of security and have severe consequences on the research program. Similarly, false positives could result in unnecessary anxiety, retesting, and culling. ZIRC has established a successful health-monitoring program (Murray et al. 2016). Current monitoring is predominantly based on histopathology.

ZIRC's goal is to develop and utilize PCR assays for detecting the most prevalent pathogens of laboratory zebrafish. Development of a panel of PCR assays, and its incorporation into the in-house health monitoring program, will improve the overall efficiency of disease detection. PCR-based assays are higher throughput than the use of histopathology. The use of molecular assays will enable the processing of much larger numbers of fish than can be typically achieved through histopathology. The latter technique requires time-consuming fixation, processing, sectioning, and staining of individual fish, and then relies on a specialized expert to read the tissue sections. Detection of pathogens in tissue sections is often influenced by whether or not pathogens occur within the section plane, the number of pathogenic organisms present, the surrounding tissue changes, and the expertise of the individual reading the slides. PCR is anticipated to be a more sensitive means of pathogen detection than histopathology. PCR can be used to simultaneously identify the species Mycobacterium present. This has important implications for husbandry and control efforts. Histopathology-based examination of tissue sections allows for the visualization of acid-fast bacilli and a presumptive diagnosis of Mycobacterium but cannot identify bacterial species. PCR assays may also be applied to environmental samples (biofilms, debris, water).

Information about PCR-based pathogen detection assays and the ZIRC health-monitoring program are shared with the research and user communities. Detailed protocols are made available on the Health Services website. Health monitoring and testing information are also presented at scientific meetings. Zebrafish research laboratories often have the equipment and expertise to run molecular assays but are not involved in their local fish facility's health programs. Providing the tools to perform in-house diagnostics and the experience obtained at ZIRC, the project has the potential to encourage communication and cooperation between zebrafish facility staff and research personnel and to inspire interest and investment in fish health.

ZIRC DATABASE MANAGEMENT AND ACCESS

To manage the large number of resources (fish, sperm samples, alleles, etc.), to facilitate staff and work group activities, and to monitor physical systems, ZIRC has created several databases and web-based applications. A chart of the primary functional areas where automation has been developed is shown in Figure 7.6.

When new orders for fish lines are received, relevant staff are given notice via automated e-mail, and the order detail is automatically made available for scheduling of staff and curation of the information. The order is linked to an existing laboratory or organization, or a new one is created, and a scheduled shipment date is established and coordinated with the customer. Once a scheduled shipment date is established, individuals responsible for production of orders begin processing the order late in the week prior to the week the fish are scheduled to ship. All other non-fish resources are processed soon after the order is received. When processing a fish order, relevant information about fish tanks and/or sample locations is automatically

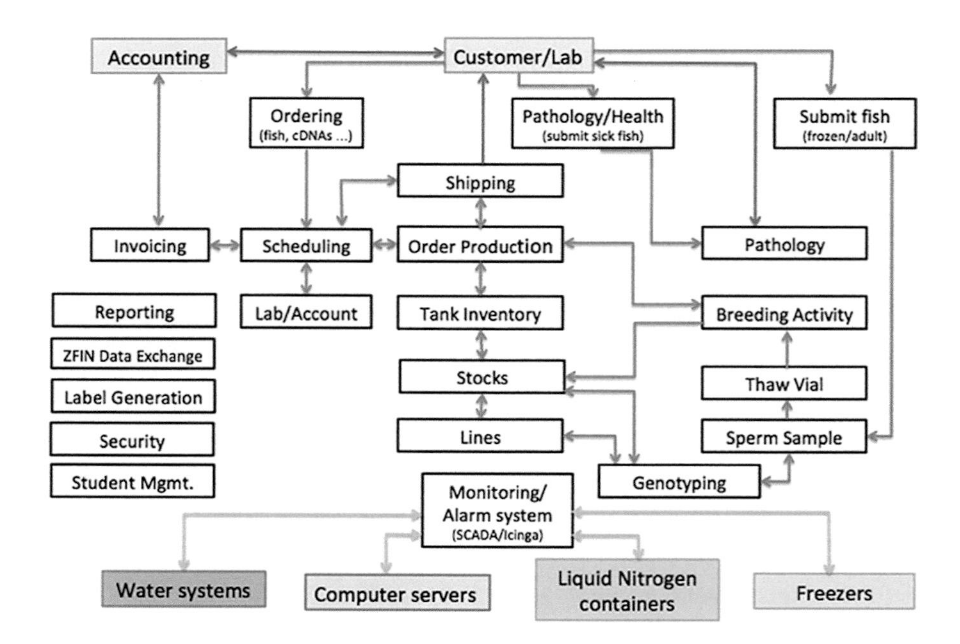

FIGURE 7.6 ZIRC's Database Applications. The chart shows functions for which applications have been created to support work group tasks, data collection and inventory, or systems monitoring.

displayed with the order item. Staff identify the stocks and tank(s) where the fish are located, or locate the frozen sperm samples to fertilize eggs to fulfill the order. Prior to shipping, staff generate an invoice from the database system and mark the order as "Shipped," which also automatically updates the count of fish in the tanks.

The database supports considerable automation and integrated information to enable staff from different work groups to carry out their tasks in a semi-automated fashion. Staff can access relevant data from the 1000+ active fish tanks and more than 80,000 frozen samples. Several workgroups can work in parallel on the same request without having to communicate updates among themselves. This improves the efficient and accurate processing of orders.

All database applications have been designed with data integrity as a major focus, applying referential integrity with browser pull-downs when possible (underlying tables of data are used to generate pull-downs), and background processes that check for data enforcement, advisory, or verification actions. However, the complexity of operations and frequent operational changes require continual enhancement of the application systems. Tracking the large number of multi-allelic live fish or frozen samples with their sample categories impacts the speed of the database at peak operating times.

Programs are run nightly (cronjobs) to review data for anomalies, reporting them to the IT workgroup. The availability of lines may be automatically altered because of changes to the number of fish in tanks and/or the number of cryopreserved sperm samples available. A download of select ZFIN data also takes place nightly and checks for updates and changes. If an allele name has changed in the ZFIN database, a process notifies staff and then applies the change to the database.

The majority of ZIRC applications are written in PHP, a server-side scripting language, where data are accessed and managed from a PostgreSQL database server. Web pages generated by PHP applications are made up of HTM and Javascript. ZIRC uses Apache2 as its web page server. Icinga, an open source computer system and network monitoring application (Benthin 2010), is used to monitor systems and notify staff, and the Ganglia software graphing tool is used to display data trends over time.

To support out PostgreSQL servers, Apache2 web servers, SCADA monitoring system, and other software servers, we rely on two rack-mounted servers. The server applications are installed onto multiple virtual servers. ZIRC also hosts several Web600 Web-Monitor Alarm devices that receive and send information connected to the many sensors used to monitor the aquaculture system, computer servers, LN freezers, and refrigerators.

COLLECTION SECURITY

The aquaculture systems are monitored for anomalies to water levels, water temperatures, pressure, pumps, and ultraviolet sterilizer operation. Computer servers are monitored for hardware and software failures and warnings, LN containers for liquid levels and temperature, and other freezers and refrigerators for temperature. If any of the monitored parameters falls outside the acceptable range, alarms are set, e-mails sent, and notices texted (SMS) to the smartphones of appropriate team members.

DISTRIBUTION OF RESOURCES

A key element of ZIRC operations is the efficient distribution of zebrafish resources to the research community. Distributions from ZIRC have increased significantly over time as a result of the growth of the research community (Figure 7.7). Here, we focus on the distribution of adults and embryos; however, similar procedures apply to supplying other products: cDNA/ESTs, monoclonal antibodies, paramecia, and *The Zebrafish Book*.

Clients can access ZIRC through the home page (https://zebrafish.org) for information on the availability of lines, fees, the Material Transfer Agreement (MTA), and to place an order. Researchers can access the ZIRC zebrafish lines inventory from several public databases including the strains catalog (https://zebrafish.org/fish/lineAll.php), ZFIN, and the Zebrafish Mutation Project site for Sanger alleles (http://www.sanger.ac.uk/sanger/Zebrafish_Zmpbrowse). The latter two sites provide direct links to specific resources in the ZIRC catalog. The ZIRC catalog is a shopping cart system through which various materials and fish strains can be selected and added to an order. During checkout, shipping, billing, grant, and principal investigator details are collected, and the MTA is provided. When an order is received, the availability of the requested lines is determined. We then reply to the customer to confirm or clarify which lines they require or to negotiate alternatives (i.e., a different allele or embryos, if adults are unavailable). At this point, a shipping date is also determined.

LICENSING, MATERIAL TRANSFER AGREEMENTS, AND SHIPPING

All resources submitted to ZIRC are deposited under an academic license, which allows for distribution to academic institutions. The ZIRC Material Transfer Agreement (MTA), under which ZIRC distributes resources, is based

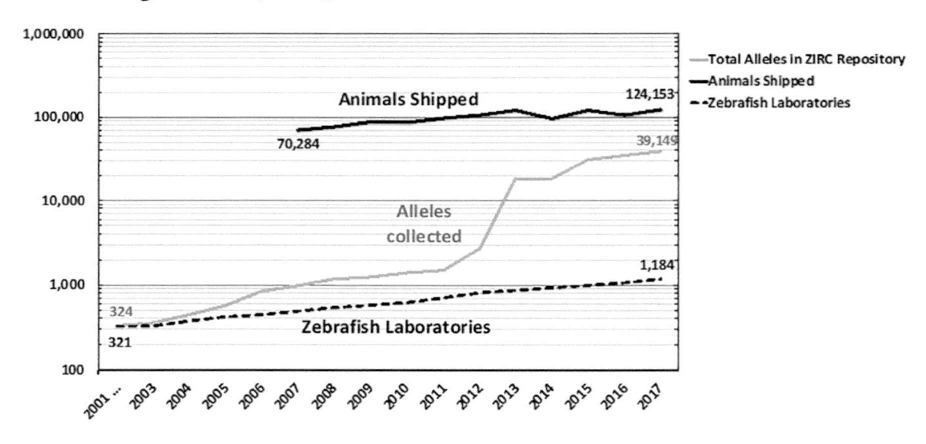

FIGURE 7.7 Annual Animal Shipments since 2007. The continuous black line (top) indicates the number of animals shipped each year; the stippled line indicates the growth of the zebrafish community measured by the number of laboratories registered at ZFIN. The grey line is the number of alleles imported since 2001 (see Figure 7.2). The comparison suggests that demands on the Resource Center have increased with the growth of the research community, not by the number of available alleles.

on the NIH Universal Biological Material Transfer Agreement (UBMTA), which provides a framework for their use, acknowledgments, and limitations (https://www.autm.net/resources-surveys/material-transfer-agreements/uniform-biological-material-transfer-agreement/). The UBMTA can be used in place of the ZIRC MTA if requested. To streamline the process, the ordering system has an electronic "click-through" MTA that all purchasers must agree to before placing an order. If requested, a hardcopy document with signatures (either the ZIRC MTA or UBMTA) is available and will supersede the electronic agreement in the ordering system. The University of Oregon Innovation Partnership Services (IPS) is available to address any legal concerns that arise with these agreements.

The resources available through ZIRC are licensed solely for academic use. Any commercial entities requesting resources from ZIRC must obtain a material-specific license from the source institution that originally provided the materials. Similarly, other stock centers requesting lines are required to obtain a distribution license agreement from the source institutions.

Due to the dynamic nature of international shipping, ZIRC does its best to stay apprised of changes to customs requirements and documentation, such as import permits, VAT exemptions, and health certificates. ZIRC supplies a customs letter indicating the species of fish, their history, disease status, and a customs statement declaring their use for biomedical research purposes only. In addition, an international invoice stating a nominal value of $5 is sent in triplicate for customs purposes. Customers are advised of the paperwork provided and informed that they may need to contact their customs office for any required licenses or permits. There is a document processing fee for international shipments. This fee varies based on the complexity of the importing country's requirements. For example, Australia requires a USDA heath certificate with associated fees that are passed on to the customer. After breeding of adults and collection of embryos for shipment, embryos are bleached (Figure 7.8), fish are packaged in culture flasks (embryos) or plastic shipping bags (adults), and placed in Styrofoam shipping boxes with packing peanuts and, depending on season, with heating pads. Customs papers are attached to the outside of the shipping box, and tracking numbers are sent with each shipment (Varga 2016).

Warranties/Guarantees and Payment

The distribution of live animals and temperature-sensitive reagents is inherently risky, especially when using regular mail courier services. ZIRC guarantees safe delivery of live animal shipments but limits losses to immediate report of loss and replacement of the specific line(s), excluding document fees, the packaging charge, and shipping cost. Hence, the risk is divided between the Resource Center and the recipients. ZIRC accepts all types of payments, and payment details are provided on each invoice. ZIRC will reference purchase orders on the invoice, but they do not constitute an order. Due to the online MTA, orders are accepted only via the website catalog.

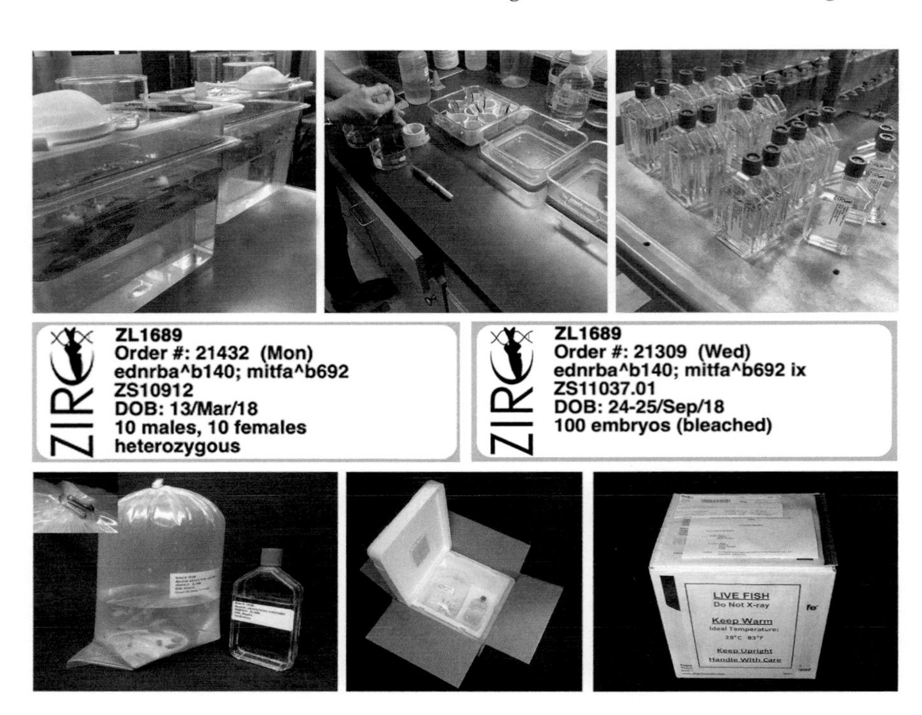

FIGURE 7.8 ZIRC Embryo and Adult Fish Shipping Procedure. Top row left: Adult wild types or carriers of genetic modifications are set up in pairs or groups in a container subdivided by a grid insert that lets fertilized eggs fall through to protect them from cannibalism by the adults. The next morning, spawning starts after lights are turned on in the facility and early embryos are collected in tea-sieves before noon. From tea-sieves, embryos are transferred to petri dishes where they are cleaned, counted, and sorted. Top row middle: At 24 hours post fertilization, eggs are surface sanitized in a dilute bleach solution and transferred into tissue culture flasks in clean water with methylene blue. Top row right: Flasks are labeled with the appropriate genotype and ordering information. Middle row: Typical shipping labels for adults (left) and embryos (right). The ZIRC line number (ZL, unique identifier), Order # and shipping day, genotype, stock number (ZS), and date of birth (DOB) are provided to clients and logged in the database for potential troubleshooting. Adults (left) are also labeled with the number and zygocity of males and females in the bag, whereas embryos are labeled with animal number and indication of completed surface sanitization. Bottom row left: If adults have been requested, they are placed into double plastic bags that are closed with a metal clip (inset). Bottom row center: Embryos in flasks and/or adults in plastic bags are placed in thick-walled Styrofoam boxes and unfilled space is buffered and isolated with water-insoluble packing peanuts. In the cold season heat bags may be taped to the inside of the box lid. Bottom row right: The cardboard box containing the Styrofoam box with animals is taped shut and the waybill, shipping labels, and (international) customs and veterinary documents are added to the outside of the box before it is picked up by a courier for transport to its destination.

REFERENCES

Ali, S., D.L. Champagne, H.P. Spaink and M.K. Richardson. 2011. Zebrafish embryos and larvae: A new generation of disease models and drug screens. *Birth Defects Research Part C: Embryo Today: Reviews* 93:115–133.

Amsterdam, A., S. Burgess, G. Golling, W. Chen, Z. Sun, K. Townsend, S. Farrington, M. Haldi and N. Hopkins. 1999. A large-scale insertional mutagenesis screen in zebrafish. *Genes & Development* 13:2713–2724.

Barman, R.P. 1991. A Taxonomic revision of the Indo-Burmese species of *Danio* Hamilton Buchanan (Pisces: *Cyprinidae*). Occasional Paper No.137, Records of the Zoological Survey of India, Calcutta, India.

Botstein, D., R.L. White, M. Skolnick and R.W. Davis. 1980. Construction of a genetic linkage map in man using restriction fragment length polymorphisms. *American Journal of Human Genetics* 32:314–331.

Chang, C.T., E.G. Colicino, E.J. DiPaola, H.J. Al-Hasnawi and C.M. Whipps. 2015. Evaluating the effectiveness of common disinfectants at preventing the propagation of *Mycobacterium* spp. isolated from zebrafish. *Comparative Biochemistry and Physiology C-Toxicology & Pharmacology* 178:45–50.

Dooley, C.M., C. Scahill, F. Fényes, R.N. Kettleborough, D.L. Stemple and E.M. Busch-Nentwich. 2013. Multi-allelic phenotyping—A systematic approach for the simultaneous analysis of multiple induced mutations. *Methods* 62:197–206.

Driever, W., L. Solnica-Krezel, A.F. Schier et al. 1996. A genetic screen for mutations affecting embryogenesis in zebrafish. *Development* 123:37–46.

Eisen, J.S. 1996. Zebrafish make a big splash. *Cell* 87:969–977.

Engeszer, R.E., L.B. Patterson, A.A. Rao and D.M. Parichy. 2007. Zebrafish in the wild: A review of natural history and new notes from the field. *Zebrafish* 4:21–40.

Ferguson, J.A., V. Watral, A.R. Schwindt and M.L. Kent. 2007. Spores of two fish microsporidia (*Pseudoloma neurophilia* and *Glugea anomala*) are highly resistant to chlorine. *Diseases of Aquatic Organisms* 76:205–214.

Golin, J., D.J. Grunwald, F. Singer, C. Walker and G. Streisinger. 1982. Ethyl nitrosourea induced germline mutations in zebrafish *Brachydanio-rerio. Genetics* 100:S27.

Grunwald, D.J. and G. Streisinger. 1992. Induction of recessive lethal and specific locus mutations in the zebrafish with ethyl nitrosourea. *Genetics Research* 59:103–116.

Haffter, P., M. Granato, M. Brand et al. 1996. The identification of genes with unique and essential functions in the development of the zebrafish, *Danio rerio. Development* 123:1–36.

Hamilton, F. 1822. *An Account of the Fishes Found in the River Ganges and Its Branches.* Edinburgh, UK: Archibald Constable and Co.

Hawke, J.P., M. Kent, M. Rogge, W. Baumgartner, J. Wiles, J. Shelley, L.C. Savolainen, R. Wagner, K. Murray and T.S. Peterson. 2013. Edwardsiellosis caused by *Edwardsiella ictaluri* in laboratory populations of Zebrafish *Danio rerio. Journal of Aquatic Animal Health* 25:171–183.

Ho, R.K. and D.A. Kane. 1990. Cell-autonomous action of zebrafish *spt-1* mutation in specific mesodermal precursors. *Nature* 348:728–730.

Howe, K., M.D. Clark, C.F. Torroja et al. 2013. The zebrafish reference genome sequence and its relationship to the human genome. *Nature* 496:498–503.

Jayaram, K.C. 1981. *The Freshwater Fishes of India, Pakistan, Bangladesh, Burma, and Sri Lanka: Handbook*, Zoological Survey of India, Calcutta Laser Graphics, Government of India Publications, Calcutta.

Kent, M.L. and J.K. Bishop-Stewart. 2003. Transmission and tissue distribution of *Pseudoloma neurophilia* (Microsporidia) of zebrafish, *Danio rerio* (Hamilton). *Journal of Fish Diseases* 26:423–426.

Kettleborough, R.N., E.M. Busch-Nentwich, S.A. Harvey et al. 2013. A systematic genome-wide analysis of zebrafish protein-coding gene function. *Nature* 496:494–497.

Kinth, P., G. Mahesh and Y. Panwar. 2013. Mapping of zebrafish research: A global outlook. *Zebrafish* 10:510–517.

Kwok, S., D.E. Kellogg, N. McKinney, D. Spasic, L. Goda, C. Levenson and J.J. Sninsky. 1990. Effects of primer-template mismatches on the polymerase chain reaction: Human immunodeficiency virus type 1 model studies. *Nucleic Acids Research* 18:999–1005.

LaFave, M.C., G.K. Varshney, M. Vemulapalli, J.C. Mullikin and S.M. Burgess. 2014. A defined zebrafish line for high-throughput genetics and genomics: NHGRI-1. *Genetics* 198:167–170.

Lawrence, C., D.G. Ennis, C. Harper, M.L. Kent, K. Murray and G.E. Sanders. 2012. The challenges of implementing pathogen control strategies for fishes used in biomedical research. *Comparative Biochemistry and Physiology Part C: Toxicology & Pharmacology* 155:160–166.

Mason, T., K. Snell, E. Mittge, E. Melancon, R. Montgomery, M. McFadden, J. Camoriano, M.L. Kent, C.M. Whipps and J. Peirce. 2016. Strategies to Mitigate a *Mycobacterium marinum* outbreak in a zebrafish research facility. *Zebrafish* 13(Suppl 1):S77–S87.

Matthews, J.L., J.M. Murphy, C. Carmichael, H. Yang, T. Tiersch, M. Westerfield and Z.M. Varga. 2018. Changes to extender, cryoprotective medium, and *in vitro* fertilization improve zebrafish sperm cryopreservation. *Zebrafish* 15:279–290.

Murray, K.N., J. Bauer, A.Tallen, J.L. Matthews, M. Westerfield and Z.M. Varga. 2011a. Characterization and management of asymptomatic Mycobacterium infections at the Zebrafish International Resource Center. *Journal of the American Association for Laboratory Animal Science* 50:675–679.

Murray, K.N., M. Dreska, A. Nasiadka, M. Rinne, J.L. Matthews, C. Carmichael, J. Bauer, Z.M. Varga and M. Westerfield. 2011b. Transmission, diagnosis, and recommendations for control of *Pseudoloma neurophilia* infections in laboratory zebrafish (*Danio rerio*) facilities. *Comparative Medicine* 61:322–329.

Murray, K.N., Z.M. Varga and M.L. Kent. 2016. Biosecurity and health monitoring at the Zebrafish International Resource Center. *Zebrafish* 13(Suppl 1):S30–S38.

Neff, M.M., J.D. Neff, J. Chory and A.E. Pepper. 1998. dCAPS, a simple technique for the genetic analysis of single nucleotide polymorphisms: Experimental applications in *Arabidopsis thaliana* genetics. *Plant Journal* 14:387–392.

Newton, C.R., A. Graham, L.E. Heptinstall, S.J. Powell, C. Summers, N. Kalsheker, J.C. Smith and A.F. Markham. 1989. Analysis of any point mutation in DNA. The amplification refractory mutation system (ARMS). *Nucleic Acids Research* 17:2503–2516.

Norris, L.J., V. Watral and M.L. Kent. 2018. Survival of bacterial and parasitic pathogens from zebrafish (*Danio rerio*) after cryopreservation and thawing. *Zebrafish* 15:188–201.

Parichy, D.M. 2015. Advancing biology through a deeper understanding of zebrafish ecology and evolution. The natural history of model organisms; *eLIFE* 4. doi:10.7554/eLife.05635.001.

Patton, E.E., P. Dhillon, J.F. Amatruda and L. Ramakrishnan. 2014. Spotlight on zebrafish: Translational impact. *Disease Models & Mechanisms* 7:731–733.

Pelegri, F. and M.C. Mullins. 2016. Genetic screens for mutations affecting adult traits and parental-effect genes. *Methods in Cell Biology* 135:39–87.

Ramsay, J.M., V. Watral, C.B. Schreck and M.L. Kent. 2009. *Pseudoloma neurophilia* infections in zebrafish *Danio rerio*: Effects of stress on survival, growth, and reproduction. *Diseases of Aquatic Organisms* 88:69–84

Sanders, J.L., V. Watral, K. Clarkson and M.L. Kent. 2013. Verification of intraovum transmission of a microsporidium of vertebrates: *Pseudoloma neurophilia* infecting the zebrafish, *Danio rerio. PLoS One* 8(9):e76064. doi:10.1371/journal.pone.0076064.

FIGURE 1.1 Three different Mexican axolotl stocks that differ in their pigment pattern. The wildtype axolotl has three different pigment cells (xanthophores, melanophores, and iridophores) that combine to yield a dark greenish coloration (left). The white axolotl (center) is homozygous for a mutated *endothelin 3* gene that results in the loss of dark melanocytes soon after hatching. The golden albino (right) has an abundance of yellow xanthophores but lacks a functional *tyrosinase* gene for melanin production. (Courtesy of Lee Thomas.)

FIGURE 2.2 Floral organ specification mutants (*ap1-1*, *ap3-1*, and *ag-1*) and Ler-0 reference strain.

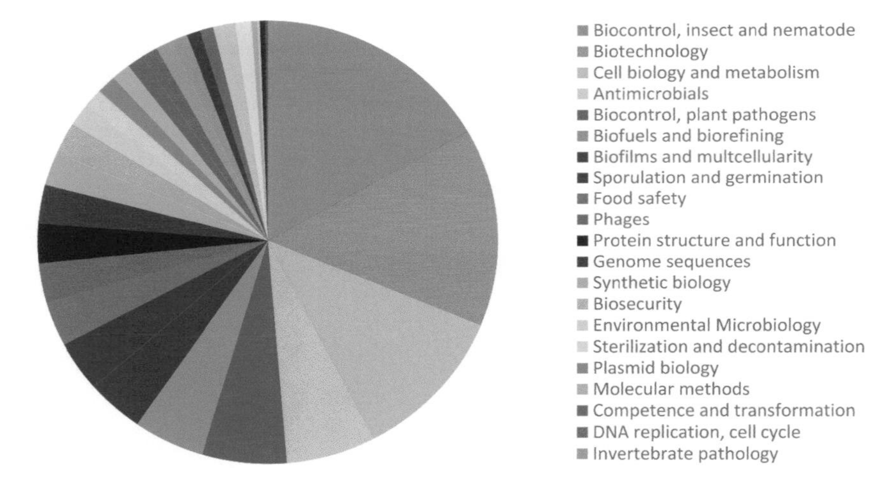

FIGURE 3.2 References citing use of Bacillus Genetic Stock Center strains from April 2013 to March 2018. Pie chart indicates the proportion of references belonging to an identified research front.

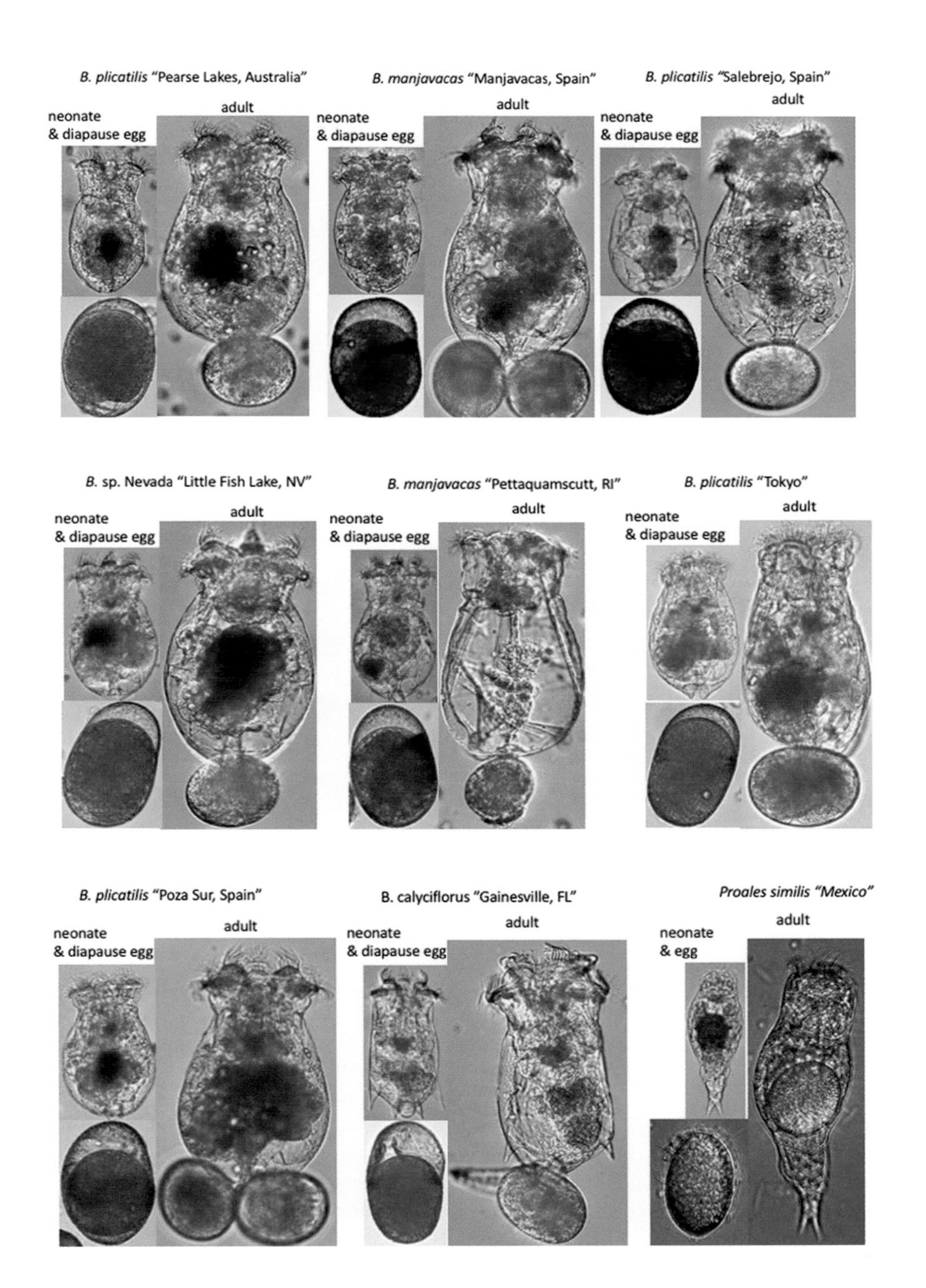

FIGURE 4.3 Photomicrographs of representative members in the *Brachionus* diapause egg collection. For each species, the large photo on the right is an adult female, the small photo in the upper left is a neonate female, and a diapause egg is pictured in the lower left.

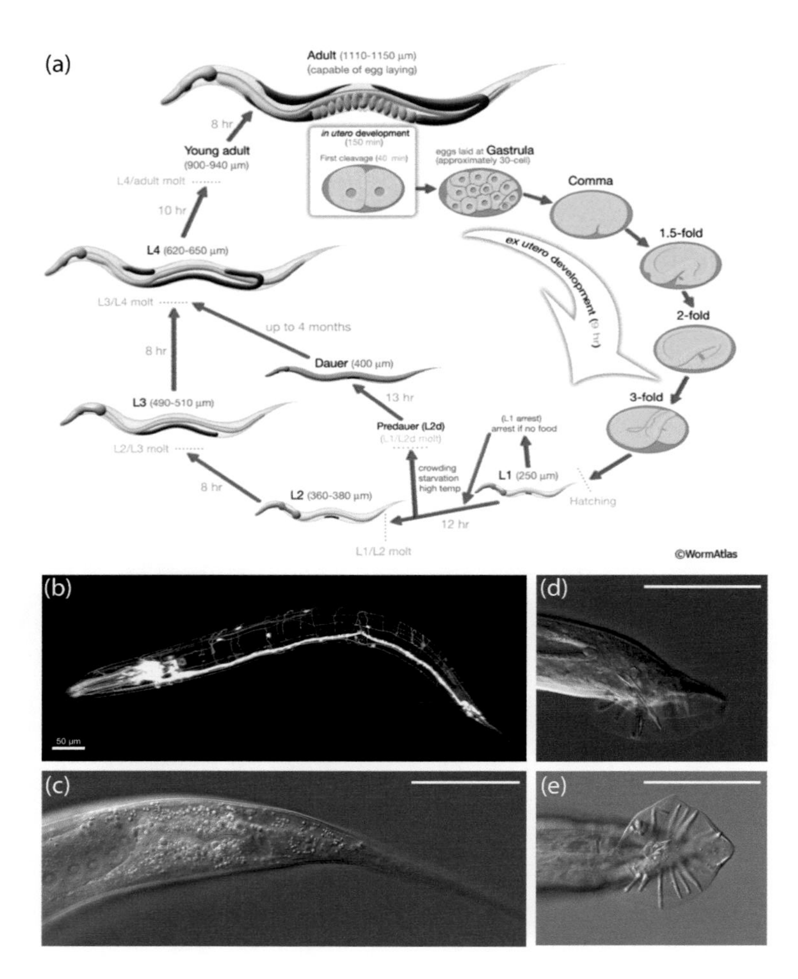

FIGURE 5.1 *Caenorhabditis elegans* life cycle, nervous system, and sexually dimorphic tail. (a) Hermaphrodite life cycle. Under ideal growth conditions, *C. elegans* will develop from a fertilized egg to an adult in just over two days. The first few hours of development occur *in utero*; the egg is then laid and embryogenesis continues until the basic body plan is complete and the first stage (L1) larva hatches. Development continues, punctuated by molts, through three additional larval stages (L2 to L4) until the sexually mature adult stage is reached. When conditions are harsh, for example due to high temperature, starvation, and/or crowding, *C. elegans* L1 larvae can enter an alternate developmental stage known as a dauer larva, which confers resistance to environmental stress and promotes survival. If conditions become favorable, dauer larvae can resume development. Reprinted with permission from WormAtlas. (b) The simple *C. elegans* nervous system is visualized by pan-neuronal expression of the GFP transgene *evIs111* [*F25B3.3::GFP* + *dpy-20*(+)]. (c, d, and e) Micrographs showing tail morphology. Scale bars are 50 μm. A hermaphrodite's tail tapers gently to a point (c), whereas an adult male tail has a fan-like copulatory structure shown in lateral (d) and ventral (e) views.

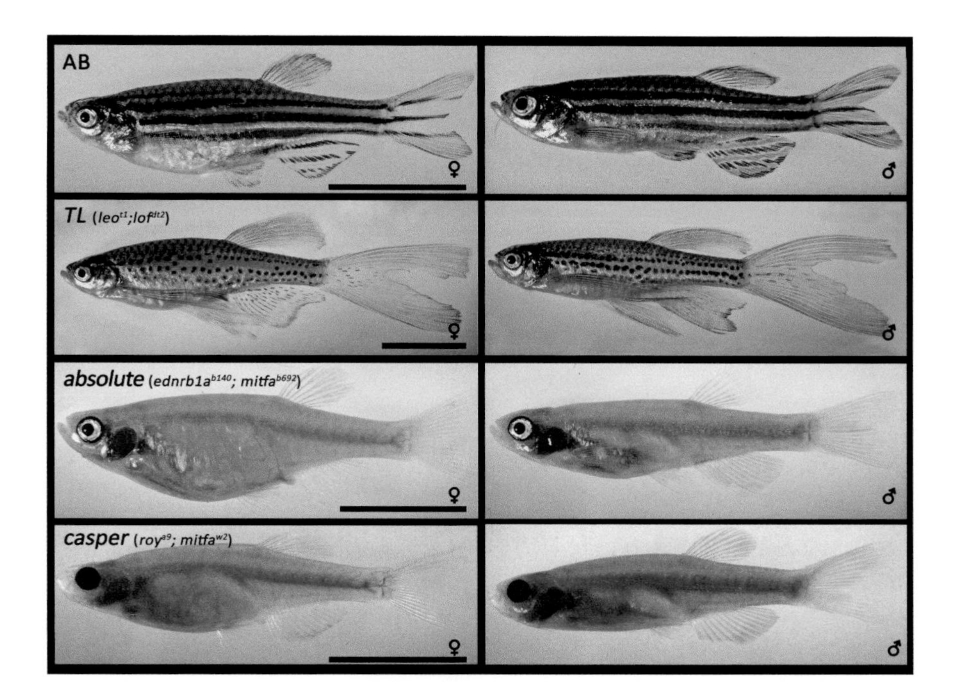

FIGURE 7.3 Adult wild-type and pigment double-mutant strains maintained at ZIRC. Top row, AB wild-type strain. Characteristic dimorphisms of the sexes include white protruding belly, yellow dorsal fin, and urogenital papilla (females) and usually slender males, with a yellowish (sometimes reddish) belly and anal fins, clear dorsal fin. Second row from top: Tüpfel-longfin (*leo*^{t1};*lof*^{dt2}) adults have dotted (dots = Tüpfel in German) pigmentation instead of stripes due to the *leopard* (*leo*) mutation. *lof*^{dt2} is a dominant homozygous viable mutation causing long fins. Third row from top: *absolute* (*ednrb1a*^{b140};*mitfa*^{b692}) double mutant fish lack melanophore, xantophore, and most iridophore cells. Iridophores are present in the iris. Bottom row: *casper (roy*^{a9};*mitfa*^{w2}) lack melanophores, most iridophore, and xantophore cells. Melanocytes are present in the eyes, which however, lack iridophores. Scale bars: 1 cm; each bar corresponds to female (left) and male (right) in the same row.

FIGURE 8.1 *Drosophila melanogaster.*

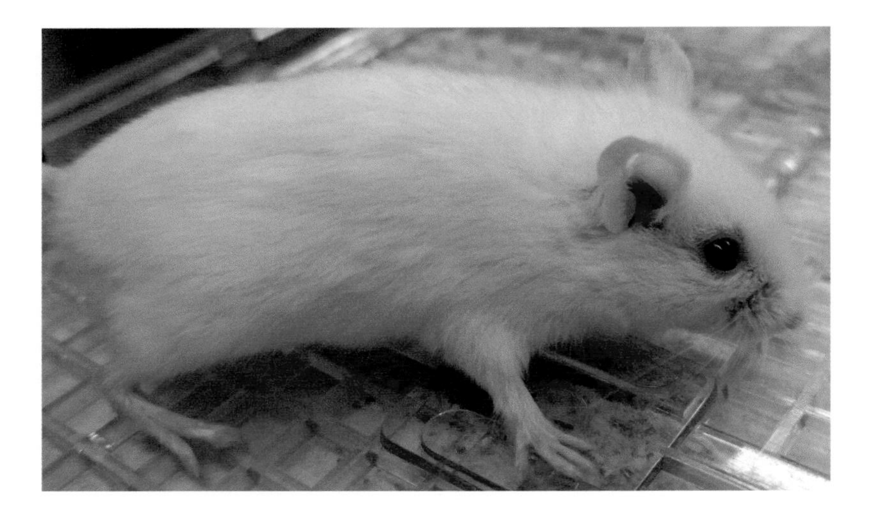

FIGURE 10.6 The albino coat color mutation in *Peromyscus maniculatus gambelii.*

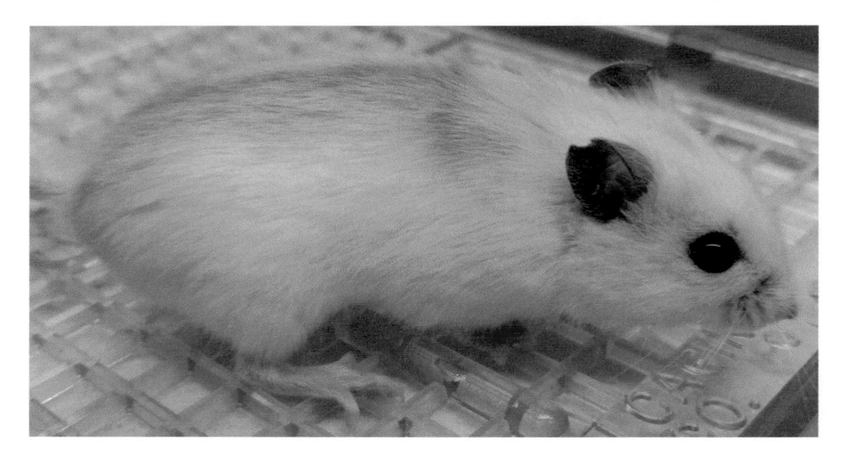

FIGURE 10.7 The dominant spot coat color mutation in *Peromyscus maniculatus bairdi.*

FIGURE 10.8 The variable white coat color mutation in *Peromyscus maniculatus bairdii.*

FIGURE 10.9 The wide-band agouti coat color mutation in *Peromyscus maniculatus bairdi*.

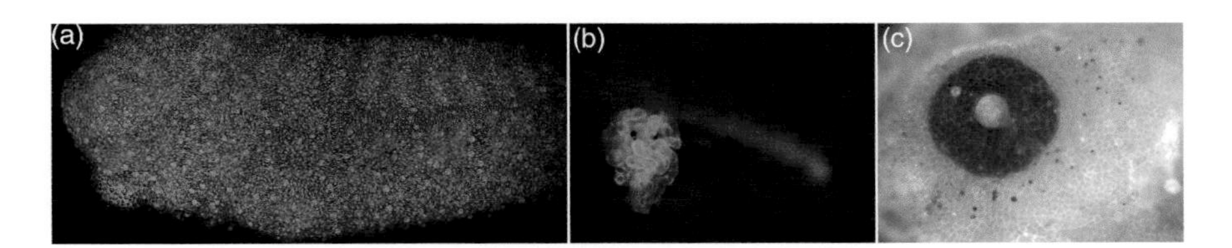

FIGURE 12.1 Transgenic *Xenopus* frogs. (a) Stage 30 *Xla. Tg(CAG:KikGR)*^{Flc} transgenic embryo where the green fluorescent KikGR is expressed ubiquitously in the embryo. Green to red fluorescent photoconversion can be achieved in specific populations of cells using a 405 nm laser. (Courtesy of Mitch Butler in the 2015 National *Xenopus* Resource Advanced Imaging Workshop.) (b) Stage 46 *Xla. Tg(Dre.cdh17:eGFP)*^{NXR} transgenic tadpole illustrating (green fluorescent protein) fluorescence in the developing *Xenopus* kidney. (c) Stage 40 *Xla. Tg(CMV:memGFP, cryga:mCherry)*^{NXR} transgenic tadpole has membrane bound green fluorescent protein expressed in all cells.

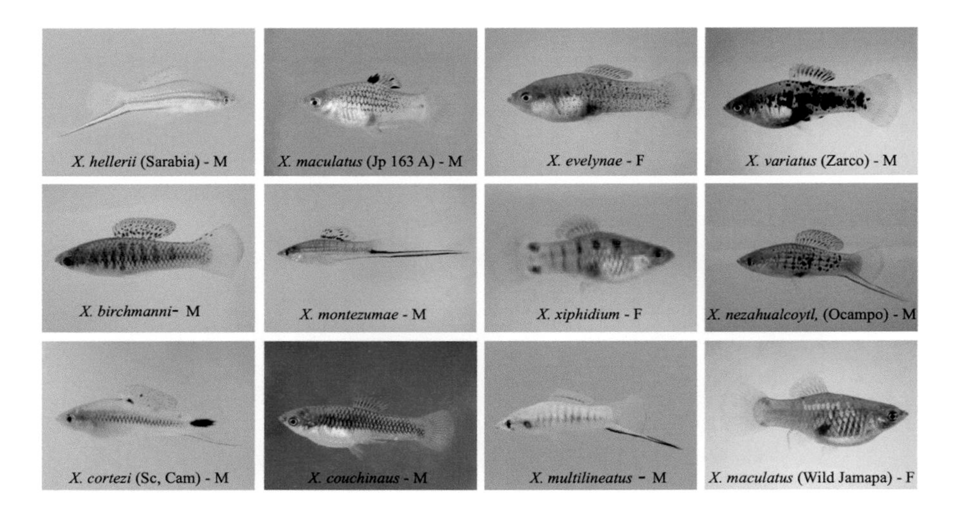

FIGURE 13.1 Examples of the varied morphology among *Xiphophorus* species.

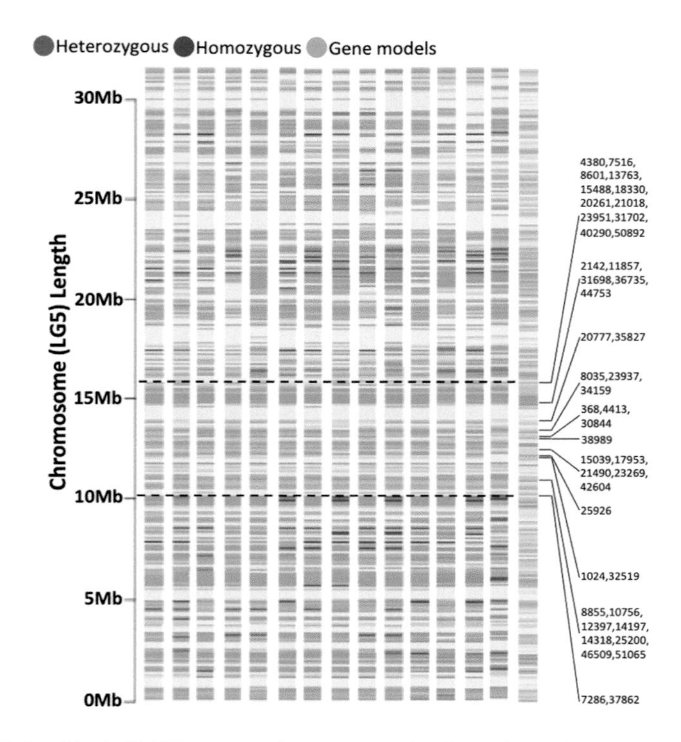

FIGURE 13.3 The *R/Diff* locus on chromosome 5. The loci on chromosome 5 that are homozygous for *X. hellerii* alleles in BC_1 hybrid progeny bearing melanoma tumors defines the *R(Diff)* locus. Each line represents a gene, and shading of the lines represents the genotype (i.e., homozygous for *X. hellerii* allele, or heterozygous for *X. maculatus* and *X. hellerii* alleles). The region between the dashed lines is 5.8 Mbp long and contains 164 gene models, and also 40 RAD-tag markers established in a previous study. This region is forwarded as the *R/Diff* locus.

Singer, A., H. Perlman, Y.-L. Yan, C. Walker, G. Corley-Smith, B. Brandhorst and J. Postlethwait. 2002. Sex-specific recombination rates in zebrafish (*Danio rerio*). *Genetics* 160:649–657.

Spagnoli, S., L. Xue and M.L. Kent. 2015. The common neural parasite *P. neurophilia* is associated with altered startle response habituation in adult zebrafish (*Danio rerio*): Implications for the zebrafish as a model organism. *Behavioural Brain Research* 291:351–360.

Streisinger, G., C. Walker, N. Dower, D. Knauber and F. Singer. 1981. Production of clones of homozygous diploid zebra fish (*Brachydanio rerio*). *Nature* 291:293–296.

Varga, Z.M. 2016. Aquaculture, husbandry, and shipping at the Zebrafish International Resource Center. *Methods in Cell Biology* 135:509–534.

Varshney, G.K., J. Lu, D. Gildea et al. 2013. A large-scale zebrafish gene knockout resource for the genome-wide study of gene function. *Genome Research* 23:727–735.

Weber, M. and L.F. De Beaufort. 1916. The Fishes of the Indo-Australian Archipelago, 3: 1-XV, 1-455, Figs. 1-214. E.J. Brill, Leiden.

Whipps, C.M., C. Lieggi and R. Wagner. 2012. Mycobacteriosis in zebrafish colonies. *Institute for Laboratory Animal Research Journal* 53:95–105.

8 The Bloomington Drosophila Stock Center *Management, Maintenance, Distribution, and Research*

Cale Whitworth

CONTENTS

Abstract: The Bloomington Drosophila Stock Center (BDSC) collects, maintains, and distributes genetically defined strains of *Drosophila melanogaster* for research and education. Over the past 32 years, the BDSC collection has grown to contain over 70,000 stocks, and in 2017 distributed more than 218,000 subcultures to ~2900 users at 982 institutions in 51 countries. The BDSC maintains and distributes only a single species, *Drosophila melanogaster*; therefore, the major focus of the program is not preservation of multiple species or genetic diversity per se, but, rather, preservation of genetic components. The purpose of this chapter is to outline the history, management practices, and research efforts of the BDSC.

INTRODUCTION

Drosophila melanogaster (Figure 8.1) has been used for over 100 years as one of the primary model organisms for genetic investigation and has contributed to many groundbreaking discoveries in foundational and applied biomedical research. Essential to this long history of success have been multiple stock centers that have distributed strains of *D. melanogaster* carrying defined genetic components.

The BDSC is a direct descendant of the *D. melanogaster* stock collection founded in the laboratory of Thomas Hunt Morgan (1933 Nobel Laureate) at Columbia University where his research established *Drosophila* as a model organism. In 1913, Calvin Bridges, a researcher in Morgan's lab, began to distribute samples of stocks to other researchers upon request. When Morgan, Bridges, and Alfred Sturtevant moved to the California Institute of Technology in 1928, the collection moved with them. As the community of *Drosophila* researchers grew, so did the Caltech collection, and in 1934 the first Caltech stock list consisting of 572 strains was published in the journal *Drosophila Information Service*. In 1948, Edward Lewis (1965 Nobel Laureate) assumed responsibility for the Caltech Drosophila Stock Center and oversaw its operations until shortly before his retirement in 1988. As Lewis's retirement approached, it became apparent that the Caltech Drosophila Stock Center would need to find a new home, and in 1986, Thomas C. Kaufman, a professor at Indiana University, volunteered to move the then 1675 stocks from Caltech to Indiana.

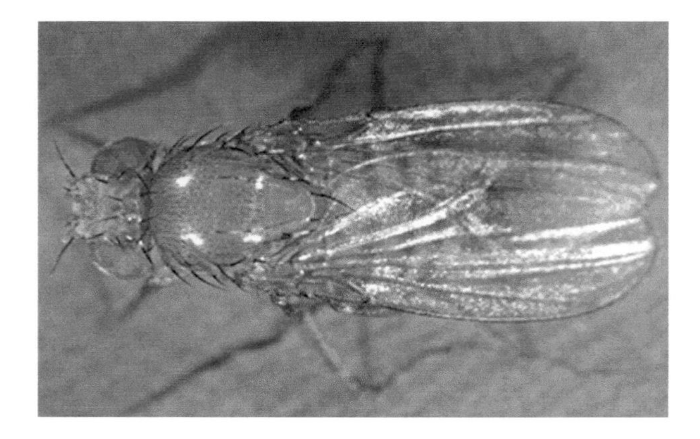

FIGURE 8.1 (See color insert.) *Drosophila melanogaster.*

A talented post-doctoral scientist in the Kaufman lab, Kathy Matthews, was hired to manage the collection, and on October 15, 1986, the newly minted Bloomington Drosophila Stock Center (BDSC) shipped its first subcultures. Many of the original strains from Morgan's lab are still in the collection, but he likely couldn't have imagined how large the stock center would grow over the next 100 plus years.

CURRENT STATUS OF THE COLLECTION

Since the 1675 stocks from Caltech were combined with ~850 stocks from the Kaufman lab to constitute the initial BDSC collection, the collection has grown to over 70,000 genetically defined strains (Figure 8.2).

The current BDSC collection contains ~72,000 unique genetic elements including alleles, chromosomal aberrations such as deficiencies and duplications, transgene insertions, and assorted other components (Table 8.1). The continual acquisition of stocks from individual scientists has allowed for gradual increases in the size of the collection, but several events have resulted in major expansions. In 1997, the Mid-America Drosophila Stock Center (Bowling Green State University) was closed and ~800 stocks were transferred to the BDSC. Between 2000 and 2003, another ~350 stocks were accessioned when the Drosophila Stock Centre in Umeå, Sweden closed. Also, in 2004 approximately 2900 stocks were generously donated by Exelixis, Inc. when the company discontinued *Drosophila* research.

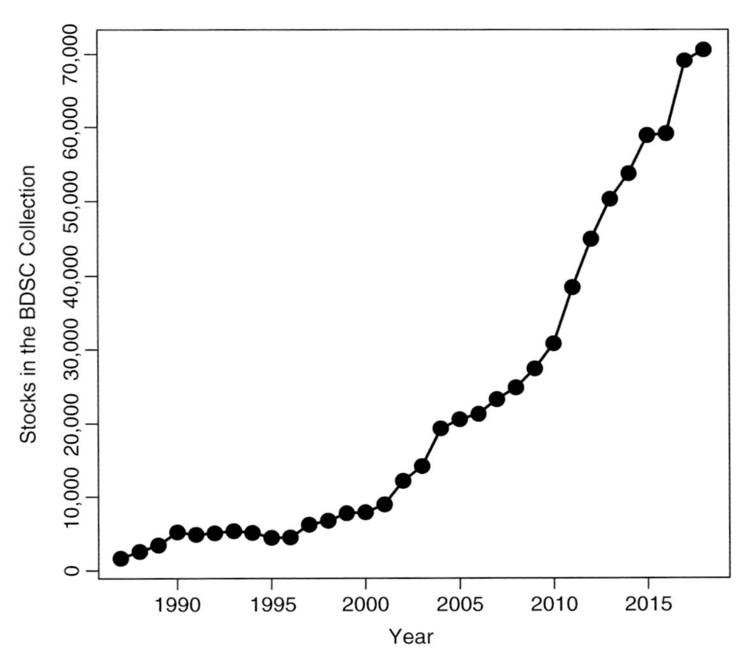

FIGURE 8.2 Stocks held per year 1987–2018. Values represent all stocks in the collection January 1 through December 31 in a given year, and as a result, stocks deaccessioned within a year are included in the year's total. The value for stocks held in 2018 was for the January 1 through November 1 period.

TABLE 8.1
Selected Categories of Stock Features/Uses

Feature or Use	Stocks
Alleles	68,161
Human disease-related	35,804
Balancer chromosomes	23,600
Binary transcriptional activation (e.g., GAL4–UAS)	21,110
Gene Disruption Project transposon insertions	20,313
RNAi transgenes	13,066
Chromosomal deletions	3,178
FRT–FLP site-specific recombination	1,907
Chromosomal duplications	1,784
Fluor-tagged proteins under native control	1,420
CRISPR–Cas9 and guide RNA transgenes	1,269
Markers (neurons, mitochondria, etc.)	747
Biosensors (Calcium voltage, apoptosis, signaling, etc.)	266
φC31 site-specific transformation technology	105

Several large-scale projects have generated and donated transgene insertion alleles including ~11,000 stocks from the Berkeley Drosophila Genome Project (BDGP; http://www.fruitfly.org/) and the Gene Disruption Project (GDP; http://flypush.imgen.bcm.tmc.edu/pscreen/index.php). The GDP has continued to donate stocks at the rate of 1100 per year. The Transgenic RNAi Project (TRiP; https://fgr.hms.harvard.edu/fly-in-vivo-rnai) began to deposit stocks in 2009 and has continued at the rate of ~1300 per year. In 2011, the Janelia Research Campus (Ashburn, Virginia, USA) began to deposit GAL4 drivers with known expression patterns generated by the FlyLight Project (Pfeiffer et al. 2008, Jenett et al. 2012, Jory et al. 2012, Manning et al. 2012, Li et al. 2014). These stocks express the yeast transcription factor GAL4 under the control of short regulatory fragments and are used to drive the expression of transgenes under the control of the GAL4 target site, UAS. Janelia eventually deposited ~7100 of these lines over a 3-year period, and also donated ~7300 stocks expressing a modified version of the GAL4 driver, termed split-GAL4, in 2017 (Tirian and Dickson 2017, Dionne et al. 2018).

As the collection has grown, so has the number of requests for stocks (Figure 8.3). For the most part, the number of stocks shipped has increased year over year since the BDSC began fulfilling requests in 1986. Many factors influence the number of shipments each year, but there is a correlation between increases in shipments and the addition of large collections of new stocks. For example, from the period of 2001–2006 a large increase in shipped samples coincided with the arrival of BDGP and GDP stocks. This was followed by a 3-year period of a more-or-less constant number or stocks shipped, which was likely explained by saturation of demand for these reagents in the community as well as less large-scale screening of these and other reagents, among other factors. The arrival of TRiP RNAi lines in 2009 and the Janelia Research Campus GAL4 lines in 2011 also corresponded with an increase in

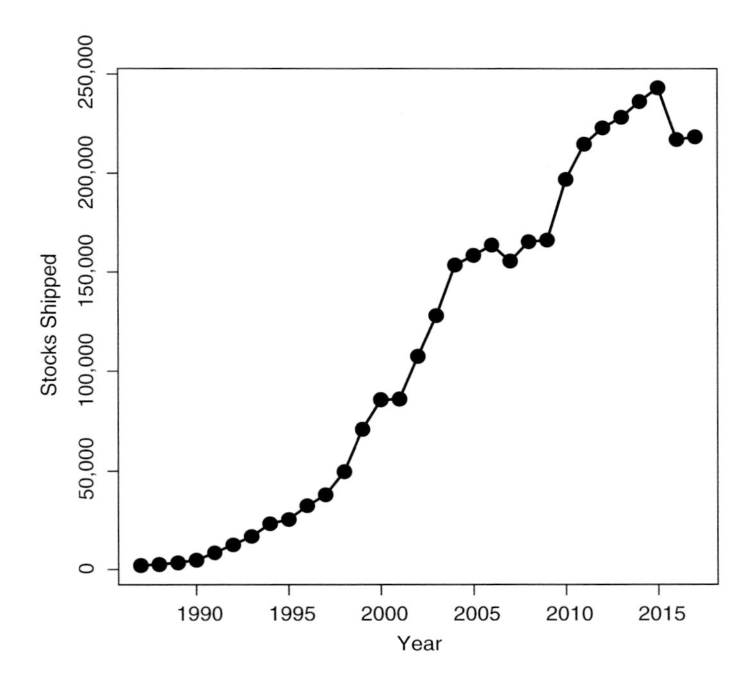

FIGURE 8.3 The total number of stocks shipped per year for the period 1987–2017. The projected number of stocks shipped in 2018 is 228,083.

samples shipped. As before, 2016–2017 saw a decrease in overall shipments as the community's demand for these and other reagents decreased.

The question of whether the collection will continue to grow remains open. With the construction of a new facility that will roughly double the storage capacity of the center, space will not likely limit the acquisition of new stocks for the foreseeable future. Funding, however, is a major constraint on growth. The size of the BDSC active user base has likely stabilized (see Figure 8.4), and, given the amount of funding available for research, increasing stock fees for an ever-growing collection would strain lab budgets. Without a major change in operations, funding, or size of the active user base, the BDSC will likely reach a steady-state number of stocks and yearly samples shipped in the near future. This will not result in stagnation of the collection. Rather, the collection's contents will continue to evolve with the needs of the research community with acquisitions of stocks carrying new tools being balanced by deaccession of stocks that have become less useful.

The active user base of the BDSC grew steadily from 1998 to 2014 (Figure 8.4) with a relatively steady state reached in 2015. Since some labs have a designated individual (usually the principal investigator or a technician) who places orders, any given "active user" may actually represent more than a single individual who utilizes the requested stocks. The user base is primarily research and teaching institutions, and users of this type accounted for 98.8% of stocks shipped in 2017 with commercial users accounting for the remainder. In 2017, each active account received a median of 34 stocks with a range of 1 to >5000 (Figure 8.5).

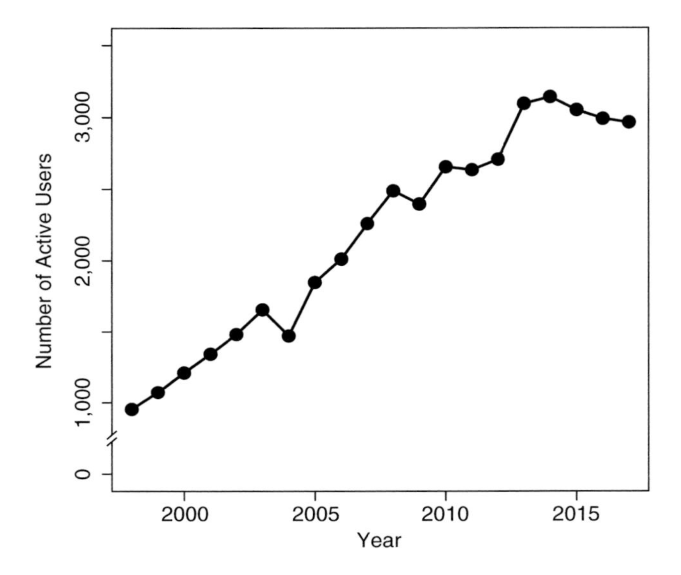

FIGURE 8.4 The number of active users per year for the period 1998–2017. Active users represent individuals who placed an order in a given year. Data from previous years are not available. The projected number of active users for 2018 is ~2800 individuals.

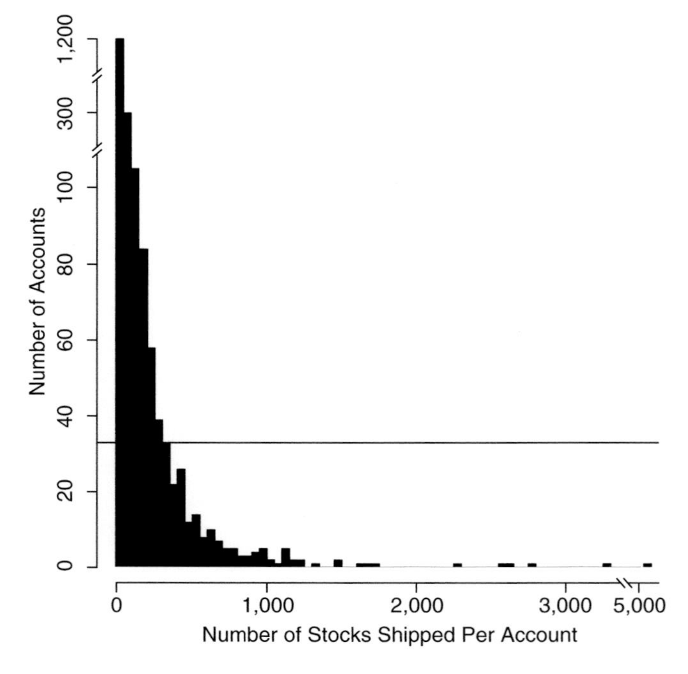

FIGURE 8.5 Binned distribution (100 bins) of the number of stocks shipped per account in 2017. The median value is indicated with a horizontal line.

FINANCIAL SUPPORT

Grant and Institutional Support

The BDSC was initially supported by a National Science Foundation (NSF) grant that moved to Indiana University in 1986 from Caltech along with the stocks. The NSF continued to fund the center for the next eight years; however, in 1995, the NIH and NSF began a period of cooperative funding for the BDSC that continued until 2014 when NSF funding priorities changed. The center currently receives financial support from the NIH as well as program income generated from fees assessed when stocks are ordered. The NIH Office of the Director funds and administers the grant with contributions from the National Institute of General Medical Sciences, the National Institute of Child Health and Human Development, and the National Institute of Neurological Disorders and Stroke.

Indiana University has housed and provided infrastructural support for the BDSC since 1986, and has also done the same for two other key resources for the *Drosophila* community: FlyBase, the primary repository of online information about *Drosophila*, since 1992, and the Drosophila Genomics Resource Center, distributors of *Drosophila*-related cell lines and DNA resources, since 2003. Indiana University's commitment to the long-term support of the *Drosophila* research community is commendable.

Program Income

Initially, all operational costs of the BDSC were covered by the NSF, thus no fees were assessed for stock orders. When the NSF and NIH began cooperatively funding the center, one requirement was the institution of fees for stock orders to generate income that could cover a portion of the center's operating expenses. To this day, an ongoing expectation of NIH program grants is continual, gradual increases in the proportion of costs covered by program income to maximize self-sufficiency. As a result, the percent of operating costs covered by program income has increased gradually since 1995, with approximately 78% of operating costs now being covered by fee income.

Stock fees are charged in a tiered system where the first five stocks in a calendar year currently cost $18, stocks 6–100 cost $7.50, and additional stocks cost $3.70 each. This fee structure was designed with two aims. First, whether an account orders 1 or 10,000 stocks per year, there are baseline costs associated with maintaining the account. These include financial and management staff who establish accounts, create invoices, and correspond with users. Stock fees are higher for the first few stocks purchased to cover these baseline costs. Second, the sliding fee schedule allows users to conduct large-scale screening projects on typical grant budgets. This promotes maximal use of the collection while also providing the most benefit to the scientific community. Handling fees for domestic and international accounts are also assessed. International accounts represent approximately 50% of annual orders, but require more effort to prepare shipments and process payments, so handling fees are set higher.

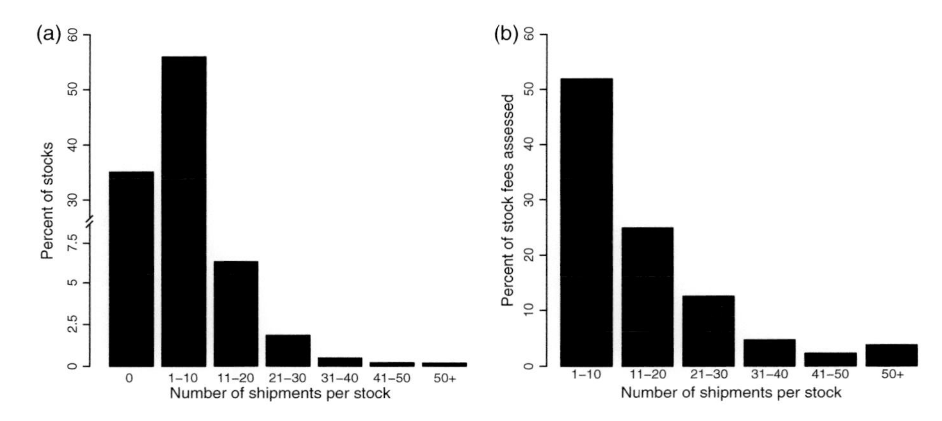

FIGURE 8.6 (a) Distribution of the percent of stocks in the collection (n = 59,126) by binned number of shipments per stock in 2017. Only stocks that were held in the collection for the entire period of January 1 through December 31, 2017 were included in this analysis. (b) Distribution of the percentage of total stock fees assessed in 2017 by binned number of shipments per stock.

As might be expected, not all stocks generate equal amounts of income. In 2017, 56% of stocks were shipped between 1 and 10 times (Figure 8.6a), but these stocks accounted for only 52% of the revenue generated from stock fees (Figure 8.6b). In contrast, the category of stocks with 21 or more shipments represented just 2.7% of the stocks shipped but accounted for nearly a quarter (23.3%) of stock fees collected. Therefore, the fees generated by high-use stocks subsidize the maintenance of low-use stocks. It is for this reason that we worry about redistribution of BDSC stocks beyond the recipient. The overall fiscal health of the BDSC and our ability to maintain important, but rarely used stocks depends on this delicate balance of high- and low-use stocks (see *Distribution Policies* below).

IMPACT OF THE BDSC

The BDSC is the sole source for >55,000 unique strains and supports a worldwide user base of ~2900 *Drosophila* researchers. Approximately 1500 publications per year acknowledge the center—although this is likely an underestimate of the actual use since many authors do not cite the BDSC as the source of stocks. The total number of publications containing the string "*Drosophila*" in their title is ~32,000 per year, and it is likely that strains and/or genetic components used in most of these publications originated here. As of July 2018, the center supported at least one principal investigator on 75% (n = 1069) of active NIH grants with "*Drosophila*" in their title or abstract. Because of the important role that stock centers play in the *Drosophila* research community, they are consistently listed as one of the highest priorities for resource funding by the fly community (https://wiki.flybase.org/mediawiki/images/e/e7/DrosBoardWP2016.pdf).

MANAGEMENT PRACTICES

BDSC management includes stock acquisition, data management, financial management, website development and maintenance, and research. Each of these areas is managed by the scientific co-directors of the center, who generally focus on one or two of these areas but are sufficiently familiar with all areas to provide redundancy. In addition, the BDSC currently employs two full-time scientists who perform quality control of stocks as well as execute the research program, one financial manager, one lab manager who processes orders and manages the stockkeeping staff, and 55 full- and part-time stockkeepers (24.7 full-time equivalents).

ACQUISITION/DEACCESSION

Stock Selection

The BDSC is not an archival collection and must balance the usefulness of stocks with budgetary constraints. Each stock is evaluated on a case-by-case basis with consideration given to current and future potential uses of the genetic components present in the stock. The acquisition policy states that the BDSC accepts "stocks of broad or exceptional interest to current and/or future research." These include stocks that:

- Represent genetic tools of use to a broad range of research projects
- Contain well-characterized mutations affecting known structures, organs, or processes
- Carry transposon insertion alleles of annotated genes that are not already represented in the collection by characterized mutations
- Carry mapped but otherwise uncharacterized visible, lethal, or sterile complementation groups
- Are special purpose strains that are not easily regenerated such as certain chromosomal aberrations and marker combinations

All stocks must be stable without active, manual selection. Due to the costs associated with stock maintenance (~$13.40/stock/year in 2017), we do not acquire or we actively cull stocks that are of limited use such as those of interest to only a few users, of purely speculative value, of a relatively uncharacterized nature, or of only historical interest. Relatively few stocks are maintained simply to provide general genetic variability/diversity.

Most stocks in the collection were deposited as part of large-scale projects from groups like the BDGP, GDP, TRiP, and the Janelia Research Campus (see *Current Status of the Collection* above). These four groups alone accounted for 67% of the 68,385 stocks in the collection at the end of 2017. Another valuable source of stocks is individual researchers submitting lines that are requested often. The BDSC scientific staff will also solicit donations of stocks identified in the literature, at conferences, or suggested by users.

Conditions of Deposition

Consistent with the longstanding ethos in the fly community of open exchange of stocks, all stocks donated to the BDSC are available to instructors or scientists at teaching or research institutions as well as scientists doing for-profit research. We no longer accept stocks where donors wish to impose restrictions on distribution such as limiting stocks to not-for-profit research (see *Distribution Policies* section below for more information).

Quarantine and Characterization of Stocks

All stocks entering the collection pass through a quarantine process and are checked for expected phenotypic markers as well as general health. Since most stocks have phenotypic markers for tracking genetic components, examination of appropriate markers when a stock arrives is a reliable and affordable method of stock validation.

One of the biggest concerns when receiving new stocks is the potential for mite infestation. Most mites discovered in *Drosophila* cultures are relatively harmless and sustain themselves on fungi or bacteria found in the culture (Ashburner et al. 2005). While these mites are not predatory toward *Drosophila* directly, they can generate large populations in a culture and outcompete flies for resources. Of serious concern are predacious mites that feed on embryonic, larval, and/or adult flies. Infestations of these mites can be rapidly detrimental to fly cultures. To screen for and protect from mite infestations, the BDSC has a strict quarantine procedure for each stock that arrives. As with all stocks in the collection (see *Maintenance* below), quarantine cultures are maintained in trays lined with cheesecloth treated with benzyl benzoate, a chemical that acts as a mite repellent. The quarantine process involves placing flies on fresh food for two consecutive days (i.e., "quick flips") in an effort to rid the stock of potential mites as the generation time of most mites is longer than that of flies, and mites tend to stay in the original culture when flies are transferred to a new container. These quick transfers occur in a room intended for the initial processing of stocks. After the two quick transfers, the flies are placed on fresh food again, and this "founder" culture is transferred to another room intended for cleaner storage of the flies for the remainder of the quarantine process. The founder culture is flipped into a new container 2–4 days later. These two cultures will each be transferred to fresh food 3 weeks later to make a total of four cultures. The original shipment culture is maintained until the stock enters the collection in case of failures in establishment of new cultures.

When the founder culture is 5 weeks old, it is checked for mites as well as bacterial and/or fungal contamination. Any stock found to have mites is taken back to step 1 of the quarantine process, and the steps are repeated until the stock is free of mites. Bacterial contamination is addressed by treatment with a mixture of penicillin and streptomycin, and fungal contamination is treated by close monitoring and rapid transfers until the contamination is eliminated. If the stock is healthy and mite-free, it is checked for expected phenotypic markers by a member of the scientific staff, and, once validated, the four cultures are split into groups of two and tagged to allow for easy identification of the stock. The two sets of stocks are then transferred to the main or backup copy of the collection for long-term maintenance.

Deaccessioning

Stocks that have either lost critical genetic components due to stock breakdown or have become genetically contaminated, too unhealthy to maintain, or less useful to the community are routinely deaccessioned. Quality control for stocks is a challenge for a large, living collection especially when cryopreservation is not a viable option (*Maintenance*). Stock health problems are generally identified through routine surveillance. Stock breakdown or contamination can also be caught when a stock's health is compromised; however, the majority of breakdown and contamination problems are identified by BDSC users who report an average of 3–4 issues per month. All reports are addressed immediately, and, if both the main and backup copies of a stock have problems, then the stock is deaccessioned from the collection. For high-value stocks that are deaccessioned through this process, center scientists contact the stock's donor or users who recently requested the stock and ask for a replacement sample. This new sample is treated as a new addition to the collection and is given a new stock number since its provenance is different from the original stock.

Deaccessioning of stocks based on decreased usefulness to the community is a time- and labor-intensive process since, by removing a stock from the collection, the BDSC is effectively preventing future researchers from acquiring the stock. Each stock is considered on a case-by-case basis, and the decision to deaccession is influenced by multiple factors. First, the *Drosophila* community provides indirect input on the usefulness of a stock via the number of requests received. While this is a useful metric, it is by no means the only factor taken into consideration since some rarely used stocks are nonetheless extremely valuable. BDSC scientists have a high degree of expertise in genetics as well as knowledge of contemporary research trends and use this information to evaluate stocks for deaccession. Factors such as the allelic strength, publication history, phenotype, and availability at other stock centers are also taken into consideration. Except in rare cases, a stock is not considered for deaccession without being available to the community for at least 3 years. BDSC users are notified of upcoming deaccessions via newsletter and email, and stocks chosen for deaccession are listed on the website for at least 6 weeks before the removal from the collection. Direct input received from users may affect the decision to deaccession a stock.

One measure of effective accession and deaccession practices is the percentage of the collection that is ordered in a given year. In 2017, 64% of stocks accessioned before 2017 and 84% of stocks accessioned before 2014 were requested at least once. This high across-the-board use shows that the overall collection is useful to the community. There is some room for deaccession of the ~16% of stocks that are not ordered as frequently, although, as stated earlier, some valuable reagents are inherently low-use.

Stock Maintenance

Stocks are maintained as live cultures in 10 × 2 cm cylindrical glass vials that must be established once every 14 days by transferring adult flies to fresh medium.

To protect against stock loss from genetic breakdown, contamination, or accident (such as a broken culture vial), two copies of each stock are maintained by different stockkeepers. Independent copies of each stock are never mixed or used to rescue another copy without approval of a BDSC scientist, and records are maintained of such events. Each copy of a stock consists of two culture vials that are separated in age by 14 days. For stocks that are heavily requested by users, up to six vials will be maintained in the main collection. As a result, the BDSC currently holds approximately 300,000 active cultures at any given time for the ~70,000 stocks. Stocks are maintained on metal shelving in trays lined with benzyl benzoate-treated cheesecloth. Each tray carries ~60 culture vials and is stored in one of three climate-controlled rooms held at 22°C.

At two-week intervals, both copies of the entire collection are transferred to fresh food by a staff of ~55 full- or part-time stockkeepers. This is a labor-intensive process requiring ~3600 hours of effort each month. Approximately 45 minutes are required to retrieve a tray of culture vials from the collection storage room, acquire fresh food vials and rayon (used to plug the open end of the vial), transfer the tray of ~60 cultures to fresh food, deliver the discarded culture vials to the appropriate location, and return the tray to the storage room. As a routine part of transferring stocks, stockkeepers also spot check a single culture from each tray for mite contamination, and any unhealthy cultures are segregated by the stockkeeper for special attention. If additional care is needed, stocks are passed to a stockkeeper or scientist specializing in the care of weak or unhealthy stocks. Depending on the nature of the problem, stocks may be treated with antibiotics or scientists may outcross them to rescue important components in a more vigorous genetic background. Stocks fluctuate in vigor and health, so back-up copies are used regularly.

The main and backup copies of the collection are currently maintained in the same building albeit physically separated by six floors and multiple firewalls as well as having separate air handling systems. While this provides some protection from potential disasters like fire and flood, the BDSC is constructing a separate facility across campus to house the backup copy of the collection and provide space for stockkeepers.

In discussions of *Drosophila* stock maintenance, cryopreservation is often mentioned as a solution to the costs of large-scale live culture. There is currently no method that can be used for efficient, efficacious cryopreservation of *Drosophila*. While the actual storage of cryopreserved samples would likely be inexpensive compared to live culture, the costs of preparing samples for storage and resuscitating them post-thaw would be expensive and time-consuming, and would likely eliminate the cost savings. Nevertheless, a robust and reliable cryopreservation method would be ideal for maintenance of a third copy of the collection for disaster planning and would perhaps eliminate permanent deaccession of stocks and allow for accession of stocks that the BDSC would not normally acquire. The most likely scenario would be that commonly ordered stocks would be maintained as live cultures with cryopreserved backups, and less-used stocks would be cryopreserved. The inherent cold sensitivity of *Drosophila* embryos combined with a short window in embryonic development where freezing is possible (Steponkus et al. 1990, Mazur et al. 1992) make improvements to current embryo cryopreservation protocols difficult, although the NIH has supported efforts to encourage development of robust protocols. We would modify our practices should there be an experimental breakthrough, but devising new cryopreservation protocols is beyond the BDSC's research capacities.

Food Preparation and Dishwashing

As with stockkeeping, dishwashing and food preparation for the ~300,000 vials used per month is a huge undertaking. The collection is maintained in glass vials that are reused after each stock transfer. Used glass vials are autoclaved to eliminate living flies and to melt the agar-based food. Rayon plugs are removed by hand, the vials are soaked in an enzymatic detergent, run through a lab dishwasher with a high-alkaline and high-bleach cleaner followed by distilled water rinses, and finally dried in drying closet. The cleaned vials are delivered to the media preparation kitchen where the vials are manually transferred to trays that hold 100 vials in a standard 10 × 10 array. Fly food (recipe and detailed protocol can be found at https://bdsc.indiana.edu/information/recipes/bloomfood.html) is prepared in an 80 gallon electric steam-jacketed kettle fitted with a peristaltic pump and dispensed into 100 vials at a time using the Droso-Filler™ (Genesee Scientific, 59–168).

Problem Stocks

When an issue with a stock is discovered, both the main and backup copy of the stock are examined for phenotypic markers that should correlate with genetic elements in the stock. If a problem is confirmed, then scientists remove the stock from circulation and address the issue via genetic tests or work with the user and/or donor to resolve the problem. This can include genetic crosses to test for predicted phenotypes, PCR of genetic components, or other approaches specific to a particular stock. Problems with stocks are documented (see *Data Management* below) so that issues can be tracked over time. The existence of the independently maintained backup collection is essential to help protect against the loss of stocks due to contamination or breakdown.

DATA MANAGEMENT

BDSC scientists spend considerable effort entering, manipulating, and extracting data from a relational database that houses information falling into two main categories: data related to stocks and data related to ordering. The purpose of maintaining these records is to allow the center's small staff to track and report detailed stock and ordering information. Maintaining accurate and useful data provides users with the most up-to-date information possible.

Each stock and its constituent genetic components, donor(s), potential uses, etc. are cataloged when a stock enters the collection. These data are displayed on the stock browsing pages and the searches on the BDSC website (https://bdsc.indiana.edu) and are provided to users when they order stocks. In order to maintain accurate and up-to-date records at the BDSC, considerable importance is placed on data integration with FlyBase (http://flybase.org). FlyBase is the primary source for information on fly genetics and serves as the information repository for, among other topics, genetic components, phenotypes, and reagents in Drosophila. Seamless integration with FlyBase allows users to identify useful stocks via FlyBase and to find detailed information about stocks listed on the BDSC website. It is BDSC policy that every genetic component in each stock is documented in FlyBase. In most cases, during the ~6-week quarantine process of new stocks discussed above, a member of the scientific staff collaborates with the donor of a stock and FlyBase curators to create entries that contain the origin, publication history, and other pertinent information for each genetic component. If a component in a new stock is not already in FlyBase, then BDSC scientists will request curation of the relevant publication(s) or will provide unpublished information from the donor in a personal communication. The integration between BDSC and FlyBase allows users to transition between the two sites when planning experiments, and to use the sophisticated FlyBase search tools. There is also a considerable cost savings to the *Drosophila* community as the BDSC can rely on FlyBase as the repository of critical details about stock components, and thereby avoid duplication of efforts in providing this valuable information.

Because discoveries necessitate changes in genetic symbols and genome annotation, FlyBase is continually updating entries. A key aspect, then, of coordinating data between FlyBase and the BDSC is keeping stock genotypes up to date with the latest symbols from FlyBase. The BDSC has worked with FlyBase programmers to develop parsing software that "reads" a genotype, attempts to associate each component with a FlyBase object (allele, insertion, etc.), and proposes changes to BDSC stock nomenclature. BDSC scientists then use this information to propagate the updates to BDSC genotypes. It can be a time-consuming process, but maintaining clear and consistent nomenclature across the two platforms helps make the relationships between stocks and the related data and literature associated with stock components explicit and unambiguous.

Data related to fees, payments, invoices, shipping/billing address/methods, account users, and their order histories are maintained for each transaction processed by the BDSC (see *Distribution*). Users are able to access these records by logging in to their account on the BDSC website, which displays up-to-date information by accessing ordering and account database tables.

DISTRIBUTION

Orders for stocks are placed through an online ordering system found at https://bdsc.indiana.edu. Users must log in to order stocks, and each user is associated with an account (usually at the level of an individual lab) that can have multiple users. Order cycles are closed on Monday and Thursday at noon, and between these times, users can add stocks to their order. Order processing involves semi-automated database manipulations that eventually result in the production of quotes, account statements, packing lists, stock information sheets, address labels, shipping information, and customs documentation. The website is then updated to reflect the status of the order as to whether it is being prepared for shipment or requires payment depending on the account type of the user.

Orders are prepared for shipment in two steps. First, the requests for stocks in a given ordering cycle are divided among specialized stock keeping staff who, in addition to their regular stock keeping duties, prepare samples for shipment. To prepare a sample, a stock keeper will transfer 10–20 adult flies to fresh food in a plastic, cylindrical shipping vial. Second, orders are packaged into individual boxes for shipment. In 2017, the BDSC shipped an average of 4412 samples in 276 shipments each week.

Distribution Policies

Most stocks survive shipment, but the BDSC cannot guarantee stocks will arrive alive. Any stock that dies in transit is replaced once without incurring stock or handling fees, and, if the shipment costs are less than $5, shipping charges are also waived. If the replacement stock dies in transit, then the user must place a new order for the stock(s) and pay all required fees.

The BDSC places a single condition on the use of stocks once they arrive to the user: users are not allowed to provide stocks to individuals outside of their parent institution. Unless explicitly given permission, a user agrees to this redistribution policy on behalf of themselves, all users of their account, and anyone in their organization allowed access to the stocks when they request an account and/or order stocks. This policy is aimed to prevent disruption of the sale of high-use stocks (i.e., those most likely to be redistributed) supporting the maintenance of low-use stocks, as doing so would severely undermine the BDSC financial structure and thus harm the overall quality of this community resource.

Regulatory Compliance

One important aspect of managing a large collection is to be aware of and advocate for changes in regulations that affect the international exchange of strains. The BDSC invested considerable effort in working with the United States Postal Service (USPS) and the U.S. Department of State to amend the Universal Postal Convention allowing for shipment of *Drosophilids* via international mail. As a result, it is now legal to mail *D. melanogaster* throughout the world. These regulations cover only the transit of the samples and do not cover import/customs regulations. We rely on users to obtain and provide any import/customs paperwork that must accompany the shipment once it arrives at the border. In order to import flies into the United States, individuals need appropriate documents from the United States Department of Agriculture (USDA) to accompany shipments. The BDSC has worked closely with the USDA to provide step-by-step explanations for scientists filling out online USDA forms (https://bdsc.indiana.edu/information/permits/index.html).

RESEARCH

All of the BDSC's research efforts have focused on building or characterizing stocks with the goal of adding value to the collection. Some research efforts have been undertaken with research funds included in the NIH grant supporting BDSC operations, while other efforts have been funded by independent grants. Below are three examples of research projects completed by the BDSC.

Chromosomal Deficiencies

In 1998, BDSC co-director Kevin Cook was awarded a NIH R24 grant to create chromosomal deficiencies with molecularly defined breakpoints. These valuable reagents are vital for mapping mutations, identifying genetically interacting loci, and characterizing alleles. Having a set of deficiencies that cover as much of the genome as possible and finely subdivide chromosomal regions had been a longstanding goal of the *Drosophila* research community. The deficiencies generated under this grant combined with deletions generated by similar projects elsewhere now provide ~99 euchromatic coverage and breakpoints spaced apart by a median of seven genes (Parks et al. 2004, Ryder et al. 2004, Cook et al. 2012). They also allowed the assembly of a second-generation "Deficiency Kit" providing maximal genomic coverage with the fewest stocks. These reagents remain some of the most ordered in the stock collection.

Mapping Mutations

Under the research component of the current NIH grant funding BDSC operations, molecularly defined deficiencies and single-gene mutations have been used in genetic crosses to characterize two different subsets of stocks. One subset carries transposon insertions associated with recessive sterile or lethal mutations that may be caused by the transposon insertion itself or may be caused by second-site mutations that arose during transposon mobilization (i.e., "hit-and-run" mutations). Attributing the phenotypes directly to transposon insertion makes these stocks more valuable because it provides functional information about nearby genes. The second subset of stocks carries alleles in complementation groups that have not been mapped to single genes or a small genomic region. The majority of these stocks are maintained by the BDSC because they have interesting phenotypes: primarily lethality or sterility but also interesting morphological phenotypes involving eyes, wings, bristles, etc. Mapping these complementation groups precisely allows researchers to associate them with sequence-defined genes. These efforts showed that transposon insertions are directly responsible for mutating 62 genes to recessive lethality or sterility, and mapped 77 complementation groups to small groups of closely linked genes and mapped seven complementation groups to single genes (Kahsai et al. 2016, Kahsai 2018).

Characterization of Balancer Chromosomes

One essential component of the *D. melanogaster* genetic toolkit is the balancer chromosome. These chromosomes carry multiple inversions that prevent crossing over between them and their homologs during meiosis. Most balancers also carry alleles or insertions that give easily recognized visible phenotypes. Balancers allow for long-term maintenance of mutations without selecting for the mutation every generation. While extremely important for *Drosophila* genetic experimentation and stock keeping, these chromosomes had not been characterized molecularly. Kevin Cook collaborated with Danny Miller, Scott Hawley, and colleagues to sequence balancer chromosomes from a variety of stocks (Miller et al. 2016a, 2016b, 2018). They were able to determine the exact genomic coordinates of the inversion breakpoints and characterize the molecular lesions in many marker mutations. They were also able to define chromosomal regions between inversion breakpoints that were subject to rare double crossovers in long-term culture and thereby better explain why some balanced stocks lose mutations spontaneously. This work should prove useful to all fly geneticists as they select balancer chromosomes best suited to particular purposes.

SUMMARY

As a community resource, the BDSC has been a provider of the fundamental reagents used in *Drosophila* research for 32 years. The center's successes can be attributed to effective management practices that have allowed for continued growth on a disciplined budget. The establishment of a successful cost recovery program has allowed the collection to go well beyond the "eventual growth to 5000 [strains]" proposed in the BDSC's first grant. Strong data management practices have allowed for reporting of accurate stock and order information on which the plans for future acquisitions and fees are based. The center has established systems that allow staff to handle the rapidly changing nature of contemporary research—including nimble and fast-paced accessioning and deaccessioning of stocks while still preserving valuable stocks from the century-old history of *Drosophila* genetics. With the continued support of the *Drosophila* community, Indiana University, and the NIH, the BDSC will continue to serve the *Drosophila* community for many years to come.

ACKNOWLEDGMENTS

This chapter is dedicated in memory of Kathy Matthews whose stewardship and impassioned commitment guided the BDSC from its inception. With her passing, the *Drosophila* community lost a true hero. I am indebted to Kevin Cook, Annette Parks, Sam Zheng, and Thom Kaufman for critiques and comments on this work. The BDSC is supported by NIH grants P40OD018537 and R21OD026525.

REFERENCES

Ashburner, M., K.G. Golic and R.S. Hawley. 2005. *Drosophila: A Laboratory Handbook*. Cold Spring Harbor, NY: Cold Spring Harbor Laboratory Press.

Cook, R.K., S.J. Christensen, J.A. Deal, et al. 2012. The generation of chromosomal deletions to provide extensive coverage and subdivision of the *Drosophila melanogaster* genome. *Genome Biology* 13:R21. doi:10.1186/gb-2012-13-3-r21.

Dionne, H., K.L. Hibbard, A. Cavallaro, J.C. Kao and G.M. Rubin. 2018. Genetic reagents for making split-GAL4 lines in *Drosophila*. *Genetics* 209:31–35.

Jenett, A., G.M. Rubin, T.T. Ngo, et al. 2012. A GAL4-driver line resource for *Drosophila* neurobiology. *Cell Reports* 2:991–1001.

Jory, A., C. Estella, M.W. Giorgianni, et al. 2012. A survey of 6300 genomic fragments for cis-regulatory activity in the imaginal discs of *Drosophila melanogaster*. *Cell Reports* 2:1014–1024.

Kahsai, L. and K.R. Cook. 2018. Mapping second chromosome mutations to defined genomic regions in *Drosophila melanogaster*. *G3 (Bethesda)* 8:9–16.

Kahsai, L., G.H. Millburn, and K.R. Cook. 2016. Phenotypes associated with second chromosome P element insertions in *Drosophila melanogaster. G3* (Bethesda) 6:2665–2670.

Li, H.H., J.R. Kroll, S.M. Lennox, et al. 2014. A GAL4 driver resource for developmental and behavioral studies on the larval CNS of Drosophila. *Cell Reports* 8:897–908.

Manning, L., E.S. Heckscher, M.D. Purice, et al. 2012. A resource for manipulating gene expression and analyzing cis-regulatory modules in the Drosophila CNS. *Cell Reports* 2:1002–1013.

Mazur, P., K.W. Cole, J.W. Hall, P.D. Schreuders and A.P. Mahowald. 1992. Cryobiological preservation of Drosophila embryos. *Science* 258:1932–1935.

Miller, D.E., K.R. Cook, A.V. Arvanitakis and R.S. Hawley. 2016a. Third chromosome balancer inversions disrupt protein-coding genes and influence distal recombination events in *Drosophila melanogaster. G3* (Bethesda) 6:1959–1967.

Miller, D.E., K.R. Cook, E.A. Hemenway, et al. 2018. The molecular and genetic characterization of second chromosome balancers in *Drosophila melanogaster. G3* (*Bethesda*) 8:1161–1171.

Miller, D.E., K.R. Cook, N.Y. Kazemi, et al. 2016b. Rare recombination events generate sequence diversity among balancer chromosomes in *Drosophila melanogaster. Proceedings of the National Academy of Sciences (USA)* 113:E1352–E1361.

Parks, A.L., K.R. Cook, M. Belvin, et al. 2004. Systematic generation of high-resolution deletion coverage of the *Drosophila melanogaster* genome. *Nature Genetics* 36:288–292.

Pfeiffer, B.D., A. Jenett, A.S. Hammonds, et al. 2008. Tools for neuroanatomy and neurogenetics in Drosophila. *Proceedings of the National Academy of Sciences (USA)* 105:9715–9720.

Ryder, E., F. Blows, M. Ashburner, et al. 2004. The DrosDel collection: A set of P-element insertions for generating custom chromosomal aberrations in *Drosophila melanogaster. Genetics* 167:797–813.

Steponkus, P.L., S.P. Myers, D.V. Lynch, et al. 1990. Cryopreservation of *Drosophila melanogaster* embryos. *Nature* 345:170–172.

Tirian, L. and B. Dickson. 2017. The VT GAL4, LexA, and split-GAL4 driver line collections for targeted expression in the Drosophila nervous system. bioRxiv. doi:10.1101/198648.

9 The Fungal Genetics Stock Center Supporting Foundational and Emerging Model Systems

Kevin McCluskey

CONTENTS

Abstract: The Fungal Genetics Stock Center (FGSC) has validated, preserved, and distributed strains of filamentous fungi for research and development for more than 55 years. Beginning with wild types and classical mutants of *Neurospora crassa*, and soon thereafter of *Aspergillus nidulans*, the holdings at the FGSC have grown increasingly larger in number and in diversity. The FGSC collection now maintains and distributes over 23,000 mutant strains that include both classical and gene deletion strains, in addition to 3000 wild-type strains obtained from nature and from laboratory research programs. Additional resources available from the FGSC include more than 800 plasmids, gene libraries, and 2900 and 6400 gene deletion mutants of *Candida albicans* and *Cryptococcus neoformans*, respectively. Since its founding in 1960, the FGSC has shipped over 45,000 individual strains as well as over 700,000 strains in arrayed sets to more than 5000 recipients around the world.

INTRODUCTION

Filamentous fungal model organisms have had a major impact on research over many years. This is directly attributable to the enormous impact of fungi as plant and animal pathogens, industrial organisms, and their use in food production and processing. Filamentous fungal model organisms have been instrumental in facilitating many of the seminal discoveries of the classical genetics era. In the early 1920s, and into the 1940s, fungi were studied because they were easily managed eukaryotic organisms that had life cycles that could be completed in a few weeks in a laboratory setting and on defined medium (Davis and Perkins 2002, Pontecorvo 1946). Mating type, among the first genes identified in fungi, was shown to segregate one to one in asci of *Neurospora* and to be independent of the albino trait (Dodge 1939). The ability to conduct sexual crosses among mutagenized auxotrophic strains of *Neurospora* (Beadle and Tatum 1941) led to the first demonstration of genetic control of a biochemical process in a eukaryote. The significance of this work was recognized by the awarding of the 1958 Nobel Prize for Physiology or Medicine (Horowitz 1991, Strauss 2016) to George Beadle and Edward Tatum (shared with Joshua Lederberg). The recognition of the impact of these findings, and the fact that the scientists who conducted the research had moved on to the study of other systems or to administrative responsibilities, led directly to the establishment of the Fungal Genetics Stock Center (FGSC).

HISTORY OF THE FGSC

Over its history, the FGSC has been hosted by a variety of academic institutions. First established at Dartmouth College in 1960, the FGSC moved to California State College at Humboldt in 1970 when the director relocated there. In 1985, the collection moved again to the University of Kansas Medical Center (KUMC). It remained at KUMC until 2006 when it moved to the University of Missouri—Kansas City. When the National Science Foundation (NSF) changed its support for living collections, the FGSC moved to its present home at Kansas State University where it is supported by the College of Agriculture as part of the Kansas State 2025 Plan.

Strains in the FGSC collection were initially genetic mutants and related wild types of *Neurospora* and *Aspergilus nidulans*. It was believed, at the time, that one to two thousand strains would be sufficient to represent the diversity of *Neurospora* mutants that was then limited to 120 loci. Similarly, it was estimated that 350 unique *A. nidulans* stocks would be adequate. Although this original intent was narrow, genetic strains of related fungi including *N. tetrasperma*, *N. discreta*, and *N. intermedia* as well as *Sordaria* and *A. niger* were included in the collection, as were wild-type strains that had been subjected to at least a preliminary genetic analysis. The type of genetic analysis that would qualify a strain as wild type included formal mating type or vegetative compatibility testers (Garnjobst and Wilson 1956). Other model systems such as *Ustilago maydis*, *Sordaria fimicola*, and *Schizophyllum commune* were in use at the time, as well as various *Fusarium* strains. However, these were used in only a small number of laboratories and were not included in the original plans. By 1995, the FGSC had grown to include over 35 species, although there were several for which only five or fewer strains were held.

THE FGSC COLLECTION HOLDINGS

At the present time, the FGSC holds strains from 129 species (although a few are interspecific hybrids). Of these 129 species, 52 are represented by only a single strain. Among the genera that are most highly represented, 17 *Neurospora* species account for 22,234 strains while 2064 strains are from 9 *Aspergillus* species. There are 38 species of *Fusarium* totaling 620 strains. Strains in the FGSC collection represent both reference wild types as well as defined mutants. The reference wild types include strains collected from various environments, inbred strains from laboratory genetics programs, mating-type or vegetative compatibility strains, and heterokaryon testers.

In keeping with its historical emphasis on genetic mutants, many of the FGSC holdings are unique. However, some strains in the FGSC collection are cross-referenced with strains in other collections. For example, the reference genome strain for *N. crassa* is FGSC 2489. This strain is also identified as American Type Culture Collection (ATCC) MYA-4614. The FGSC holds 143 strains that have corresponding ATCC accession numbers. Forty-one strains are cross-referenced with strains maintained by the United Agricultural Research Service's Northern Regional Research Laboratory in Peoria, IL. Sixteen strains also have corresponding CBS accession numbers. Current interest in whole genome sequencing and analysis has encouraged the deposit of materials that historically were not included in the FGSC. This has resulted in an expansion of its mandate and collection activities.

Reference Genome Strains

FGSC 2489 (*N. crassa*) is also known as 74-OR23-1V A. This alternative number refers to its Oak Ridge (OR) lineage and indicates that it is a mating type (*mat*) *A* (A), vegetative (V) re-isolate that was selected for high fertility and for isogenicity with mating partner FGSC 4200, 74-ORS-6a (Mylyk et al. 1974). These strains have historically been valuable in cytological studies of chromosome pairing and the formation of well-ordered and highly pigmented asci. The pedigree of the reference genome strain is known, and representative strains have been evaluated for nucleolus characteristics, genetic polymorphisms in heterokaryon incompatibility (*het*), nucleolus satellite (*sat*), temperature sensitivity, and spreading colonial morphology (*scot*) (Newmeyer et al. 1987). Twelve strains in the pedigree of the reference genome have had their genomes sequenced via a collaboration between the US Department of Energy's Joint Bioenergy Institute and the FGSC. The contributions of the polymorphisms detected in those data are being evaluated (McCluskey 2015b, McCluskey et al. 2011). The common genetic background is also referred to as St. Lawrence (SL) to acknowledge the role of Patricia St. Lawrence (UC Berkeley) in breeding standard strains for mutagenesis and cytology (St. Lawrence 1954). The *N. crassa* genome was sequenced using Sanger chemistry at the Broad Institute (Galagan et al. 2003) and genomic libraries from the FGSC were used in assembling it (Kelkar et al. 2001, Orbach 1991). The actual strain from which DNA was obtained was a version of 74-OR23-1V A that had been acquired from Stanford University and maintained separately at the University of Oregon. Because of its unique history, the strain from which genomic DNA was actually isolated has the accession number FGSC 9013. Subsequent sequencing and mutant generation were conducted in bona fide clones of FGSC 2489 (Collopy et al. 2010, Colot et al. 2006, Dunlap et al. 2007, Lambreghts et al. 2009, Park et al. 2011).

Other *Neurospora* strains based on the reference genotype include 2454 classical mutants that were either generated in the common background (SL or OR), or repeatedly backcrossed into the SL background. Among these strains, 113 were backcrossed three times, 61 were backcrossed four times, 30 were backcrossed five times, 81 were backcrossed six times, and 66 were backcrossed seven times. This level of backcrossing was intended to provide near-isogenic strains (Leslie 1981) of both mating types for genetic, physiological, and cell biology studies.

The reference genome strain for *A. nidulans* was deposited with the FGSC in 1962 from the Glasgow collection started by Guido Pontecorvo (University of Glasgow) where it was known as G00 (Pontecorvo et al. 1953, Yuill 1939). When it was agreed that the FGSC would accept *Aspergillus* strains, a decision was made to prefix accession numbers with the character "A." Strain FGSC A4 was used in the genome sequencing (Galagan et al. 2005) program and is generally considered to be prototrophic. It is *veA*$^{+}$ (velvet) and has a wild-type morphology. Many strains in use within the *Aspergillus* research community are *veA* and are therefore free from light regulation of complex biochemical processes (Kim et al. 2002b). Strain FGSC A4 was also used as a source of RNA to generate the cDNA libraries that were used to develop Expressed Sequence Tags (Kupfer 1999) and also as the source of DNA for ordered and chromosome-specific sub-libraries (Brody et al. 1991). The *Aspergillus* research community is divided among researchers studying either *A. nidulans*, *A. niger*, *A. flavus*, *A. parasiticus*, *A. oryzae*, or increasingly, *A. fumigatus*. Thus, shared resources for each are more difficult to develop than for only *N. crassa* where most work is carried out with a single organism. This is similar to the situation with *Fusarium*. Although the FGSC holds 620 *Fusarium* strains, 217 are Restriction Fragment Length Polymorphsm (RFLP) mapping strains for *F. graminarium* (Jurgenson et al. 2002) and 122 are for RFLP mapping in *F. moniliforme* (Xu and Leslie 1996).

Mutant Strains

FGSC mutant strains are named for their specific defect. Hence, a strain that does not produce pigments is called "albino" and the genes which often encode proteins for the biosynthesis of pigments are known for the trait that is disrupted. The gene deletion mutants of *Neurospora* each carry one mutation, marked with a hygromycin resistance gene (Colot et al. 2006). Classical mutant strains may contain one or more markers, and specialized strains have been developed that carry multiple markers. The so-called alcoy strains of *N. crassa* simplified mapping of classical traits by having multiple unlinked translocations tagged with auxotrophic or morphological markers. As such, novel markers could be scored directly without having to transfer the progeny strain (Perkins 1991). Alcoy strains were limited in their ability to map mutations or translocations and so strains with multiple markers, including readily visible markers such as color mutations as well as those near the centromere (Perkins 1972, 1990), were developed. These have been utilized extensively for genetic and RFLP mapping (Simmons et al. 1987).

In 2007, the FGSC began accepting arrayed sets of gene deletion mutants of *Cryptococcus neoformans*. Soon thereafter, the FGSC accepted arrays of gene deletion mutants of *Candida albicans*. These resources grew incrementally. The FGSC now holds 47 plates of *C. neoformans* mutants and 33 plates of *C. albicans* mutants. While individual deposits over many years have made a significant impact on the FGSC's holdings, coordinated deposits such as the *N. crassa* gene deletion set have had a major impact on FGSC holdings (Table 9.1).

Other Collaborator Donated Strains

More than 13,000 strains were deposited with the FGSC from 2005 to 2014. The FGSC accepted this large number of accessions as a collaborator on an NIH-sponsored project to generate gene deletion mutants for every gene identified in the assembled genome sequence (Colot et al. 2006). The support received enabled the

TABLE 9.1
Growth of the Fungal Genetics Stock Center Collection as Evidenced by the Number of Items Formally Accessioned into the Collection by Year and Type

	Year							
Type of Accession	**2010**	**2011**	**2012**	**2013**	**2014**	**2015**	**2016**	**2017**
Neurospora	605	497	551	920	288	51	14	133
Aspergillus	170	200	259	29	187	15	1	24
Other strains	5	10	11	239	1	10	25	70
Plasmids	22	19	22	69	21	25	7	6

FGSC to accession the knock-out strains through collaboration with the depositing laboratories. Strains were provided by the depositing laboratories in a format that allowed their rapid preservation in desiccated and cryopreserved formats at the FGSC. Because most of these have no visible phenotype, the FGSC did not carry out routine validation prior to accessioning them, a deviation from historical practices. In addition to the terms and responsibilities outlined in the original grant, the FGSC also assembled an ordered set of mutants that have been distributed to researchers in the US, Asia, Europe, and South America.

In 2003, the FGSC accepted a deposit of 253 strains of *Schizophyllum commune* when the laboratories that had characterized them were facing multiple retirements (Raper and Fowler 2004). Among these are representatives of the different mating-type loci identified through multiple years of study in an effort to understand self-recognition in fungi. Included among these were auxotrophs and morphological mutants characterized as being involved in mating (Fowler et al. 2001, Raper and Raper 1973, Raper et al. 1965). The FGSC also accepted a series of mutants of *Coprinus cinereaus* (*Coprinopsis cinerea*) (Stajich et al. 2010) in 2013. These were mostly auxotrophs that require para-amino benzoic acid (PABA), adenine, and/or tryptophan in the context of mutations (Kües 2000) at the mating type locus. Since *Coprinus* exhibits synchronized meiosis, it has found continued use in studies to understand how development is synchronized (Kües et al. 2016). *Coprinus* is also a fungus with a biotechnological impact (Sugano et al. 2017), the production of laccases (Ducros et al. 1998).

Non-accessioned Strains and Resources

The FGSC has accepted resources, including strains and reagents, over many years without accessioning them. These resources are not subject to existing quality control, but rather are accepted on an as-is basis and are provided to requestors without cost (Simione et al. 2012). Among the fungal strains accepted in this manner are lyophilized stocks originally prepared in the Tatum laboratory at Yale or Stanford (Barratt 1986). These have not been frequently requested, but have been utilized to test viability and to establish a long-term viability record for lyophilized conidia (McCluskey 2000a). The FGSC also holds, without accessioning, wild *Neurospora* strains provided by D.D. Perkins (Turner et al. 2001), *ad-3* mutant *N. crassa* strains from F.J. de Serres (de Serres 1989), *Ustilago maydis* auxotrophs (Perkins 1949), and a collection of mutants and wild-type strains of the chytrid fungus *Allomyces* (Olson 1984). While these are not formally curated, they are available for distribution and are listed on the FGSC website.

The FGSC has historically accepted select reagents and molecular resources. Donors have provided the FGSC with stocks of pluronic prill, the reverse agar that is solid at room temperature and liquid when chilled (Seifert 1994), the selectable agent Ignite, and the cell wall synthesis inhibitor PolyoxinB (Din and Yarden 1994). The FGSC provides these to requestors at no cost. The FGSC has an inventory of glass "race tubes" (Ryan et al. 1943) and provides them to clients at no additional cost, thereby lowering the cost of participation for labs that do not require large numbers of the glassware.

MOLECULAR RESOURCES AT THE FGSC

In 1989, in addition to accepting fungal strains, the FGSC began accepting cloning vectors and cloned genes.

Plasmids

The first plasmid accessioned into the collection was pJR2 which carries a full-length copy of glutamate dehydrogenase and was used to restore an "amination deficient" strain to prototrophy (Kinsey and Rambosek 1984). This plasmid has been distributed 37 times and the article describing it cited 119 times (Google Scholar, accessed 2/2/18). The plasmid collection has grown rapidly and currently includes 811 unique plasmids that are used in a variety of studies (Table 9.2). Since the establishment of the plasmid collection, 4854 plasmids have been sent to FGSC clients. Cloning, genetic, and genomic technologies have advanced since the plasmid collection was initially established. Not all types of plasmids have been often utilized. Plasmids used for transformation and gene expression have been widely distributed. Many clones were useful only for RFLP mapping (Botstein et al. 1980) and these were rarely used. While the FGSC continues to hold, manage, and distribute plasmids, the economy of scale at Addgene, a dedicated plasmid distribution center (Kamens 2014), enables them to provide a higher level of service than generally available at the FGSC, and so deposits of plasmids into the FGSC, collection have decreased since 2015.

Gene Libraries

Among the diverse non-accessioned materials, gene libraries have been highly utilized. These libraries include both genomic or cDNA clones, as well as ordered genome libraries. The FGSC currently holds 33 cDNA and 21 ordered genome libraries. More than 1000 libraries have been distributed to clients around the world. While 748 of these libraries were pools of phage particles or plasmid DNA, 252 were ordered libraries distributed in 96-well plates. These latter libraries were typically

TABLE 9.2
Types and Numbers of Plasmids in the Fungal Genetics Stock Center

Type of Plasmid	Number of Plasmids	Target Organisms(s)
Transformation	174	Numerous
Cloned gene	129	Numerous
RFLP[a]	275	*Fusarium, Magnaporthe*
Expression	14	*Neurospora, Aspergillus*
Tag[b] (GFP, RFP, Flag)	154	Numerous

[a] Restriction fragment length polymorphism.

[b] GFP—green fluorescent protein. RFP—red fluorescent protein.

distributed frozen on dry ice, but in extraordinary circumstances, the FGSC was able to grow the library on agar-solidified medium simplifying international shipment.

Collaboration among researchers resulted in the identities of cosmids carrying identified genes, and these have been posted annually on the FGSC website and as part of the FGSC catalog. *Neurospora crassa* and *Aspergillus nidulans* libraries were ultimately used in the assembly of their respective genomes and for many years the genome coordinates of the clones from these library sets were published on their respective genome websites. Many of these libraries were much utilized for diverse purposes. Library pMoCosX, one of the foundational libraries for *N. crassa* (Orbach 1991, Orbach 1994), was distributed to 56 clients and cited 28 times in Google Scholar (accessed January 26, 2018). This cosmid library was utilized to orient genetic maps (Bowring and Catcheside 1995), to identify transposable elements (Ramussen et al. 2004), and for cloning genes by complementation (Dieterle et al. 2010, Kim et al. 2002, Kinney et al. 2009, McCluskey et al. 2007, Wiest et al. 2008). A similar library of *A. nidulans* genomic DNA is comprised of clones in two unique vectors, pLorist2 and pWE15. This set was distributed 43 times and has received 1889 citations in Google Scholar (accessed January 26, 2018). This library was notably used to identify novel polyketide synthetases (Yu and Leonard 1995), histone deacetylase genes (Graessle et al. 2000), and as the basis for chromosome-specific sets and a physical map of the genome (Prade et al. 1997). Many of these resources became archival after the respective genomes were sequenced and published (Galagan et al. 2003, 2005). For several years, the FGSC accepted and held genome libraries for the sequencing of various fungi including *Magnaporthe grisea* and *Fusarium graminearum*. These libraries were never utilized and were destroyed when the FGSC relocated in 2014.

In contrast, many of the cDNA libraries held at the FGSC have been widely utilized. For example, the 24 h cDNA library of *A. nidulans* was distributed 89 times and used in many diverse projects. The sequencing of cDNAs in an Expressed Sequence Tag library identified over 8000 genes (Kupfer et al. 1997) presaging the information from the assembled genome. The self-replicating genome libraries of *A. nidulans* (Osherov et al. 2000) were also very popular having been distributed a total of 109 times. The article that describes these libraries has been cited 63 times. These citing articles have in turn been cited over 2200 times (Google Scholar, accessed January 26, 2018) demonstrating the multiplication of the impact generated using FGSC resources.

Gene Deletion Cassettes

In addition to gene libraries, the FGSC holds a set of gene-deletion cassettes for *A. nidulans* as DNA samples in a 96-well format. These cassettes are provided to FGSC clients as 10 ul of DNA plus an aliquot of pooled gene-specific PCR primers. Of the approximately 10,000 cassettes provided to date, 440 samples have been sent to 33 different recipients, although some unique cassettes were sent to more than one recipient. There was the expectation that cassette recipients would deposit the deletion strain into the FGSC collection, but this has not come to pass. In general, most researchers have generated knock-out cassettes in their own laboratories using fusion PCR (Dohn et al. 2018).

PRACTICES FOR PRESERVATION

The FGSC is fortunate that most of the materials in its care are readily and easily preserved for many years by desiccation. This allows most freely sporulating strains to be stored on anhydrous silica gel or as freeze-dried (lyophilized) spores (Perkins 1962, Raper and Alexander 1945). These techniques are widely utilized and highly robust. The current holdings include over 20,000 strains preserved on silica gel and 10,000 preserved by lyophilization. The FGSC presently holds 848 silica gel stocks and 1174 sets of lyophilized spores that were prepared in the 1960s, and these were last tested for viability in the early 2000s. Silica gel is a very convenient format as it allows one to repeatedly reactivate a strain by transferring a few grains to appropriate culture medium. It does require that stocks be kept desiccated and the FGSC maintains dry conditions by keeping silica gel stocks in screw-cap glass tubes in desiccated refrigerators or within desiccated boxes.

CRYOPRESERVATION

Both lyophilization and storage on silica gel require survival of desiccation. Hence, the FGSC began a significant investment in cryopreservation equipment and installed −80 freezers, as well as liquid nitrogen (LN) storage, beginning in the 1970s (Figure 9.1). These were expanded in the 1990s. This equipment has been used to support the maintenance and distribution of morphological mutant strains, plasmids, arrayed gene libraries, and arrayed sets of gene deletion mutants. The FGSC assessed the impact of cryopreservation at −80 in glycerol and milk on *N. crassa* conidia and observed 90% lethality following one freeze-thaw cycle, and smaller decreases in subsequent freeze-thaw cycles (McCluskey et al. 2006). While the viability of conidia

FIGURE 9.1 Cryo-storage of fungal strains at the Fungal Genetics Stock Center.

was low, the vast numbers of cells meant that this was a practical approach to storing large numbers of gene deletion mutants, even with their potential heterogeneity.

The FGSC currently stores 7753 specimens at −80°C and/or over (vapor phase) LN with the difference being whether cryopreservation is the primary storage or a backup for other formats. *Escherichia coli* strains carrying plasmids are stored at −80°C with a backup of the plasmid DNA stored at −20°C. While considered to be archival, copies of ordered genome libraries in *E. coli* hosts are still stored cryopreserved at −80°C in 25% glycerol. This format has been used for all genome libraries held, copied, and distributed by the FGSC (McCluskey 2011). A recent study of the preservation of *Fusarium* isolates at the FGSC, and their copy at the USDA National Laboratory for Genetic Resources Preservation (NLGRP, Fort Collins, CO), found little differences in viability of strains stored in 25% glycerol and 3.5% nonfat milk at −80°C versus strains stored over LN (Webb et al. 2018).

No strains at the FGSC are maintained under sterile water or oil, or by serial transfer, although the so-called cell-wall deficient "slime" strains of *N. crassa* were routinely transferred in the years immediately following their introduction into the FGSC. These strains are now stored in glycerol at −80°C and over LN, or as sheltered heterokaryons (Emerson 1963, Selitrennikoff 1978). In the past, some strains were cryopreserved using DMSO as the cryoprotectant (Wilson 1986). While this technique is used elsewhere, the ability to use glycerol simplifies preservation as it can be used for most fungi and also for preserving *E. coli* carrying plasmids (Feltham et al. 1978).

Quality Control

As a genetic collection, for many years the FGSC validated the genotype of strains. When the majority of strains were auxotrophic or physiological, this genetic testing was accomplished by growing a strain on supplemented and non-supplemented medium and verifying the response to the appropriate supplement. Physiological mutants, including fertility, morphological, or temperature-sensitive mutants, were validated by crossing, observing colony characteristics, or growing at an elevated temperature. The FGSC last tested all *Neurospora* strains up to number 4555 and all active strains (defined as strains that had been deposited or requested within the last 10 years) above number 4555 in 2005. Similarly, all *Aspergillus* strains, including spore preparations stored over silica gel or as lyophilized spore preparations, were last tested in 2006. This testing was accomplished by starting each strain from silica gel or lyophilized spore stocks on supplemented medium for strains with nutritional requirements. These strains were then tested for growth without the specific supplements. While many collections provide lyophilized or cryopreserved spores to clients, the FGSC has always sent living cultures, and this represents a further level of quality control; each strain is examined for viability and, superficially, for taxonomy every time a strain is requested by a client.

DISTRIBUTION PRACTICES AND POLICIES

Many strains in the FGSC collection have been highly utilized over the years. Distribution data reflects trends in research and, as such, provide insight not directly available through other resources (Table 9.3). For example, strains were widely utilized

TABLE 9.3
Distribution of Resources from the Fungal Genetics Stock Center over Periods of Time (1960 Through 2018)

Period	Individual Strains	Arrayed Strains	Plasmids	Gene Libraries
1960–1985	32,672	NA[a]	NA[a]	NA[a]
1986–1997	14,491	NA[a]	1,205	400
1998–2018	29,444	>700,000	4,854	596

[a] Not available.

in high school and undergraduate curricula throughout the 1960s and into the 1980s. While it is impossible to predict what strains will be valuable to the community in the future, it is well established that genetic resources saved for one purpose may be (and frequently are) used for one or more other purposes. As an example, FGSC strains carrying the osmotic-2 (*os-2*) marker were used sparingly over many years for genetic mapping. In 1999, this mutation was shown to confer resistance to phenylpyrrole fungicides. Subsequently, strains carrying classical and gene deletion mutations at this locus rapidly became the most widely requested mutants (McCluskey and Plamann 2008).

Among the FGSC holdings, only twelve strains have been distributed 100 or more times since modern data keeping was implemented (Table 9.4). Similarly, 9638 strains have been distributed ten or fewer times and 4765 strains having been sent out only once. An estimate of strain distribution carried out to support the 1975 FGSC grant proposal showed that only 57 out of 1000 strains were never distributed. Since 1998, 1295 *Aspergillus* strains and 5481 *Neurospora* strains were not distributed although they are maintained as archival (McCluskey 2011).

Frequently requested strains have a number of unifying characteristics. They are either reference genome strains, strains used as mating-type testers, as transformation recipients, or for use in the classroom. Among the most highly utilized *N. crassa* strains, auxotrophic mutants have a special role. The 1-234-723 allele of *his-3* is widely used as a target for heterologous gene replacement (Aramayo and Metzenberg 1996). Similarly, strains that are mutant at the *mus-51* or *mus-52* loci are deficient in non-homologous integration and so show higher efficiency of gene replacement (Ninomiya et al. 2004). These were used as the recipient in the *N. crassa* systematic gene deletion project (Colot et al. 2006). The fluffy strain that has been distributed 100 times is used as a mating type tester that does not produce asexual spores by virtue of a defect in a zinc finger transcription factor gene (Bailey and Ebbole 1998, Perkins et al. 1989). Mutants in the arginine biosynthetic pathway (*arg*) are used in classroom exercises demonstrating biochemical pathway characterization (Stine 1973), and these are routinely requested by various undergraduate institutions.

The FGSC has been successful in distributing material in an arrayed format. These include the *N. crassa* gene deletion set (Colot et al. 2006) that has been distributed to 36 clients in the US, Europe, and Asia, over 1200 plates of *C. neoformans* mutants, and 1300 plates of *C. albicans* mutants, to 200 unique clients. Distribution

TABLE 9.4
Strains of *Neurospora* and *Aspergillus* Distributed >100 Times in the Last 20 Years (1998–2018) by the Fungal Genetics Stock Center

FGSC #	Species	Designation	Number Distributions	Reference
2489	*N. crassa*	74-OR23-1V A[a]	359	Mylyk and Threlkeld (1974)
A4	*A. nidulans*	type (*veA*+)[b]	236	Pontecorvo et al. (1953)
4200	*N. crassa*	74-ORS-6a	180	Käfer and Fraser (1979)
1459	*N. crassa*	*arg-1*	145	Newmeyer (1957)
66	*N. crassa*	*arg-2*	136	Mitchell et al. (1952)
A1100	*A. fumigatus*	AF293[c]	129	Pain et al. (2004)
262	*N. crassa*	STA4[d]	120	Perkins (2004)
90	*N. crassa*	*arg-4*	117	Davis (1979)
987	*N. crassa*	74-OR23-1A[e]	108	Case et al. (1965)
6103	*N. crassa*	*his-3*	106	Webber (1965)
9718	*N. crassa*	Δ*mus-51::bar+*	101	Colot et al. (2006)
4317	*N. crassa*	*fl* (fluffy)	100	Perkins et al. (1989)

[a] ATCC MYA-4614.
[b] ATCC 38163.
[c] ATCC MYA-4609.
[d] ATCC 14692.
[e] ATCC 24698.

to foreign destinations is complicated by the need to ship on dry ice. In special cases (and for an additional fee) the FGSC staff will replicate arrayed sets onto agar solidified medium to allow shipment at ambient temperature.

FEES

Fees charged at the FGSC impact strain distribution. For many years, the FGSC did not charge fees for strains and began implementing a cost-recovery process only when required to do so. Fees for FGSC strains are especially problematic for international distribution where a large valuation can trigger import taxes or fees. A recent situation where botanic specimens were sent to Australia with a very low declared value led to the destruction of irreplaceable material (Stokstad 2017). While fungal collections have typically shared strains in the past, recent requests for reference genome strains from other collections have been declined, an acknowledgment that collections (in general) have entered a revenue-centric era.

IMPACT OF THE FGSC

Among the major impacts of the FGSC on the research communities is its ability to provide living material with a common genetic background. In 1976, the *Neurospora* community adopted the St. Lawrence/Oak Ridge background as the standard lineage for genetic manipulation. After this, most simple mutant strains in the FGSC collection were backcrossed into the reference genome background. The impact of this common background is seen in gene and genome sequencing projects and builds upon shared resources from the molecular genetics era (McCluskey and Plamann 2004). The FGSC has employed a metric to assess impact called the h-index that measures the number of citations to published work. While this metric is more commonly used to document individual impact, it is equally applicable to an institution such as the FGSC. Using this metric, the FGSC has an impact of 141, which translates to 141 articles each of which have been cited 141 or more times. Overall, these articles that acknowledge use of FGSC strains have been cited over 100,000 times (https://scholar.google.com/citations?user=EZgw2Z0AAAAJ&hl=en accessed 4/4/18).

METADATA

The FGSC acquires and maintains data as described in the WFCC and per the Organization for Economic Co-operation and Development best practice guidelines. For all strains, the FGSC maintains taxonomic data (genus, species) as well as depositor name, date, affiliation, and any trivial names that refer to the strain in publications or in other collections. Wild-type strains have associated geographic origin information, although for most FGSC strains latitude and longitude are not available and historical place names are all that was provided by the depositor. Mutant strains have different data requirements, and so in addition to taxonomic and origin data, strain history (genetic background), mutant locus, and method of mutagenesis are available. For all strains, citation to published research is requested (Table 9.5).

The FGSC was established before the era of digital computers and the original database was maintained on paper in 3-ring binders. Beginning in the 1980s a digital database was established and maintained in the program Dbase. Migration to a more widely deployed and web-enabled platform in 1998 made the FGSC database

TABLE 9.5
Data Maintained on All Strains, Mutant Strains, and Wild-Type Strains at the Fungal Genetics Stock Center

All Strains	Mutant Strains[a]	Wild-Type Strains[a]
Genus	Genotype	Country of origin
Species	Allele designation/locus	Location
Accession no.	Linkage group of marker	
External accession no.	Linked markers	
Mating type	Mutagen	
Deposit date	Genetic history	
Depositor name and contact	Nutritional requirements	
Published citations		
Preservation dates		
Test dates		

[a] In addition to data maintained for all strains.

available to customers online. Through collaboration with the authors of the published *Neurospora* compendia (Perkins et al. 1982, 2000) as well as the online e-compendium (http://www.bioinf.leeds.ac.uk/~gen6ar/newgenelist/genes/gene_list.htm) extensive data on mutant phenotypes and the specific mutations underlying each phenotype was incorporated into the database (McCluskey 2000b).

RESEARCH AT THE FGSC

Due to the nature of the support received from the NSF, research activities at the FGSC were initially largely restricted to collection improvement. Moreover, the transition from classical to molecular to genomics studies has influenced research activities. In the classical genetics era, detection of genetic anomalies such as translations or duplications represented the overlap of strain characterization and collection quality assurance. In 1975, 8% of strains were replaced for reasons of instability, secondary mutations, or chromosome abnormalities (Barratt and Ogata 1980).

In the modern data era, research at the FGSC has engaged students and technicians for a broader impact. Student projects are as diverse as the diversity of the materials in the collection. Individual studies have sought to characterize germination of asexual spores in response to ethylene gas and to examine the viability of *Neurospora* through multiple cycles of freezing and thawing (McCluskey et al. 2006) in anticipation of preparing arrayed sets of gene deletion mutants of *N. crassa*. Due to the availability of strains with multiple mutations at a single locus, several projects have re-examined mutations at previously identified mutant loci such as including *ad-8* (Wiest et al. 2012) and *trp-3* (Baker et al. 2015, Wiest et al. 2013). These studies largely validated the results of the fine-scale mapping studies of the classical genetics era.

The FGSC collection includes a large number of temperature sensitive (TS) lethal markers that are relatively straightforward to identify by complementation of the TS phenotype with wild-type DNA. These were used to demonstrate that the TS lethal mutant known as unknown-16 (*un-16*) was a point mutation in the ribosomal protein S9 encoding gene (McCluskey et al. 2007), that *un-4* was a mitochondrial import protein (Wiest et al. 2008), that *un-10* was a mutation in the eukaryotic translation initiation factor 3, subunit B (Kinney et al. 2009), and that *un-7* was an allele of *png-1* and both were in a gene responsible for shuttling proteins into the proteasome for recycling (Dieterle et al. 2010). The FGSC has also, more recently, used whole genome sequencing and *in silico* comparative analysis to identify the mutated gene in a group of classical mutants with diverse phenotypes including aberrant colony morphology, lack of sexual reproduction, and perethecial pigment production (McCluskey et al. 2011). The success of this project led to a larger scale analysis of over 500 strains with mutant phenotypes that have not, as yet, been found to be associated with an open reading frame in the annotated genome.

CONCLUSIONS

Research in fungal biology in the widest sense benefits from the availability of quality controlled genetic resources (McCluskey 2013, 2015b), and sharing of materials enhances research productivity in every area (Furman and Stern 2011,

Furman et al. 2010). The FGSC is a unique resource. Fungal collections, including fungal type strains at the Westerdijk Institute (Netherlands) collection, industrial and agricultural strains at the USDA/ARS/NRRL collection and the ARS Entomopathogen collection, and biodiversity collections such as the Phaff Yeast collection and the World Phytophthora Resource typically emphasize wild-type strains of limited genetic exploitation.

The FGSC has been closely tied to the research community, and it has benefitted when the community was most active. Because work with filamentous fungal model organisms was so very successful, the FGSC expanded in breadth as technical advances brought more research systems into a modern context (Roche et al. 2014). The ability to work directly in economically important systems has led to a decline in utilization of materials from the FGSC collection. Examples include human pathogens like *A. fumigatus*, Candida or *Cryptococcus*, agricultural pathogens and post-harvest contaminants like *Fusarium*, various Aspergilli, and diverse plant pathogens, as well as fungi used in industrial and pharmaceutical biotechnology, means that the FGSC has needed to expand its scope to serve its constituency. Reductions in support for fungal genetics research at the NSF has limited the ability of the FGSC to attract grant support. Ironically the developments in biotechnology, and especially high throughput DNA sequencing that allow high-level research in what were once considered intractable systems, has meant that fewer laboratories are working with model systems.

Moreover, as research moves beyond questions of primary metabolism, cell biology, and population genetics, other systems provide valuable insight not available through work with models like *N. crassa* and *A. nidulans*. As the growth and increased diversity at the Fungal Genetics Conference has reflected the nature of research with fungi, the FGSC has worked to maintain the irreplaceable materials that led to the biochemical genetics revolution.

ACKNOWLEDGMENTS

This is publication XYX from the Kansas Agricultural Experiment Station.

REFERENCES

Aramayo, R. and R. Metzenberg. 1996. Gene replacements at the his-3 locus of *Neurospora crassa*. *Fungal Genetics Newsletter* 43:9–13.

Bailey, L.A. and D.J. Ebbole. 1998. The fluffy gene of *Neurospora crassa* encodes a Gal4p-type C6 zinc cluster protein required for conidial development. *Genetics* 148:1813–1820.

Baker, S.E., W. Schackwitz, A. Lipzen et al. 2015. Draft genome sequence of *Neurospora crassa* strain FGSC 73. *Genome Announcements* 3:e00074–00015.

Barratt, R.W. 1986. Stocks from Tatum *Neurospora* collection. *Fungal Genetics Newsletter* 33:49–58.

Barratt, R.W. and W.N. Ogata. 1980. *Neurospora* stock list. Tenth revision. *Neurospora Newsletter* 27:39–128.

Beadle, G.W. and E.L. Tatum. 1941. Genetic control of biochemical reactions in *Neurospora.Proceedings of the National Academy of Sciences(USA)* 27:499–506.

Botstein, D., R.L. White, M. Skolnick and R.W. Davis. 1980. Construction of a genetic linkage map in man using restriction fragment length polymorphisms. *American Journal of Human Genetics* 32:314–331.

Bowring, F.J. and D.E. Catcheside. 1995. The orientation of gene maps by recombination of flanking markers for the am locus of *Neurospora crassa. Current Genetics* 29:27–33.

Brody, H., J. Griffith, A.J. Cuticchia, J. Arnold and W.E. Timberlake. 1991. Chromosome-specific recombinant DNA libraries from the fungus *Aspergillus nidulans. Nucleic Acids Research* 19:3105–3109.

Case, M., H. Brockman and F. De Serres. 1965. Further information on the origin of the Yale and Oak Ridge wild-type strains of *Neurospora crassa. Fungal Genetics Reports* 8:31.

Collopy, P.D., H.V. Colot, G. Park et al. 2010. High-throughput construction of gene deletion cassettes for generation of *Neurospora crassa* knockout strains. *Methods in Molecular Biology* 638:33–40.

Colot, H.V., G. Park, G.E. Turner et al. 2006. A high-throughput gene knockout procedure for *Neurospora* reveals functions for multiple transcription factors. *Proceedings of the National Academy of Sciences (USA)* 103:10352–10357.

Davis, R.H. 1979. Genetics of arginine biosynthesis in *Neurospora crassa. Genetics* 93:557–575.

Davis, R.H. and D.D. Perkins. 2002. Timeline: *Neurospora*: A model of model microbes. *Nature Reviews Genetics* 3:397–403.

de Serres, F.J. 1989. X-ray-induced specific locus mutations in the ad-3 region of two-component heterokaryons of *Neurospora crassa*. III. Genetic fine structure analysis of the ad-3 and immediately adjacent genetic regions by means of complementation tests. *Mutatation Research* 211:89–102.

Dieterle, M.G., A.E. Wiest, M. Plamann and K. McCluskey. 2010. Characterization of the temperature-sensitive mutations un-7 and png-1 in *Neurospora crassa. PLoS One* 5(5):e10703.

Din, A.B. and O. Yarden. 1994. The *Neurospora crassa* chs-2 gene encodes a non-essential chitin synthase. *Microbiology* 140:2189–2197.

Dodge, B.O. 1939. Some problems in the genetics of the fungi. *Science* 90:379–385.

Dohn, J.W., A.W. Grubbs, C.E. Oakley and B.R. Oakley. 2018. New multi-marker strains and complementing genes for *Aspergillus nidulans* molecular biology. *Fungal Genetics and Biology* 111:1–6. doi:10.1016/j.fgb.2018.01.003.

Ducros, V., A.M. Brzozowski, K.S. Wilson et al. 1998. Crystal structure of the type-2 Cu depleted laccase from *Coprinus cinereus* at 2.2 Å resolution. *Nature Structural and Molecular Biology* 5:310–316.

Dunlap, J.C., K.A. Borkovich, M.R. Henn et al. 2007. Enabling a community to dissect an organism: Overview of the *Neurospora* functional genomics project. *Advances in Genetics* 57:49–96.

Emerson, S. 1963. Slime, a plasmodioid variant of *Neurospora crassa. Genetica* 34:162–182.

Feltham, R., A.K. Power, P.A. Pell and P. Sneath. 1978. A simple method for storage of bacteria at −76°C. *Journal of Applied Microbiology* 44:313–316.

Fowler, T.J., M.F. Mitton, L.J. Vaillancourt and C.A. Raper. 2001. Changes in mate recognition through alterations of pheromones and receptors in the multisexual mushroom fungus *Schizophyllum commune. Genetics* 158:1491–1503.

Furman, J.L., F. Murray and S. Stern. 2010. More for the research dollar. *Nature* 468:757–758.

Furman, J.L. and S. Stern. 2011. Climbing atop the shoulders of giants: The impact of institutions on cumulative research. *American Economic Review* 101:1933–1963.

Galagan, J.E., S.E. Calvo, K.A. Borkovich et al. 2003. The genome sequence of the filamentous fungus *Neurospora crassa. Nature* 422:859–868.

Galagan, J.E., S.E. Calvo, C. Cuomo et al. 2005. Sequencing of *Aspergillus nidulans* and comparative analysis with *A. fumigatus* and *A. oryzae. Nature* 438:1105–1115.

Garnjobst, L. and J.F. Wilson. 1956. Heterocaryosis and protoplasmic incompatibility in *Neurospora crassa. Proceedings of the National Academy of Sciences(USA)* 42:613–618.

Graessle, S., M. Dangl, H. Haas et al. 2000. Characterization of two putative histone deacetylase genes from *Aspergillus nidulans. Biochimica et Biophysica Acta* 1492:120–126.

Horowitz, N.H. 1991. Fifty years ago: The *Neurospora* revolution. *Genetics* 127:631–635.

Jurgenson, J., R. Bowden, K. Zeller, J. Leslie, N. Alexander and R. Plattner. 2002. A genetic map of *Gibberella zeae* (*Fusarium graminearum*). *Genetics* 160:1451–1460.

Käfer, E. and M. Fraser. 1979. Isolation and genetic analysis of nuclease halo (nuh) mutants of *Neurospora crassa. Molecular and General Genetics* 169:117–127.

Kamens, J. 2014. The Addgene repository: An international nonprofit plasmid and data resource. *Nucleic Acids Research* 43(D1):D1152–D1157.

Kelkar, H.S., J. Griffith, M.E. Case et al. 2001. The *Neurospora crassa* genome: Cosmid libraries sorted by chromosome. *Genetics* 157:979–990.

Kim, H., R.L. Metzenberg and M.A. Nelson. 2002a. Multiple functions of *mfa-1*, a putative pheromone precursor gene of *Neurospora crassa. Eukaryotic Cell* 1:987–999.

Kim, H.-S., K.-Y. Han, K.-J. Kim, D.-M. Han, K.-Y. Jahng and K.-S. Chae. 2002b. The *veA* gene activates sexual development in *Aspergillus nidulans. Fungal Genetics and Biology* 37:72–80.

Kinney, M., A. Wiest, M. Plamann and K. McCluskey. 2009. Identification of the *Neurospora crassa* mutation un-10 as a point mutation in a gene encoding eukaryotic translation initiation factor 3, subunit B. *Fungal Genetics Reports* 56:6–7.

Kinsey, J.A. and J.A. Rambosek. 1984. Transformation of *Neurospora crassa* with the cloned am (glutamate dehydrogenase) gene. *Molecular and Cellular Biology* 4:117–122.

Kües, U. 2000. Life history and developmental processes in the basidiomycete *Coprinus cinereus. Microbiology and Molecular Biology Reviews* 64:316–353.

Kües, U., S. Subba, Y. Yu et al. 2016. Regulation of fruiting body development in *Coprinopsis cinerea*. In *Science and Cultivation of Edible Fungi: Mushroom Science* IXX, ed. J.J.P. Baars and A.S.M. Sonnenberg, pp. 318–322. Coatesville, PA: International Society for Mushroom Science.

Kupfer, D.M. 1999. Development, analysis and use of an expressed sequence tag database from the multicellular asomycete, *Aspergillus nidulans*. PhD dissertation. Norman, OK: University of Oklahoma.

Kupfer, D.M., C.A. Reece, S.W. Clifton, B.A. Roe and R.A. Prade. 1997. Multicellular ascomycetous fungal genomes contain more than 8000 genes. *Fungal Genetics and Biology* 21:364–372.

Lambreghts, R., M. Shi, W.J. Belden et al. 2009. A high-density single nucleotide polymorphism map for *Neurospora crassa. Genetics* 181:767–781.

Leslie, J.F. 1981. Inbreeding for isogeneity by backcrossing to a fixed parent in haploid and diploid eukaryotes. *Genetics Research* 37:239–252.

McCluskey, K. 2000a. Long term viability of *Neurospora crassa* at the FGSC. *Fungal Genetics Newsletter* 47:110.

McCluskey, K. 2000b. A relational database for the FGSC. *Fungal Genetics Newsletter* 47:74–78.

McCluskey, K. 2011. From genetics to genomics: Fungal collections at the Fungal Genetics Stock Center. *Mycology* 2:161–168.

McCluskey, K. 2013. Biological resource centers provide data and characterized living material for industrial biotechnology. *Industrial Biotechnology* 9:117–122.

McCluskey, K. 2015a. Boosting research and industry by providing extensive resources for fungal research. In *Gene expression systems in fungi: Advancements and applications*, ed. M. Schmoll, and C. Dattenböck, 34–50. Berlin, Germany: Springer.

McCluskey, K. 2015b. New resources, research, and progress at the Fungal Genetics Stock Center. *Fungal Genetics Reports* 62S:34.
McCluskey, K. and M. Plamann. 2004. 10th Fungal Genetics Stock Center catalogue of strains. *Fungal Genetics Newsletter* 51 (supplement).
McCluskey, K. and M. Plamann. 2008. Perspectives on genetic resources at the Fungal Genetics Stock Center. *Fungal Genetics Reports* 55:15–17.
McCluskey, K., A. Wiest and S.A. Walker. 2006. The effect of repeated freeze-thaw cycles on cryopreserved *Neurospora crassa* samples. *Fungal Genetics Newsletter* 53:37.
McCluskey, K., S.A. Walker, R.L. Yedlin, D. Madole and M. Plamann. 2007. Complementation of un-16 and the development of a selectable marker for transformation of *Neurospora crassa*. *Fungal Genetics Newsletter* 54:9–11.
McCluskey, K., A. Wiest, I.V. Grigoriev et al. 2011. Rediscovery by whole genome sequencing: Classical mutations and genome polymorphisms in *Neurospora crassa*. *G3* 1:303–316.
Mitchell, M.B., T. Pittenger and H. Mitchell. 1952. Pseudo-wild types in *Neurospora crassa*. *Proceedings of the National Academy of Sciences (USA)* 38:569–580.
Mylyk, O., E. Barry and D. Galeazzi. 1974. New isogenic wild types in *N. crassa*. *Neurospora Newsletter* 21:24.
Mylyk, O.M. and S.F. Threlkeld. 1974. A genetic study of female sterility in *Neurospora crassa*. *Genetics Research* 24:91–102.
Newmeyer, D. 1957. Arginine synthesis in *Neurospora crassa:* Genetic studies. *Microbiology* 16:449–462.
Newmeyer, D., D.D. Perkins and E.G. Barry. 1987. An annotated pedigree of *Neurospora crassa* laboratory wild types, showing the probable origin of the nucleolus satellite and showing that certain stocks are not authentic. *Fungal Genetics Newsletter* 34:46–51.
Ninomiya, Y., K. Suzuki, C. Ishii and H. Inoue. 2004. Highly efficient gene replacements in *Neurospora* strains deficient for nonhomologous end-joining. *Proceedings of the National Academy of Sciences (USA)* 101:12248–12253.
Olson, L.W. 1984. *Allomyces*- a different fungus. *Opera Botanica* 73:1–96.
Orbach, M. 1991. The Orbach/Sachs cosmid library of *N. crassa* DNA sequences (pMOcosX). *Fungal Genetics Newsletter* 38:97.
Orbach, M.J. 1994. A cosmid with a HyR marker for fungal library construction and screening. *Gene* 150:159–162.
Osherov, N., J. Mathew and G.S. May. 2000. Polarity-defective mutants of *Aspergillus nidulans*. *Fungal Genetics and Biology* 31:181–188.
Pain, A., J. Woodward, M.A. Quail et al. 2004. Insight into the genome of *Aspergillus fumigatus*: Analysis of a 922 kb region encompassing the nitrate assimilation gene cluster. *Fungal Genetics and Biology* 41:443–453.
Park, G., H.V. Colot, P.D. Collopy et al. 2011. High-throughput production of gene replacement mutants in *Neurospora crassa*. *Methods in Molecular Biology* 722:179–189.
Perkins, D. 1972. Linkage testers having markers near centromere. *Fungal Genetics Reports* 19: 21.
Perkins, D., B. Turner, V. Pollard and A. Fairfield. 1989. *Neurospora* strains incorporating fluffy, and their use as testers. *Fungal Genetics Reports* 36:64.
Perkins, D.D. 1949. Biochemical mutants in the smut fungus *Ustilago maydis*. *Genetics* 34:607–626.
Perkins, D.D. 1962. Preservation of *Neurospora* stock cultures with anhydrous silica gel. *Canadian Journal of Microbiology* 8:591–594.
Perkins, D.D. 1990. New multicent linkage testers for centromere-linked genes and rearrangements in *Neurospora*. *Fungal Genetics Reports* 37:19.
Perkins, D.D. 1991. *Neurospora* alcoy linkage tester stocks with group VII marked, and their use for mapping translocations. *Fungal Genetics Reports* 38:83.

Perkins, D.D. 2004. Wild-type *Neurospora crassa* strains preferred for use as standards. *Fungal Genetics Reports* 51:7–8.

Perkins, D.D., A. Radford, D. Newmeyer and M. Bjorkman. 1982. Chromosomal loci of *Neurospora crassa. Microbiological Reviews* 46:426–570.

Perkins, D.D., A. Radford and M.S. Sachs. 2000. *The Neurospora Compendium: Chromosomal Loci*. Cambridge, MA: Academic Press.

Pontecorvo, G. 1946. Genetic systems based on hetero caryosis. In *Cold Spring Harbor Symposia on Quantitative Biology*. Cold Spring Harbor, NY: Cold Spring Harbor Laboratory Press.

Pontecorvo, G., J. Roper, L. Chemmons, K. MacDonald and A. Bufton. 1953. The genetics of *Aspergillus nidulans. Advances in Genetics* 5:141–238.

Prade, R.A., J. Griffith, K. Kochut, J. Arnold and W.E. Timberlake. 1997. In vitro reconstruction of the *Aspergillus* (= *Emericella*) *nidulans* genome. *Proceedings of the National Academy of Sciences* (*USA*) 94:14564–14569.

Ramussen, J., A. Taylor, L.-J. Ma, S. Purcell, F. Kempken and D. Catcheside. 2004. Guest, a transposable element belonging to the Tc1/mariner superfamily is an ancient invader of *Neurospora* genomes. *Fungal Genetics and Biology* 41:52–61.

Raper, C.A. and T.J. Fowler. 2004. Why study *Schizophyllum*? *Fungal Genetics Reports* 51:30–36.

Raper, C.A. and J.R. Raper. 1973. Mutational analysis of a regulatory gene for morphogenesis in Schizophyllum. *Proceedings of the National Academy of Sciences (USA)* 70:1427–1431.

Raper, J.R., D.H. Boyd and C.A. Raper. 1965. Primary and secondary mutations at the incompatibility loci in *Schizophyllum. Proceedings of the National Academy of Sciences (USA)* 53:1324–1332.

Raper, K.B. and D.F. Alexander. 1945. Preservation of molds by the lyophil process. *Mycologia* 37:499–525.

Roche, C.M., J.J. Loros, K. McCluskey and N.L. Glass. 2014. *Neurospora crassa*: Looking back and looking forward at a model microbe. *American Journal of Botany* 101:2022–2035.

Ryan, F.J., G. Beadle and E. Tatum. 1943. The tube method of measuring the growth rate of *Neurospora. American Journal of Botany* 30:784–799.

Seifert, K.A. 1994. A novel method of growing fungi for DNA extraction. *Fungal Genetics Reports* 41:79–80.

Selitrennikoff, C. 1978. Storage of slime strains. *Fungal Genetics Newsletter* 25:16.

Simione, F.P., R.H. Cypess, M. Wigglesworth and T. Wood. 2012. Managing a global biological resource of cells and cellular derivatives. In *Management of Chemical and Biological Samples for Screening Applications*, ed. M. Wigglesworth and T. Wood, pp. 143–164. Hoboken, NJ: John Wiley & Sons.

Simmons, J., P. Chary and D. Natvig. 1987. Linkage group assignments for two *Neurospora crassa* catalase genes: The Metzenberg RFLP mapping kit applied to an enzyme polymorphism. *Fungal Genetics Reports* 34:55.

St. Lawrence, P. 1954. The association of particular linkage groups with their respective chromosomes in *Neurospora crassa*. PhD thesis, New York: Columbia University.

Stajich, J.E., S.K. Wilke, D. Ahrén et al. 2010. Insights into evolution of multicellular fungi from the assembled chromosomes of the mushroom *Coprinopsis cinerea* (*Coprinus cinereus*). *Proceedings of the National Academy of Sciences (USA)* 107:11889–11894.

Stine, G.J. 1973. *Laboratory Exercises in Genetics*. London, UK: Macmillan Publishing Company.

Stokstad, E. 2017. Botanists fear research slowdown after priceless specimens destroyed at Australian border. *Science*. doi:10.1126/science.aal1175.

Strauss, B.S. 2016. Beadle and Tatum and the origins of molecular biology. *Nature Reviews Molecular Cell Biology* 17:266. doi:10.1038/nrm.2016.42.

Sugano, S.S., H. Suzuki, E. Shimokita, H. Chiba, S. Noji, Y. Osakabe and K. Osakabe. 2017. Genome editing in the mushroom-forming basidiomycete *Coprinopsis cinerea*, optimized by a high-throughput transformation system. *Scientific Reports* 7:1260. doi:10.1038/s41598-017-00883-5.

Turner, B.C., D.D. Perkins and A. Fairfield. 2001. *Neurospora* from natural populations: A global study. *Fungal Genetics and Biology* 32:67–92.

Webb, K.M., G. Holman, S. Greene and K. McCluskey. 2018. Frozen fungi: Cryogenic storage is an effective method to store *Fusarium* cultures for the long-term. *Annals of Applied Biology* (in press).

Webber, B.B. 1965. Genetical and biochemical studies of histidine-requiring mutants of *Neurospora crassa*. IV. Linkage relationships of hist-3 mutants. *Genetics* 51:275–283.

Wiest, A., D. Barchers, M.H. Eaton, R.R. Schnittker and K. McCluskey. 2013. Molecular analysis of intragenic recombination at the tryptophan synthetase locus in *Neurospora crassa*. *Journal of Genetics* 92:523–528.

Wiest, A., A.J. McCarthy, R. Schnittker and K. McCluskey. 2012. Molecular analysis of mutants of the *Neurospora* adenylosuccinate synthetase locus. *Journal of Genetics* 91:199–204.

Wiest, A., M. Plamann and K. McCluskey. 2008. Identification of the *Neurospora crassa* mutation un-4 as the mitochondrial inner membrane translocase subunit tim16. *Fungal Genetics Reports* 56:37–39.

Wilson, C. 1986. FGSC culture preservation methods. *Fungal Genetics Reports* 33:47.

Xu, J. and J.F. Leslie. 1996. A genetic map of *Gibberella fujikuroi* mating population A (*Fusarium moniliforme*). *Genetics* 143:175–189.

Yu, J.-H. and T.J. Leonard. 1995. Sterigmatocystin biosynthesis in *Aspergillus nidulans* requires a novel type I polyketide synthase. *Journal of Bacteriology* 177:4792–4800.

Yuill, E. 1939. Two new *Aspergillus* mutants. *Journal of Botany* 77:174–175.

10 The *Peromyscus* Genetic Stock Center

Amanda Havighorst, Vimala Kaza and Hippokratis Kiaris

CONTENTS

Abstract: Animals of the genus *Peromyscus* are an uncommon, but useful, animal model with superficial similarity to the more commonly used laboratory mouse. The Peromyscus Genetic Stock Center (PGSC) serves as a repository for a number of different wild-type stocks of *Peromyscus* animals originating from varied habitats and localities. Current holdings include a variety of species in addition to coat-color and disease model mutants. Unlike laboratory mice, animals at the PGSC are bred to maintain genetic diversity, making them a unique and valuable resource for many different fields of study. Animals are shipped both nationally and internationally in support of scientific research.

INTRODUCTION TO *PEROMYSCUS*

Peromyscus is a genus of new world mice that is distributed across North America and into Central America (Kirkland and Layne 1989). In many places, they are the most abundant mammal found within their range. Members of the genus occupy most terrestrial habitats including alpine regions, deserts, grasslands, both temperate and tropical forests, wetlands, and beaches (Dewey and Dawson 2001, Kirkland and Layne 1989, Whitaker and Hamilton 1998). As a genus, *Peromyscus* includes more than 50 species, which can be further divided taxonomically into 200 subspecies, each of which is uniquely adapted to the environment in which it lives (Kirkland and Layne 1989). These nocturnal mice are cricetids, more closely related to hamsters than to standard laboratory mice (Romanenko et al. 2007).

The wide use of animals of this genus in biomedical research necessitates their breeding in captivity. Unlike laboratory mice, *Peromyscus* is difficult to inbreed. This leads to unique requirements and strategies for colony management that promote, rather than eliminate, genetic diversity (Dewey and Dawson 2001, Havighorst et al. 2017). Rather than being bred for docility, *Peromyscus* colonies have retained their wild characteristics and behaviors (Havighorst et al. 2017). However, like standard laboratory mice, *Peromyscus* are easily maintained in captivity in conditions similar to those in which laboratory mice are kept (Dewey and Dawson 2001). Though they do best in opaque rather than transparent cages, *Peromyscus* thrive on standard laboratory mouse chow and can be kept in the same type of bedding (Dewey and Dawson 2001). *Peromyscus* has been used in many different types of research over a long period of time and has contributed to many different fields of research. Such studies include genetics, hybridization, speciation, parasitology, physiology, adaptation and evolution, ecology, animal behavior, epigenetics, a limited number of pathological studies involving aging, metabolic disorders, tumorigenesis, and others (Dewey and Dawson 2001, Havighorst et al. 2017, Kirkland and Layne 1989, Shorter et al. 2012).

HISTORY OF THE PGSC

The PGSC was established in 1985 by Dr. Wallace D. Dawson. However, the University of South Carolina was not the first location to house colonies of captive deer mice. Laboratory colonies of deer mice were established as early as 1915, and other colonies have also existed at the University of Oregon, the University of Michigan, Ohio State University, among others. The very first of the large colonies of *Peromyscus* was established by Francis Sumner at the Scripps Institution at La Jolla, California (Bedford and Hoekstra 2015). Animals from this colony were later given to Dr. Lee R. Dice at the University of Michigan. Animals from some other institutions were abandoned due to the cost of animal upkeep and a lack of funding (Bedford and Hoekstra 2015). In the early 1980s, the National Academy of Sciences concluded that the loss of such colonies was a loss of potentially valuable resources. In response, the National Science Foundation (NSF) made funds available for the upkeep of colonies of non-traditional animal models. Since 1985, the NSF, with minimal interruptions, has continuously supported the PGSC. In 1998, the PGSC also received funding from the National Institute of Health due to the animals' usefulness as a model for the study of Lyme disease and hantavirus. The founding animals of the *P. maniculatus bairdii* colony that are kept at the PGSC today originally came from the University of Michigan colony (Bedford and Hoekstra 2015).

THE PGSC TODAY

The *Peromyscus* Genetic Stock Center (PGSC) is a unique repository and genetic stock center for *Peromyscus*. The PGSC is located at the University of South Carolina. Stocks at the PGSC are maintained to preserve their wild characteristics and genetic diversity, making them an alternative to laboratory mice. The wild characteristics of these animals are preserved in the colonies at the PGSC due to the lack of inbreeding. These characteristics are useful for modeling diverse populations (Havighorst et al. 2017). The analysis of behaviors and traits inherent to *Peromyscus* species can therefore provide information in a manner consistent with the observation of wild animals (Bedford and Hoekstra 2015).

In addition to its mission of maintaining and supplying experimental animals, the PGSC has also played a role in the preservation of the endangered subspecies *Peromyscus polionotus trissyllepsis*—the Perdido Key beach mouse (Greene et al. 2016). Captive-bred individuals maintained at the stock center were successfully reintroduced to their native habitat in 2007, thus demonstrating that beach mice can be bred in captivity and their populations replenished using captive-bred mice (Greene et al. 2016).

STOCKS MAINTAINED AT THE PGSC

Throughout its operation, the PGSC has maintained and distributed a varying number of animals and species. Logistically, materials to be acquired and maintained were determined based on the demand for their utilization and the financial support available to the PGSC to attain and maintain its operational objectives. The PGSC strives to maintain the maximum number of species without compromising the characteristics (genetic diversity or integrity) of those species, as required by the user community.

Peromyscus Species

Currently, five major species are maintained for which stocks exist. All photos, unless otherwise noted, are of female animals approximately 3 months of age.

Peromyscus maniculatus (PM)

Two subspecies of *P. maniculatus* are kept at the PGSC.

1. *Peromyscus maniculatus bairdii* (BW). Also called the prairie deer mouse, these animals are smaller than other deer mice, with an average weight of 15.45 g in the wild (Svihla 1935) (Figure 10.1). A unique characteristic of these animals is that they can be cross-bred with *Peromyscus polionotus*, rendering them ideal for genetic studies. These animals have been used in this manner for studies of genetic imprinting and reproductive isolation (Dewey and Dawson 2001). Crossing female *P. maniculatus* with male *P. polionotus* leads to offspring which are smaller than their parents, while crossing in the other direction leads to offspring which are larger (Dewey and Dawson 2001).
2. *Peromyscus maniculatus sonoriensis* (SM2). Also called the Sonoran deer mouse, these animals are another subspecies of *P. maniculatus* originally collected from White Mountain, CA, in 1995 (Careau et al. 2011) (Figure 10.2). The PGSC population is descended from 50 wild-caught animals (Careau et al. 2011). Because it was collected from a high-altitude region, this subspecies, when compared with low-altitude-adapted BW mice, represents a valuable resource for studying high-altitude adaptation (Dewey and Dawson 2001). On average, *P. m. sonoriensis*, with an average weight of 20–24 g, is larger than other subspecies of deer mouse.

FIGURE 10.1 *Peromyscus maniculatus bairdii* (the prairie deer mouse).

FIGURE 10.2 *Peromyscus maniculatus sonoriensis* (the Sonoran deer mouse).

Peromyscus eremicus (EP)

The cactus mouse, descended from animals collected from Tucson, AZ, in 1993, is a desert-dwelling species of the subgenus *Haplomylomys* (Avise et al. 1974, Bedford and Hoekstra 2015). There are currently fifteen recognized subspecies of *P. eremicus*. One of these is kept at the PGSC (Veal 1979).

Peromyscus leucopus (LL)

The white-footed mouse (Figure 10.3) is the species that is currently the most widely used for experimental studies. This is due to its relevance to the study of Lyme disease. *Peromyscus leucopus* has been recognized as a natural reservoir for the spirochete

FIGURE 10.3 *Peromyscus leucopus* (the white-footed mouse).

Borrelia burgdorferi (Anderson 1989). The stock maintained at the PGSC traces its origins to 38 wild ancestors caught between 1982 and 1985 near Linville Falls, NC. Reportedly, this is the longest lived of the *Peromyscus* species with a maximum recorded lifespan of 7.9 years (Dice 1937). This renders it ideal for aging studies (Weigl 2005, Zhao et al. 2015). In many ways, the appearance of the white-footed mouse is similar to *P. maniculatus*, but *P. leucopus* is distinguished by a lack of tail hair and protruding eyes (Feldhamer et al. 1983, Whitaker and Hamilton 1998).

Peromyscus polionotus subgriseus (PO)

The oldfield mouse (Figure 10.4) at the PGSC, descended from 21 wild ancestors caught prior to 1952 in Ocala National Forest, FL, is the smallest species of *Peromyscus* with a weight range of 8–9 g (Lacy et al. 1996, Whitaker and Hamilton 1998). Oldfield mice are native to the southeastern United States and tend to favor open areas such as sandy fields, beaches, and cotton and corn fields (Whitaker and Hamilton 1998). Those subspecies of *P. polionotus* that live in coastal areas are called "beach mice" rather than oldfield mice (Whitaker and Hamilton 1998). *Peromyscus polionotus* is monogamous, and males exhibit paternal care of their young (Gubernick and Teferi 2000, Whitaker and Hamilton 1998). This characteristic makes them suitable for behavioral studies.

Peromyscus californicus insignis (IS)

The California mouse (Figure 10.5) colony at the PGSC is descended from animals caught in the Santa Monica Mountains, CA, over a period of several years (1979 to 1987) (Bedford and Hoekstra 2015). The California mouse,

FIGURE 10.4 *Peromyscus polionotus subgriseus* (the oldfield mouse).

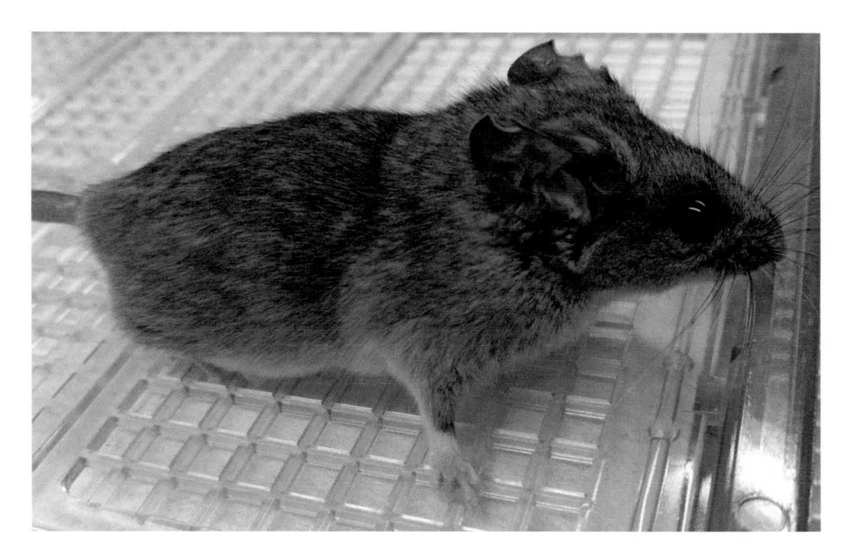

FIGURE 10.5 *Peromyscus californicus insignis* (the California mouse).

weighing an average of 45 g, is one of the largest species of *Peromyscus* (Kirkland and Layne 1989). Like *P. eremicus*, this species belongs to the subgenus *Haplomylomys* (Avise et al. 1974). *Peromyscus californicus* is also monogamous and provides paternal care (Gubernick and Teferi 2000). This species is used for studies of these behaviors.

Coat Color Mutants

In the past, *Peromyscus* was extensively used in the study of coat color genetics and physiology, as well as behavioral changes associated with coat color. Unfortunately, many of these stocks were culled due to funding limitations and only a few coat color mutations are currently available as live animals at the PGSC. A select few of these are described here. Many other coat color mutations exist in the form of cryopreserved sperm samples.

Albino

Albinism in deer mice, as in the standard laboratory mouse, obeys Mendelian genetics and is recessive (Castle 1912). These animals lack pigmentation in the skin and eyes due to an absence of tyrosinase activity (Castle 1912, Vrana et al. 2014). Albino deer mice have been known since 1909 when an albino deer mouse was wild-caught in Michigan (Castle 1912). The mutation represented in the PGSC arose spontaneously in 1919 in captive stock of *P. maniculatus. gambelii* (Figure 10.6).

FIGURE 10.6 (See color insert.) The albino coat color mutation in *Peromyscus maniculatus gambelii.*

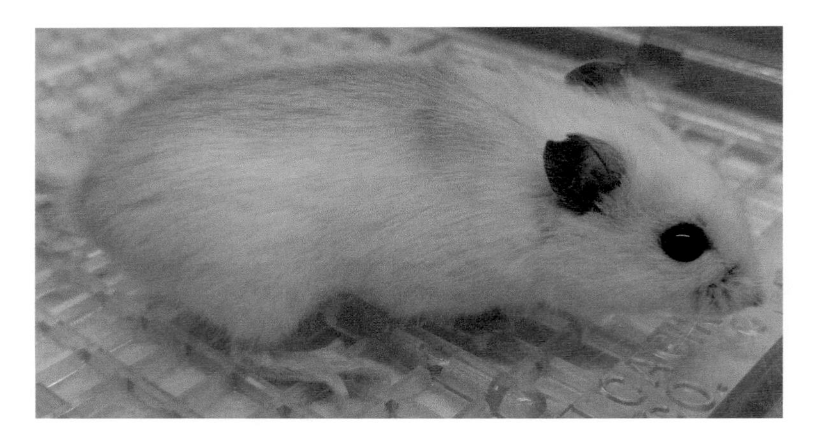

FIGURE 10.7 (See color insert.) The dominant spot coat color mutation in *Peromyscus maniculatus bairdi.*

Dominant Spot (DSP)

First collected in 1928 from a wild population of *P. maniculatus bairdi*, this dominant mutation leads to an unpigmented spot on the forehead and at the end of the tail, with variably expressed spotting over the rest of the body (Figure 10.7).

Tan Streak (TNS)

The tan streak mutation arose in a group of wild-caught *P.m. nubiterrae* from Macon County, NC. This phenotype becomes apparent a few days after birth, with affected

FIGURE 10.8 (See color insert.) The variable white coat color mutation in *Peromyscus maniculatus bairdii.*

animals being pale in color. By weaning, it is fully expressed. Affected animals are nearly completely white, with a pale tan streak along the dorsal line (Figure 10.8). Some animals have a shorter streak which only extends partway down the back, while in others, it extends to the base of the tail (Wang et al. 1993).

Variable White (WHT)

First appearing in a group of captive-bred *P.m. bairdii*, this semi-dominant lethal mutation is believed to arise from a defect in neural crest cell migration due to deafness associated with the phenotype. These animals may also exhibit ataxic or circling behavior, with the intensity of the behavior varying with the expression of the coat color phenotype (Cowling et al. 1994).

Wide-Band Agouti (WBA)

These animals overexpress the agouti gene, leading to a longer yellow hair band, resulting in a coat color that is more yellow than the *P.m. bairdii* wild type (Figure 10.9). The length of this band can be influenced by feeding the animal a methyl-donor-enriched diet, which leads to a heightened variability in the width of the band (Shorter et al. 2012).

Disease Models

A number of stocks are kept which exhibit phenotypes resembling various diseases. While not as varied as transgenic laboratory mouse models, all deer mouse disease models are spontaneous mutations which arose within wild populations or stock center colonies.

FIGURE 10.9 **(See color insert.)** The wide-band agouti coat color mutation in *Peromyscus maniculatus bairdi*.

Epilepsy (EPL)

The EPL stock, derived from BW deer mice, possess a mutation (likely in a single gene) that produces audiogenic epilepsy that is triggered by high-pitched sounds, such as jingling keys (Figure 10.10). Seizures manifest in the form of tics, waltzing, convulsions, and other abnormal movements. This trait is only observed in younger animals, as adults tend to become deaf by the age of 4 months (Chance and Yaxley 1949).

Juvenile Ataxia (JTX)

This mutation arose in a population of four blonde (bl) colored *P. maniculatus bairdii*, a coat color mutant which the stock center no longer maintains. Juvenile

FIGURE 10.10 *Peromyscus maniculatus bairdii* carrying the gene for epilepsy.

ataxia mice between 15 and 30–45 days old exhibit an ataxic gait, are unable to right themselves, and have difficulty climbing. The ataxia fades in older animals and they become capable of moving normally (Van Ooteghem 1983).

Cataract Webbed (CW)

This recessive mutation of *P. maniculatus*, first noticed in 1951, is characterized by the development of cataracts and syndactyly of the soft tissues. The age at which cataracts develop varies widely, ranging from 5 weeks to over 1 year. Similarly, the degree of webbing varies widely. There is no association between the severity of webbing and the likelihood of developing cataracts. Animals may also present with other issues, including retinal dysplasia (Anderson and Burns 1979).

Boggler (BGL)

Boggler deer mice, originating from *P. maniculatus blandus* and first noted in 1943, have pronounced tremors and an uncoordinated gait which produces awkward “boggling” movements. Boggler deer mice present with either tremors and staggering together, or neither. In some cases, boggler deer mice may have seizures similar to that characteristic of the EPL stock (Barto 1955).

Alcohol Dehydrogenase Null (ADHNN)

These mice, descended from *P. maniculatus bairdii*, have a naturally occurring recessive mutation that results in the lack of a key liver enzyme (alcohol dehydrogenase) that metabolizes ethanol (Dewey and Dawson 2001). Ethanol elimination in these animals occurs at approximately half the rate of normal animals (Burnett and Felder 1980, Norsten et al. 1989).

MAINTENANCE OF STOCK CENTER COLONIES

Colonies of *Peromyscus* are maintained to preserve their wild characteristics and genetic diversity. All animals of each stock kept at the PGSC are pedigreed, and their lineages can be traced back to the founding members of their respective colonies. Matings generally consist of pairs (one male, one female) rather than trios (one male, two females). The selection of mating pairs is semi-random; animals are selected to avoid brother-sister matings except in the case of inbred stocks, but no other selection occurs in the mating process. Males are left in the cages with their young, as they generally do not harm them. Males of *P. californicus* and *P. polionotus* provide paternal care (Gubernick and Teferi 2000). Litters are left with their parents until they are approximately 25 days old (35 in the case of *P. californicus*), at which time they are weaned, assigned a number, and their ears notched for later identification. Animals are placed in cages separated by sex, with six animals per cage (three in the case of *P. californicus*, due to their large size).

Animals are housed in opaque 12” × 9” × 6” polypropylene cages, with wire mesh tops for feeding and watering. The top of each cage is micro barrier filtered to prevent the spread of infectious agents that potentially may be present in specific animals. All stocks are currently closed, meaning no animals are wild-collected and added to our existing stocks, further reducing the likelihood of illness.

RECORD KEEPING AND DATABASE MANAGEMENT

Records of all animals in the stock center are kept both as paper copies and in an electronic database. Upon weaning, animals are assigned numbers from a book in which the animals' gender, birth date, weaning date, and parent numbers are recorded. In addition, each mating has a mating record sheet on which the litter size, number weaned, the number of the first animal weaned from each litter, and the number of males and females from each litter are recorded. All of this information is also entered into a database in which each animal is categorized according to its status (alive, dead, culled for genotype, culled for sickness, used in an experiment, or sold). This strategy facilitates tracing an individual animals' pedigree for many generations and back to the original foundation stock from which they were derived.

MATERIALS PROVIDED

PGSC ships animals worldwide. In addition, the center also provides tissues, other biological materials, and is able to perform histopathological evaluation of tissues, on request. Protocols such as the administration of special diets or supplements, the harvesting of embryos, or the set up of timed pregnancies have been developed by the PGSC in collaboration with requesting investigators. These protocols help to better meet the needs of individual researchers.

CONCLUSIONS

The PGSC represents a unique resource in the world of animal research. The center maintains animals similar in form to standard laboratory mice, but whose potential in research has not yet been fully exploited. The outbred nature of *Peromyscus* makes it not only a valuable tool for ecological and behavioral research, but a potentially interesting model for the study of human diseases as its genetic variation more accurately reflects that of the diverse human population (Dewey and Dawson 2001, Havighorst et al. 2017). At present, many of the disease models maintained by the PGSC are not fully characterized. Thus, they remain underutilized despite their potential.

REFERENCES

Anderson, J.F. 1989. Ecology of Lyme disease. *Connecticut Medicine* 53:343–346.

Anderson, R.S. and R.P. Burns. 1979. Cataract-webbed *Peromyscus*: I. Genetics of cataract in *P. maniculatus*. *Journal of Heredity* 70:27–30.

Avise, J.C., M.H. Smith, R.K. Selander, T.E. Lawlor and P.R. Ramsey. 1974. Biochemical polymorphism and systematics in the genus *Peromyscus*. V. Insular and mainland species of the subgenus *Haplomylomys*. *Systematic Biology* 23:226–238.

Barto, E. 1955. Boggler, an inherited abnormality of the deermouse (*Peromyscus maniculatus*), characterized by a tremor and a staggering gait. PhD Dissertation, Ann Arbor: University of Michigan.

Bedford, N.L. and H.E. Hoekstra. 2015. The natural history of model organisms: *Peromyscus* mice as a model for studying natural variation. *Elife* 4:e06813.

Burnett, K.G. and M.R. Felder. 1980. Ethanol metabolism in *Peromyscus* genetically deficient in alcohol dehydrogenase. *Biochemical Pharmacology* 29:125–130.

Careau, V., D. Thomas, F. Pelletier et al. 2011. Genetic correlation between resting metabolic rate and exploratory behaviour in deer mice (*Peromyscus maniculatus*). *Journal of Evolutionary Biology* 24:2153–2163.

Castle, W.E. 1912. On the origin of an albino race of deermouse. *Science* 35:346–348.

Chance, M.R. and D.C. Yaxley. 1949. New aspects of the behaviour of *Peromyscus* under audiogenic hyper-excitement. *Behaviour* 1:96–105.

Cowling, K., R.J. Robbins, G.R. Haigh, S.K. Teed and W.D. Dawson. 1994. Coat color genetics of *Peromyscus*: IV. Variable white, a new dominant mutation in the deer mouse. *Journal of Heredity* 85:48–52.

Dewey, M.J. and W.D. Dawson. 2001. Deer mice: "The *Drosophila* of North American mammalogy." *Genesis* 29:105–109.

Dice, L.R. 1937. Variation in the wood-mouse, *Peromyscus leucopus novebora-censis*, in the northeastern United States. *Occasional Papers of the Museum of Zoology, University of Michigan* 232:1–32.

Feldhamer, G.A., J.E. Gates and J.H. Howard. 1983. Field identification of *Peromyscus maniculatus* and *P. leucopus* in Maryland: Reliability of morphological characteristics. *Acta Theriologica* 28:417–423.

Greene, D.U., J.A. Gore and M.A. Stoddard. 2016. Reintroduction of the endangered Perdido Key beach mouse (*Peromyscus polionotus trissyllepsis*): Fate and movements of captive-born animals. *Florida Scientist* 1:1–3.

Gubernick, D.J. and T. Teferi. 2000. Adaptive significance of male parental care in a monogamous mammal. *Proceedings of the Royal Society of London B: Biological Sciences* 267:147–150.

Havighorst, A., J. Crossland and H. Kiaris. 2017. *Peromyscus* as a model of human disease. *Seminars in Cell and Developmental Biology* 61:150–155.

Kirkland, G.L. and J.N. Layne. 1989. *Advances in the study of Peromyscus (Rodentia).* Lubbock, TX: Texas Tech University Press.

Lacy, R.C., G. Alaks and A. Walsh. 1996. Hierarchical analysis of inbreeding depression in *Peromyscus polionotus*. *Evolution* 50:2187–2200.

Norsten, C., T. Cronholm, G. Ekström, J.A. Handler, R.G. Thurman and M. Ingelman-Sundberg. 1989. Dehydrogenase-dependent ethanol metabolism in deer mice (*Peromyscus maniculatus*) lacking cytosolic alcohol dehydrogenase. Reversibility and isotope effects in vivo and in subcellular fractions. *Journal of Biological Chemistry* 264:5593–5597.

Romanenko, S.A., V.T. Volobouev, P.L. Perelman et al. 2007. Karyotype evolution and phylogenetic relationships of hamsters (Cricetidae, Muroidea, Rodentia) inferred from chromosomal painting and banding comparison. *Chromosome Research* 15:283–298.

Shorter, K.R., J.P. Crossland, D. Webb, G. Szalai, M.R. Felder and P.B. Vrana. 2012. *Peromyscus* as a mammalian epigenetic model. *Genetics Research International* 2012:179159. doi:10.1155/2012/179159.

Svihla, A. 1935. Development and growth of the prairie deermouse, *Peromyscus maniculatus bairdii*. *Journal of Mammalogy* 16:109–115.

Van Ooteghem, S.A. 1983. Juvenile ataxia—A new behavioral mutation in the deermouse. *Journal of Heredity* 74:201–202.

Veal, R. and W. Caire. 1979. *Peromyscus eremicus*. *Mammalian Species* 8:1–6.

Vrana, P.B., K.R. Shorter, G. Szalai et al. 2014. *Peromyscus* (deer mice) as developmental models. *Wiley Interdisciplinary Reviews: Developmental Biology* 3:211–230.

Wang, L.R., J.P. Crossland and W.D. Dawson. 1993. Coat color genetics of *Peromyscus*. II. Tan streak—A new recessive mutation in the deer mouse, *P. maniculatis*. *Journal of Heredity* 84:304–306.

Weigl, R. 2005. *Longevity of Mammals in Captivity; From the Living Collections of the World*. Stuttgart, Germany: Kleine Senckenberg-Reihe.

Whitaker, J.O. and W.J. Hamilton. 1998. *Mammals of the Eastern United States*. Ithaca, NY: Cornell University Press.

Zhao, X., Y. Ueda, S. Kajigaya et al. 2015. Cloning and molecular characterization of telomerase reverse transcriptase (TERT) and telomere length analysis of *Peromyscus leucopus*. *Gene* 568:8–18.

11 The *Tetrahymena* Stock Center

A Versatile Research and Educational Resource

Donna Cassidy-Hanley, Eduardo Orias, Paul Doerder and Theodore Clark

CONTENTS

Abstract: As a typical ciliated protozoan, *Tetrahymena* cells compartmentalize their germline and somatic genetic information into two different nuclei within the same cell. This compartmentalization permits experimental manipulation not possible in other systems and has resulted in fundamental discoveries in cell and molecular biology. The optimal way to preserve the germline genetic integrity of *Tetrahymena* cell lines during laboratory storage is by maintaining the cells frozen at liquid nitrogen temperature. The *Tetrahymena* Stock Center (TSC) was originally established to preserve and distribute valuable strains of wild type and mutant *Tetrahymena*. More than 2200 strains representing dozens of species and a variety of mutant and genetically modified cell lines are currently available to the scientific research community at TSC. As TSC has grown, it has leveraged its resources to maintain a *Tetrahymena* Genome Database and to provide a large array of valuable research products and services to the research community.

The TSC carries out a proactive research and educational community outreach targeting K-12 and college undergraduate students, including a community-based *Tetrahymena* barcoding research project. As the use of its resource grows, the TSC will continue its mission to support existing resources while identifying new and changing needs and evolving the best means to serve the research and educational communities.

INTRODUCTION TO *TETRAHYMENA*

Tetrahymena, a ciliated protozoan commonly found in fresh-water lakes, ponds, and streams, is a highly evolved single-celled eukaryote with a genetic complexity rivaling that of metazoans (Figure 11.1). The very first "animal-like" cell to be grown in axenic culture (Lwoff 1923), *Tetrahymena* has served as an important model for studies of eukaryotic cellular and molecular biology since the 1950s (Nanney and Simon 2000). The use of *Tetrahymena* as a genetic system began in earnest with the collection of natural isolates of defined mating type by Alfred M. Elliot and co-workers from ponds around Woods Hole, Massachusetts; Vermont; and Michigan. Like other ciliates, *Tetrahymena* is characterized by nuclear dimorphism. The polyploid somatic macronucleus (MAC) controls cell function, while the micronucleus (MIC), which is transcriptionally silent except for a brief burst of transcription during an early stage of conjugation (Martindale and Bruns 1983, Sugai and Hiwatashi 1974) provides the genetic blueprint for the next sexual generation. This unique genetic organization is the basis for the development of *T. thermophila* as a powerful genetic system that allows examination of genetic phenomena not possible in other systems (Karrer 2000).

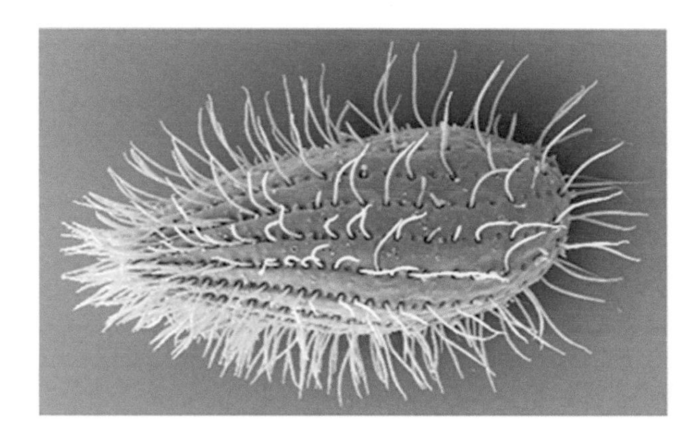

FIGURE 11.1 *Tetrahymena thermophila*. (SEM courtesy of Dr. Aswati Subramanian.)

Tetrahymena thermophila, the species most commonly used for molecular and genetic research, can quickly and easily be propagated vegetatively in simple proteose peptone-based media, but can also be readily induced to enter a sexual cycle (conjugation) by starvation in non-nutritive buffer (Karrer 2012). In the presence of sufficient food, reproduction normally occurs through binary fission during which the MIC divides by mitosis, the MAC divides by amitotic binary fission, and the cell divides by transverse cytokinesis. When sexually mature cells are starved and in the presence of complementary mating types, the sexual process (conjugation) occurs. During the temporary, partial anterior fusion of two cells, the MIC of each parental cell undergoes meiosis and one mitotic division to give rise to gametic nuclei, followed by nuclear exchange and reciprocal fertilization. The diploid zygote nucleus gives rise to a new MIC and MAC in each conjugant, and the old MAC is destroyed. Though rare among ciliates, the amicronucleate condition (i.e., cells lacking the MIC) is common in *Tetrahymena*, with some species found only as amicronucleate forms. Amicronucleates are almost always incapable of conjugation and are obligatorily asexual in the sense that they have no germline genome to contribute to sexual offspring.

Tetrahymena species typically do not form resting cysts and, prior to perfection of cryopreservation methods, cells had to be maintained in vegetative culture, often in bacterized Cerophyll, a grass infusion medium (Nanney 1953). With repeated refeeding and serial transfer, *Tetrahymena* cell lines can be maintained indefinitely, but they eventually become infertile because of genetic deterioration of the MIC as evidenced by an inability when mated to complete conjugation and form viable progeny (Simon and Nanney 1979). The ability to form mating pairs is not sufficient to indicate fertility, which must be assayed by the formation of new macronuclei and the generation of sexually immature progeny that are incapable of forming mating pairs for 30–70 fissions after pair separation.

The loss of fertility during prolonged serial subculture is observed regardless of the type of culture medium. Nanney's solution, used to maintain fertility in stocks grown in continuous live culture, periodically induces conjugation between appropriate strains allowing for the selection of the most fertile offspring. This enabled the establishment of a breeding program that resulted in the generation of many of the various inbred strains commonly used today. However, the constant selective breeding to maintain genetically useful stocks is labor and time intensive and requires technical expertise. Fortunately, the development of cryopreservation techniques that allow *Tetrahymena* to be frozen in liquid nitrogen (LN) has eliminated the need for periodic rounds of inbreeding and regular subculture to maintain fertile stocks.

TETRAHYMENA SYSTEMATICS

Tetrahymena belongs to the ciliate order Tetrahymenida that presently contains five families and lesser known genera, including *Glaucoma*, *Dexiostoma*, and *Colpidium*. Though the present taxonomy of the order is relatively stable, it has not always been so, resulting in much confusion as evidenced in the older (particularly pre-1960) literature. The genus *Tetrahymena*, for instance, was once known as *Leucophrys*, and several *Tetrahymena* species were assigned to the genera *Glaucoma* and *Lambornella*. Complicating the taxonomy was the long-time failure to recognize cryptic biological species as such by giving them the same Latin names (Nanney and McCoy 1976). Thus, nearly a dozen sexually isolated, bone fide species were misleadingly lumped under the name *Tetrahymena pyriformis*.

Morphologically, most *Tetrahymena* species are indistinguishable from each other. Based on mating compatibility, micronucleate species were first called varieties (*T. pyriformis* variety 1, *T. pyriformis* variety 2, etc.) and later the word "variety" was replaced with the word "syngen." The syngen designation was intended to call attention to separate gene pools but was more cumbersome than useful. This inspired a long search for molecular features that would allow the identification of species without reference to living strains, particularly as breeding strains are laborious to maintain and breeding tests fail to identify sexually immature isolates, the numerous amicronucleates, and selfers. Finally, with the advent of the first molecular means (isoenzyme mobility) to distinguish among species (Borden et al. 1973a, 1973b, Nanney and McCoy 1976), Latin names were assigned to the syngens. In recognition of comparable molecular differences, Latin names were also assigned to previously described amicronucleate (asexual) strains, all also indistinguishable by morphology. These names remain in use. The genus *Tetrahymena* is particularly species-rich, with new species constantly being identified by molecular analyses. They fall into three small subunit ribosomal DNA (SSUrRNA) clades, "australis," "borealis" and "paravorax," with a slight majority belonging to the "borealis" clade.

TETRAHYMENA ECOLOGY

Most *Tetrahymena* species were initially collected as free-living cells feeding on unknown species of bacteria. Though some species have been collected from both ponds and streams, there is evidence of habitat preference (Doerder 2018). For example, the best studied species *T. thermophila* is found primarily in small ponds, whereas *T. canadensis*, and most species in the related genera *Dexiostoma* and *Glaucoma*, were isolated primarily from streams. Some *Tetrahymena* species (i.e., *T. rostrata* and *T. glochidiophila*) have been found as both free-living organisms and as parasites (of slugs and mussels, respectively). Some *Tetrahymena* species are known only as parasites, infecting primarily aquatic invertebrate and some vertebrate hosts (Lynn and Doerder 2012). Researchers wishing to collect *Tetrahymena* and members of related genera, as for example in the educational outreach program, should look in any fresh water source, including moist soil. *Tetrahymena* is more likely to be present in samples taken from bottom sediments or among emergent vegetation, or that include decomposing animals. Other micro-habitats, such as surfaces and rapid flows, should not be avoided. Because infection rates are low, searches for parasites typically require large numbers of samples and, in some instances, permits.

The biogeography of *Tetrahymena* is poorly understood (Doerder 2018, Simon et al. 2007). In addition to the question of the number of species alluded to above, there are also major unresolved questions with respect to distribution. Some species are globally distributed and are found in multiple biogeographic zones. Others, such as *T. thermophila*, give evidence of restricted distribution consistent with endemism (Zufall et al. 2013). As another seeming paradox, *T. thermophila* has so far been found only in eastern North America, whereas its currently closest relative *T. malaccensis* has been found only in Malaysia. *Tetrahymena elliotti*, a close relative of both, has a global distribution. Systematic global collecting of *Tetrahymena* is unlikely to occur any time soon. Hence, the use of citizen science based educational outreach in which students collect water samples and identify *Tetrahymena* offers the best long-term chance of obtaining sufficient data to test hypotheses regarding endemism and to identify new species. The TSC's ability to cryopreserve new isolates will facilitate population studies and improve the general understanding of *Tetrahymena* ecology and biogeography.

TETRAHYMENA THERMOPHILA

Two model *Tetrahymena* species, *T. pyriformis* and *T. thermophila*, have been used for the most acclaimed scientific contributions using *Tetrahymena*. Early research featured *T. pyriformis*, an amicronucleate species that is still commonly used in toxicological studies (Sauvant et al. 1999, Xue et al. 2006). However, *T. pyriformis* cannot be used for germline genetic studies. Therefore, after the mid-1950s when the first *Tetrahymena* genetic locus was identified (Nanney et al. 1955), *T. thermophila* replaced *T. pyriformis* as the most extensively used *Tetrahymena* species for genetic, molecular, cellular, developmental, and population studies. Indeed, *T. thermophila* may currently be the most thoroughly investigated ciliate. *Tetrahymena thermophila* grows at higher temperatures (up to 41°C) than other *Tetrahymena* sp. (typically 33°C–38°C), which speeds up experimental work and facilitates the isolation of useful temperature-sensitive, conditional lethal mutants.

Tetrahymena thermophila can be grown on bacteria (*Klebsiella pneumonia* or *Pseudomonas aeruginosa*) by cilia-mediated filter feeding and subsequent phagocytosis, and also axenically (i.e., via no other living organism) in a complex beef-derived broth, or in a totally chemically defined synthetic medium composed entirely of small molecules (growth media reviewed in Cassidy-Hanley 2012, Orias et al. 2000). It can also be grown in suitably supplemented complex or defined media in the absence of a functional oral apparatus (Rasmussen and Orias 1975). This dual feeding capacity enables the survival and maintenance of useful mutants with inactive cilia which are incapable of phagocytosis and would otherwise die. Methods for efficient MIC or MAC DNA transformation that allow the use of "reverse genetics" (starting with a sequenced gene and modifying it to understand what it contributes), such as MIC or MAC gene knock-outs and knock-downs, gene substitutions, gene overexpression, and protein tagging, are very well developed (Asai and Forney 2000, Collins 2012). The *T. thermophila* MIC and MAC genomes are currently the most completely described *Tetrahymena* genomes, and this species is the only ciliate for which all 5 MIC and all 181 maintained MAC chromosomes have been co-aligned (Hamilton et al. 2016).

OTHER *TETRAHYMENA* SPECIES OF INTEREST

Though *Tetrahymena* species have much in common, various species have special features that have allowed studies not possible with *T. thermophila*. For instance, Lwoff's *T. pyriformis* cells (Lwoff 1923) are easily synchronized by heat shock. Hence, they figured prominently in the earliest studies of the eukaryotic cell cycle (Scherbaum and Zeuthen 1954), although it was *T. thermophila* that was used in the subsequent mutational analysis of the *Tetrahymena* cell cycle (Frankel et al. 1976, 1977). Cells of three species (*T. vorax*, *T. paravorax*, and *T. patula*) can form either of two types of oral apparatus (OA). Microstome cells possess the small OA universally conserved in the genus and adapted for extremely efficient filter feeding on bacteria. Macrostome cells have a very large OA which allows them to prey on cells of other *Tetrahymena* species as well as to cannibalize microstome cells of their own species (Corliss 1973). Several other *Tetrahymena* species are histophagous, i.e., parasites that feed on tissues of multicellular aquatic animals. A variety of such *Tetrahymena* species are maintained by the TSC.

THE SCIENTIFIC CONTRIBUTIONS OF *TETRAHYMENA*

Tetrahymena has had a significant impact on many areas of biological research and continues to contribute to the understanding of basic biological mechanisms, including many critical to human health and disease. A series of landmark developments (Table 11.1) has expanded the use of *Tetrahymena* as a model organism for the investigation of genetic, developmental, cell physiological, and ultrastructural processes (reviewed in Asai and Forney 2000, Collins 2012, Gall 1986, Hill 1972). These studies provided numerous insights into fundamental biological processes, including Nobel Prize winning work on telomeres and telomerase (Chan et al. 2017, Kain 2009) and catalytic RNA (Cech 1990, Sengupta et al. 2016). *Tetrahymena* has also been utilized in seminal work in molecular biology, epigenetics, and nutrition, including chromatin and post-translational histone modifications (Papazyan et al. 2014, Wang et al. 2017, Xiong et al. 2016), small RNAs (Mochizuki et al. 2002, Mochizuki and Gorovsky 2004), genome rearrangement (Chalker and Yao 2011, Yao et al. 2014), genome stability (Akematsu et al. 2017,

TABLE 11.1
Landmark Studies in the Development of *Tetrahymena thermophila* as a Model Organism

Area of Study	Date	Subject
Cytogenetics	1889	Description of nuclear events of conjugation in a *Tetrahymena*-like species (Maupas 1889)
Nutrition	1923	Axenic growth in sterile beef broth (Lwoff 1923)
Taxonomy	1940	New genus *Tetrahymena* (four membranelles) established, based on unique ciliary organization of the oral apparatus (Furgason 1940)
Nutrition	1951	First chemically-defined synthetic medium for growing *Tetrahymena* (Kidder and Dewey 1951)
Sex	1952	Identification of mating types (Elliott and Gruchy 1952)
Genetics	1953	First Mendelian genetic locus reported (Nanney and Caughey 1953)
Genetics	1959	Generation of inbred strains by successive sib-matings (Nanney 1959)
Genetics	1963	One-step generation of whole-genome homozygous strains by genomic exclusion (Allen 1963); pronuclear exchange block (Orias et al. 1979); pronuclear fusion block (Hamilton et al. 1988)
Preservation	1967	First live *Tetrahymena* cells successfully frozen in liquid nitrogen (Simon and Whang 1967)
Genetics	1971	First laboratory mutants generated by mutagenesis (Carlson 1971)
Nutrition	1975	Development of media that supports growth in the absence of phagocytosis (Rasmussen and Orias 1975)
Cell biology	1976	First use of laboratory mutations to dissect a *Tetrahymena* cell mechanism (Frankel et al. 1976)
Genetics	1983	Generation of stable nullisomic deletion strains to facilitate genetic mapping (Bruns et al. 1983)
Molecular genetics	1986	Successful DNA transformation by microinjection (Tondravi and Yao 1986), electroporation (Brunk and Navas 1988), and particle bombardment (Cassidy-Hanley et al. 1997)
Molecular genetics	1994	Development of vector-mediated gene replacement and knock-out methods (Gaertig et al. 1994)
Genetics	1995	Development of genome-wide germline (Lynch et al. 1995) and somatic (Longcor et al. 1996) genetic mapping
Genomics	2006	Sequencing and assembly of MAC genome (Eisen et al. 2006)
Molecular genetics	2014	First gene identified by "forward genetics," i.e., starting with a mutant phenotype of interest (Galati et al. 2014)
Genomics	2016	Sequencing, assembly, and publication of MIC genome; co-alignment of the 5 MIC and 181 maintained MAC chromosomes (Hamilton et al. 2016)

Howard-Till et al. 2013, Sandoval et al. 2015, Shodhan et al. 2017), ribosomes and ribosomal RNAs (Rabl et al. 2011, Wilson and Doudna Cate 2012), microtubules and microtubule modification (Pucciarelli et al. 2012, Thazhath et al. 2002), cytoskeletal motors (Suryavanshi et al. 2010, Wilkes et al. 2008, Wood et al. 2007), stimulus-dependent regulated secretion (Sparvoli et al. 2018, Turkewitz 2004), epigenetics (Chalker et al. 2013), phagocytosis and transmembrane transport of small molecules (Jacobs et al. 2006), and many other areas (Asai and Forney 2000, Collins 2012). In recognition of its contributions to basic research, *T. thermophila* was one of the first eukaryotic model organisms chosen for full genome sequencing by the Human Genome Research Institute at the National Institutes of Health (NIH).

THE *TETRAHYMENA* STOCK CENTER COLLECTION

History

Prior to creation of the TSC in 2005, the large number of the native and genetically modified *Tetrahymena* strains that had been amassed over decades were held in different labs and maintained either under minimal growth conditions (as slowly dividing cultures at room temperature) or in LN freezers after the development of successful protocols for the freezing of *Tetrahymena* (Bruns and Cassidy-Hanley 2000, Cassidy-Hanley et al. 1995, Flacks 1979,Orias et al. 2000, Simon 1972, Simon and Schneller 1973). Strains propagated by serial sub-culture generally exhibit germline senescence over time unless periodically subjected to mating and generation of a new germinal nucleus (Simon and Nanney 1979). This is a time-prohibitive and tedious exercise, especially when regenerating functional heterokaryons. The development of a successful technique for freezing cultures in LN greatly facilitated the maintenance of genetically healthy strains in many labs. However, secure preservation of critical strains was still jeopardized by improper freezing and thawing or catastrophic loss due to freezer failure and lack of secondary freezer backup. Commonly, strains were shared by transfer from lab to lab, increasing the risk of cross-contamination, mislabeling (Borden et al. 1973c), or strain mix-up with each transfer. For many archival strains, descriptions of genotype and phenotype were only available in laboratory notebooks and informally transferred among users. Important strains were endangered by changing research priorities, local space considerations, and retirements. The TSC was created to address these and other issues by providing a centralized site for deposition and distribution of important *Tetrahymena* strains. In 2005, the TSC was established with funding from the National Institutes of Health.

Collection Holdings

The TSC has brought together the most important archival strains from many of the largest *Tetrahymena* collections, while also serving as a safe haven for new useful strains. The TSC currently offers 2243 different strains representing 15 species of *Tetrahymena* including; *T. americanis*, *T. australis*, *T. borealis*, *T. canadensis*, *T. elliotti*, *T. glochidiophila*, *T. hegewischi*, *T. malaccensis*, *T. mobilis*, *T. pigmentosa*, *T. pyriformis*, *T. setosa*, *T. shanghaiensis*, *T. thermophila*, and *T. tropicalis*.

Most species in the TSC are represented by several clonal isolates from different geographic locations. The clonal lines of *T. borealis*, *T. malaccensis*, *T. elliotti*, and *T. thermophila*, used for genomic sequencing, are included in the collection. The collection also includes representatives of many of 36 newly named species (Doerder 2018). Approximately 90% (2015/2243) of the strains in the collection are *T. thermophila. Tetrahymena thermophila* holdings currently include highly inbred wild-type strains, naturally occurring and induced mutations, star strains, functional heterokaryons, deletion strains, single and multiple nullisomics, genetically engineered lines with well-defined gene modifications, naturally occurring genetically polymorphic strains, and panels of meiotic segregants and terminal assortants. These strains represent an extensive resource that is of great potential benefit to the biological community.

Wild-Type Strains

Starting with Elliott and students in the 1950s, investigators have isolated strains from numerous species of *Tetrahymena*. Doerder (2018) speculates that there are likely hundreds more species to be discovered. Many of the original isolates of each species were deposited in the American Type Culture Collection prior to the establishment of the TSC. More recently, many wild samples from the Doerder collection have been deposited in the TSC. For most deposits of wild strains, fertility cannot be guaranteed. However, though such stocks may have defective MICs, their MAC genomes are intact and therefore are experimentally accessible. Unfortunately, several species described in the late 1900s were not preserved in any form and hence are unavailable for further study.

Inbred Strains

In domesticating *Tetrahymena* for genetic analysis, David Nanney established several inbred strains that are, for all practical purposes, homozygous. The first strains, Strain A and Strain B, were the result of a cross between wild isolates WH6 and WH14 followed by regular rounds of sib-mating between offspring with mating type alleles *mat*A and *mat*B, respectively. By the time these strains were preserved in LN and deposited in the ATCC, they were in their 16–18th generation of inbreeding. Other inbred strains were also established. Designated as strains C, D, E, and F, each is the result of a cross of a different wild isolate to Strain B and selection for offspring with the *mat*A-like (non-*mat*B) allele (Table 2 of Allen and Gibson 1973). In addition to differing at the *mat* locus, these strains differ at other loci, thereby having fixed some of the unique genes present in the non-B parent. In addition to Strains A–F, there exist strains such as A3, B2, C3, etc. These strains are the result of crosses among various strains and subsequent inbreeding. The letter designation indicates that the strain has the same *mat* allele as the parent-lettered strain. These strains also contain various combinations of wild genes. Though Strain B has been used for most genetic analyses, particularly mutational studies, the other strains have proved useful. Variants present in Strain C3, for instance, were exploited for the genetic mapping of DNA polymorphisms to MIC and MAC chromosomes (Brickner et al. 1996, Longcor et al. 1996). The congenic strains developed by Allen and Lee (1971) constitute an alternative to inbred strains. These strains were prepared by crossing a C2 × D F_1 to strain D in 12 consecutive backcrosses, selecting in each backcross offspring possessing certain C2 genes. As a final step, homozygosis was induced by genomic exclusion. Though rarely used, these strains are available from the ATCC.

"Star" (*) Strains

"Star" (*) strains are spontaneous *T. thermophila* variants with a non-functional MIC; they can conjugate but fail to contribute any genetic information to sexually-derived progeny. Star strains have been identified in several different inbred backgrounds and include C*, A*III, A*V, B*V, and B*VII (Allen 1967a, 1967b, Allen et al. 1967, Shabatura and Doerder 1981), all of which are maintained in the TSC. Star strains possess defective (hypodiploid) MICs and are unable to form functional gametic nuclei. When cells of star strains are mated to normal cells, the pairs undergo a variant form of conjugation called genomic exclusion (Allen 1967a) that involves two successive rounds of conjugation. In Round I of mating, the diploid partner undergoes normal meiosis and unidirectional transfer of a haploid pronucleus to the star cell. This is followed by one round of MIC endoreduplication resulting in the creation of identical whole-genome homozygous diploid MICs and retention of the parental MAC in both partners. Cells separate prematurely but remain sexually mature so they are immediately capable of a second round of mating. Round II mating proceeds as in normal conjugation and gives rise to progeny with identical whole-genome homozygous MICs and MACs, provided care is taken that two exconjugants of the same pair mate with one another in Round II. The mating process can also be halted following Round I simply by refeeding the culture.

Star strains are extraordinarily useful genetic tools since Round I genomic exclusion allows for the creation of homozygous functional heterokaryons, cell lines in which the MIC and MAC have different genotypes. Because the MIC is transcriptionally silent, Round I genomic exclusion allows the production and maintenance of clones that are homozygous in the MIC for lethal constructs, including point mutations, deletions, loss of entire chromosomes, and knockouts of essential genes, and yet are fully viable due to their wild-type macronucleus.

Mutants and Genetically Modified Strains

In addition to strains mentioned above, the TSC maintains hundreds of strains harboring various induced mutations as well as specialized genetic constructs, including knock-outs, knock-ins, and macronuclear assortants. To be of continued use, these too must be cryopreserved.

Genetic Mapping Panels

Randomly amplified polymorphic DNA (RAPDs) (Williams et al. 1990) polymorphisms between inbred strains B and C3 were used for the large-scale genetic mapping of the *T. thermophila* germline genome, based on meiotic recombinant frequency (Brickner et al. 1996, Wickert and Orias 2000). The panel of 32 whole-genome homozygous F_2 meiotic segregants used for conventional MIC genetic mapping is maintained in the TSC. Mapping DNA polymorphisms to one of the somatic chromosomes maintained at 45 copies per G1 MAC is a second kind of genetic mapping, so far unique to *Tetrahymena*. The mapping is based on the genetic assortment that occurs in heterozygous MACs because of the random distribution of MAC chromosome copies during binary fission, and the co-assortment that occurs when two loci are on the same MAC chromosome (Longcor et al. 1996). The panel of 36 "terminal assortants" generated to detect co-assortment between inbred strains B and C3 polymorphisms is also maintained at the TSC.

Genetic maps generated by using these two mapping panels were useful for independently verifying the validity of MAC chromosome assemblies when the

T. thermophila MAC genome was first sequenced and assembled (Eisen et al. 2006). They were also used to narrow down the location of the mating type locus to a ~300 kb MAC DNA segment (~0.3% of the MAC genome), greatly facilitating its identification by comparative RNAseq (Cervantes et al. 2013).

Micronuclear Deletions, Nullisomics, and Unisomic Strains

The functional division of nuclear activity between MIC and MAC enhances the usefulness of *Tetrahymena* as an experimental system by creating a very plastic genetic environment that allows a multitude of genetic manipulations not possible in most other systems. MIC genes are not expressed and are thus not required for cell viability. The MIC genome can be manipulated to create a variety of heterokaryon strains containing wild-type MACs but modified MICs. These include a panel of strains homozygous for deletions of varying size on each of the MIC chromosomes, stable nullisomic strains missing both copies of one or more MIC chromosomes (Bruns and Brussard 1981, Bruns et al. 1983), and unisomic strains containing a MIC with multiple copies of only one of the five micronuclear chromosomes. Micronuclear deletion strains, nullisomics, and unisomics are fully viable when grown vegetatively since chromosomally aberrant micronuclei, even those with severe haplodeficiency, undergo normal mitosis. These strains also carry out the meiotic and mitotic events normally associated with conjugation. Since most nullisomic and deletion lines are fertile when mated to strains carrying MIC chromosomes complementary to missing regions, these strains are valuable tools for genetic mapping of mutations to MIC chromosomes using standard crosses, Southern analysis, or PCR (Cassidy-Hanley et al. 1994, 2005). Unisomics provide the potential to separate and interrogate individual MIC chromosomes in isolated micronuclei. The TSC houses a collection of single and multiple nullisomic strains, strains carrying deletions of various sizes on each MIC chromosome, and panels of inbred B and C strain unisomics containing each of the MIC chromosomes.

Strains from Diverse Tetrahymena Laboratory Collections

Foundational *T. thermophila* clones created by Dr. Peter Bruns (Cornell University), including many of the drug resistant functional heterokaryons essential for genetic analyses, are now securely maintained in TSC freezers. The irreplaceable *T. thermophila* mutant strains developed by Dr. Joseph Frankel (University of Iowa) during a career spanning nearly 50 years are also safely stored. The extensive collection of *T. thermophila* strains created by Dr. Eduardo Orias and Dr. Eileen Hamilton (University of California, Santa Barbara) has been safely transferred to TSC. These include well-characterized genetic mapping panels (meiotic segregant and terminal assortment panels) and a number of naturally occurring genetically polymorphic strains, many carrying useful genetic mutations. Important genetically modified strains engineered in the lab of Dr. Martin Gorovsky (University of Rochester) have been deposited with the TSC. A large array of geographically unique wild-type *Tetrahymena* species collected from around the country by Dr. Paul Doerder (Cleveland State University) has been added to the TSC collection. During TSC's initial acquisition phase, smaller numbers of unique strains (primarily *T. thermophila*) were also acquired from other researchers including Dr. Eric Cole

(St. Olaf College), Dr. Robert Coyne (J. Craig Venter Institute), Dr. James Forney (Purdue University), Dr. Jacek Gaertig (University of Georgia), Dr. Todd Hennessey (University of Buffalo), Dr. Piroska Huvos (Southern Illinois University), Dr. Kathleen Karrer (Marquette University), Dr. Larry Klobutcher (University of Connecticut), the late Dr. Denis Lynn (University of British Columbia), and Dr. Mark Winey (University of Colorado Boulder). This collection of *Tetrahymena* strains, amassed by many researchers over 60 years, is fundamental to the development of *Tetrahymena* as a model organism.

Plasmids

Vectors specifically designed for use in *Tetrahymena* are critical to many uses of *Tetrahymena* as a model system for molecular biology. To help meet this need, the TSC has extended its services to include storage and distribution of plasmid constructs commonly used by *Tetrahymena* researchers. A searchable list of plasmids is available on the website, linked to a description of each plasmid including vector type and source, drug-selectable markers, complete sequence, a plasmid map, as well as an online request form. Thirty-four plasmids are currently available addressing a number of different needs, including a set of Gateway tagging vectors; pXS76 (a paromomycin selectable vector targeting foreign genes to the MTT1 metallothionein locus); a high copy-number ribosomal DNA vector, pD5H8; a paromomycin selectable TAPTAG vector; 5′- and 3′-GFP tagging vectors; vectors for insertion of transgenes into the ß-tubulin-1 locus; and vectors carrying paromomycin, cycloheximide, and blasticidin resistance selection cassettes. The bacteria containing these plasmids are stored in a −80°C freezer. The plasmids are purified from freshly streaked colonies upon receipt of a request. The purified plasmid is tested and confirmed by restriction analysis, dried on filter paper, and shipped for next day delivery. The TSC proactively solicits the deposition of useful new vectors as soon as they become available.

TSC OPERATIONAL PROCEDURES

Acquisition of Strains

Currently, the TSC accepts deposition of all well-characterized mutant and genetically modified *T. thermophila* strains, and cultures of other *Tetrahymena* species from well-defined geographical locations. Strains are accepted based on their potential usefulness to the *Tetrahymena* research and teaching communities. Submissions of lab strains of *T. thermophila* must be accompanied by a brief description of each strain, including relevant germ line and somatic genotypes and phenotypes. Appropriate GPS locations of collected wild specimens must be provided, as well as any available information related to species identification. Submission instructions and fillable forms that can be e-mailed directly to the TSC are available on the website. Once received, completed forms are converted into a spreadsheet and imported directly in the TSC database, minimizing errors during the transfer of critical strain information. For large-scale deposits, a downloadable Excel file with all the required field variables already inserted is provided on the website. Help in using currently approved nomenclature is also available to facilitate strain submission. Once the submission is authorized, the donor ships cells for next day delivery to the TSC, either as live cultures in growth medium or, for frozen cells, embedded in dry ice. Donated live cells are established as growing cultures, frozen, and test thawed before incorporation into TSC freezers. Frozen cells are transferred directly into TSC freezers upon receipt and are thawed and refrozen following the first order for the strain. As soon as transfer to TSC freezers is complete, strains are listed in the online database and made publicly available.

Maintenance and Security

Long-term storage of *Tetrahymena* generally requires that growth be minimized or eliminated since extended vegetative growth can lead to strain degradation and decreased fertility as deleterious modifications accumulate within the MIC. Several techniques have been developed for freezing viable *Tetrahymena* in LN. Early freezing protocols required expensive equipment and were often not dependable (Flacks 1979, Simon 1982). The freezing protocol currently used by the TSC (Bruns et al. 2000) is based on modifications that eliminate the need for costly equipment other than a standard LN freezer, and it has greatly improved freezing efficiency and cell recovery after thawing.

Secure maintenance of all resources is critical to ensuring that the TSC continues to meet the needs of the *Tetrahymena* research community. To ensure the safe and secure maintenance of *Tetrahymena* strains in the TSC repository even in the event of a catastrophic freezer failure, a total of six vials of each strain are maintained in two LN freezers in separate secure locations. Four vials of each strain are stored in a primary storage Custom Biogenic Systems V-3000 freezer with audible alarms (vapor phase storage with 10-day static holding time) capable of storing 22,000 samples. Two additional vials of each strain are kept in one of five 6000-sample back-up LN freezers (static holding time of ~104 days) located in a separate high-security facility accessible only by keyed entry. For commonly requested or difficult to recover strains, four additional vials, distributed among two freezers, are maintained.

Usage and strain inventory levels are monitored through a database interface. When the total number of individual vials for a specific strain drops below 4, the strain is re-established in growing culture and additional vials are re-frozen. Frequently ordered clones are maintained as slow growing stock tube cultures for a maximum of 6 months to allow same day response to many orders. To maintain genetic integrity and prevent senescence of these frequently ordered strains, new stock tubes are prepared from freshly thawed vials at least once every 6 months.

Resource Distribution

The TSC distributes strains and plasmids worldwide. The TSC website includes a searchable online strain database that provides information on available strains, as well as links to other available resources and online ordering forms. Live 1.0 mL cell cultures in sterile proteose peptone growth media are shipped in 1.5 mL cryotubes for next day delivery in the United States and by international high priority delivery to all other countries. For shipment to countries outside the United States, paperwork fulfilling the import and customs requirements for the destination country must be completed by the client prior to shipment. Less frequently requested strains are thawed upon request, tested for viability and, for most strains, shipped within 5 days. All shipping charges are paid by the recipient. Since under normal shipping conditions cells can survive 2–3 weeks or longer in transit, delivery of live healthy cultures is guaranteed. Hand warmers are included in shipping containers to locations affected by cold weather to prevent cold or uncontrolled freezing damage. Once received, cells can easily be established as growing cultures simply by dilution into fresh culture media.

Strain Integrity and Reproducibility

A primary TSC goal is to enhance reproducibility and transparency of all types of research involving *Tetrahymena* by ensuring easy access to well-characterized strains through a centralized distribution center. A critical factor for ensuring reproducibility is the maintenance of genetic integrity, which is physically assured by long-term storage of cell lines in LN to avoid germline degradation and to reduce the accumulation of random mutations in either the germline MIC or somatic MAC. Equally important is consistency in strain identification among researchers in different labs and the use of a common genetic nomenclature to clearly denote what is known about the genetic make-up of each strain. Traditionally, *T. thermophila* strains have been named according to location of origin. Wild-type isolates were given a two-letter prefix based on origin (e.g., isolates collected at Woods Hole were designated WH) followed by a serial number. Research-generated strains were named using a two-letter prefix representing the location of the lab of origin followed by a strain serial number (e.g., CU428 is Cornell University strain number 428). Newly acquired wild strains are given strain designations provided by the depositor. The TSC website uses a searchable interface that provides both the archival strain names still commonly used by most researchers and a unique stock identifier. To facilitate consistency in strain identification in publications, the TSC has also joined the Research Resource Identification Initiative (RRID; Bandrowski et al. 2016) that provides universal unique identifiers for referencing specific research resources, and offers an RRID portal that acts as a central location for obtaining and referencing information regarding *Tetrahymena* strains (https://scicrunch.org/resources/Organisms/search?q=%2A&l=&facet[]=Database:TSC).

In order to simplify cross-referencing, all TSC strains have been assigned an RRID number based on the existing *Tetrahymena* strain ID number used in the TSC database and on the TSC website. The use of persistent and unique RRID identifiers by those using TSC resources is promoted on the website and in presentations and posters to help ensure reproducibility and support resource transparency. Consistent terminology to describe genotypes and phenotypes is also important to facilitate correct strain identification and ensure coherent communication among different labs. The TSC has revised the genotypic and phenotypic information for all strains to meet the current nomenclature guidelines for describing micronuclear and macronuclear genotypes and phenotypes (Allen 2000, Allen et al. 1998). Depositors of new strains are asked to provide appropriate genetic information based on current nomenclature guidelines. Additional information on gene nomenclature is provided on the TSC website (http://*Tetrahymena*.vet.cornell.edu/extras/revised_*Tetrahymena*_nomenclature.doc).

Currently, *cox*1 barcoding has become the most effective way to distinguish among *Tetrahymena* species, including species with identical SSUrRNA sequences (Chantangsi and Lynn 2008, Doerder 2018, Kher et al. 2011). The *cox*1 barcode consists of a 689-nucleotide segment of the mitochondrial cytochrome oxidase subunit 1 gene and is similar to the diagnostic barcoding region used to identify animal species. Barcodes are relatively easy to obtain, to illustrate in publications, and to archive for access by other researchers. Unique barcodes exist for all but the six named *Tetrahymena* species for which there are no available living or preserved representatives. Intraspecific *cox*1 barcode variation is typically 0%–2%, whereas interspecific variation averages about 11%. Differences of >4% are considered diagnostic of new species (Doerder 2018). Other markers, such as the histone H3/H4 intergenic region, may also distinguish among species though this has not been thoroughly investigated (for list of markers see Doerder and Brunk 2012). The TSC routinely barcodes new isolates obtained through the educational outreach program. *Cox*1 and SSU rRNA barcodes exist for designated type strains for most of the 80 *Tetrahymena* species (GenBank accession numbers are given in Doerder 2018 and Lynn and Doerder 2012).

Inventory Management

A series of Filemaker Pro relational electronic databases are used to store critical strain information, organize and track strains, and integrate strain information with an online search engine to facilitate identification, ordering, and deposition of strains. The freezer inventory database allows TSC to keep track of all strain activity, from locating existing strains to determining empty freezer spaces for new vials to integrating new deposits into the freezers. A flagging system has been implemented to indicate when the total number of vials for any given strain drops to two. The stock descriptions database contains all of the available information for each strain and is updated as new strains are added to the center. An online search engine utilizes the information in the stock description database to facilitate identification and ordering of strains.

In addition to providing safe storage and easy access to well-characterized strains, TSC offers additional important resources to the research community including the *Tetrahymena* Genome Database, fee-based micro- and macronuclear transformation services, a selection of plasmids, and prepared materials and supplies such as genomic DNA, media components, live cell cultures, and frozen cell paste. Free technical support is also available for those using TSC resources.

THE *TETRAHYMENA* GENOME DATABASE (TGD)

In 2009, the TSC assumed responsibility for the *Tetrahymena* Genome Database (http://ciliate.org/index.php/home/welcome), originally created in 2003. The TGD provides essential information on the *T. thermophila* MIC and MAC genomes, as well as the MAC genomes of *T. borealis*, *T. elliotti*, and *T. malaccensis*, through a web-based Wiki (Stover et al. 2006, 2012). Together, these resources have helped propel *Tetrahymena* into the post-genomics era and have further enhanced its utility as a tool for basic and applied research. The TGD maintains the latest annotations for both the MIC and MAC genomes, continually updating new sequence data, bioinformatics analyses, and literature curation(s). The TGD promotes community-wide gene annotation, bringing together BLAST and GBrowse interfaces, gene descriptions including common and official TTHERM gene names, and links to important new publications. The search features at TGD provide a vital link for researchers between genes and the strains available at the TSC. To optimize information flow between TGD and the TSC, gene pages in TGD and strain information in the TSC database are reciprocally linked, making it easier to identify useful strains when browsing TGD and making information about genes and their paralogs more readily available to those accessing strain information via the TSC website. To facilitate easy comparisons with other ciliate model organisms, TGD has been replicated to produce databases for newly sequenced ciliate genomes including those from *Ichthyophthirius*, *Oxytricha*, *Stylonychia*, and *Stentor.*

A variety of extremely useful genomic resources have been assembled for *T. thermophila* at the cross-referenced TGD (http://ciliate.org/index.php/) and *Tetrahymena* Functional Genomic Database (TFGD; http://tfgd.ihb.ac.cn/). These databases include searchable, annotated gene lists, BLAST search capabilities and browser links for genomic and gene sequences, gene expression data for all genes at various times during the three main states of the life cycle (vegetative growth, starvation and conjugation), RNA-seq, microarray, gene network and phosphoproteome data. These genomic resources have facilitated the use of "forward genetics" approaches in *Tetrahymena*, i.e., starting with a mutant defective for a mechanism of interest and identifying, without any preconceptions, the mutant gene, often identifying a previously unknown gene required by the mechanism.

ADDITIONAL TSC SERVICES, RESOURCES AND ACTIVITIES

DNA-Mediated Transformation Services

One of the key strengths of *Tetrahymena* as a model system is the capability to genetically transform either the MIC, which results in a heritable germline modification, or the MAC, which results in a phenotypically expressed change that is inherited during asexual replication but is not passed on to sexual progeny. In light of the overwhelming importance of reverse genetics for functional gene analysis, TSC now offers biolistic transformation services to facilitate genetic engineering. The TSC's transformation services provide easily accessible, economic biolistic transformation targeting either the MIC or the MAC, allowing anyone to generate custom made cell lines without large investments of time or resources. TSC provides support and expertise, as well as a ready source of plasmids commonly used for vector construction, but clients are responsible for the design and construction of their transformation vectors. Once the construct has been verified, TSC staff transform the target cells and identify candidate clones using appropriate drug selection. Putative transformant clones are subjected to PCR and/or genetic analysis as appropriate to confirm the presence of the transgene. If requested, transformants are subsequently propagated at increasing concentrations of the selective drug, so as to select for the complete genotypic replacement of the wild type by the transformed allele in the polyploid somatic nucleus through joint assortment and selection. A minimum of 2 somatic or 1 germline transformant clone is provided per construct. This service increases community access to research strategies like gene disruption or modification, gene knock-ins and knock-outs, overexpression of proteins, and gene tagging without the need for expensive equipment or specialized training and is of special benefit to newcomers to the field.

Prepared *Tetrahymena* Materials and Supplies

A top priority for the TSC is to expand opportunities for researchers to utilize the unique advantages of the *Tetrahymena* model system. Easy, affordable access to high purity genomic DNA, large volumes of live cell cultures, frozen cell paste, and premixed powdered media provides additional flexibility in experimental design and support the use of the *Tetrahymena* model system by the both the *Tetrahymena* research community and non-specialist researchers unfamiliar with the organism. These resources greatly expand opportunities for researchers to undertake one-time experiments or pilot studies without investing an inordinate amount of time and resources.

TSC Technical Support

Technical information is available on the website, including basic protocols and recipes for a variety of culture media, a glossary of *Tetrahymena* related terms, information about upcoming meetings, links to other related web resources, and a link to the *Tetrahymena* Genome Database website. These resources are designed to provide additional flexibility to experienced *Tetrahymena* researchers, to support those interested in small-to-mid sized pilot studies, and to offer non-experts the opportunity to economically utilize the unique benefits of the *Tetrahymena* model system as well as the generation of custom designed transformant strains, without significant investment in time or effort. In addition to these services, the integration of the TGD as a vital part of the TSC leverages the strengths of both entities to create a dynamic resource for all members of the research community.

Community-Based Barcoding

Traditional methods of species identification in ciliates are difficult, especially in light of the existence of self-mating and amicronucleate species and the fact that many species are morphologically indistinguishable. As noted previously, *cox*1 barcodes successfully identify *Tetrahymena* species (Chantangsi and Lynn 2008, Chantangsi et al. 2007, Doerder 2018, Kher et al. 2011). However, there are instances in which additional information is useful, particularly among closely related species. Useful additional sequences in this regard include the intergenic region between the histone H3 and H4 genes, and the 5.8S rRNA Internal Transcribed Spacer (ITS) region. Currently, a small database of *Tetrahymena* 5.8S regions is available, along with an even more limited H3/H4 intergenic region database (reviewed in Doerder and Brunk 2012). To expand these databases, facilitate species identification, clarify interspecies relationships, and add to knowledge on species geographic distribution, TSC has undertaken the analysis of the *cox*1, 5.8S ITS, and the H3/H4 intergenic regions in the more than 300 wild caught strains currently housed in the collection, as well as new isolates being added to the collection through the collaboration with the ASSET educational outreach program. The data generated will provide a more extensive dataset for identifying new wild strains and will help to validate current species identification.

RESOURCE SHARING

COMMUNITY OUTREACH

The *Tetrahymena* research community is congenial and highly interactive, and TSC leverages that collaborative spirit to encourage resource sharing and exchange of information and ideas. Easy community access to all TSC resources is provided on the publicly accessible website (https://*Tetrahymena*.vet.cornell.edu/). Information regarding TSC and TGD activities is also posted on the ciliate list server maintained at the University of Georgia (https://listserv.uga.edu/cgi-bin/wa?A0=ciliatemolbio-l) and shared with the community via the dedicated TSC Twitter account, @*TetrahymenaSC* and the TGD supported Twitter account, @*thermophila*. In addition, the research community is kept updated about TSC services and offerings by direct e-mails and through posters and talks presented at professional meetings including the biennial Ciliate Molecular Biology conference. TSC also works with the *Tetrahymena* Research Advisory (TetRA) Board, whose mission is to identify and promote specific initiatives that could benefit the community as a whole. The TetRA Board provides valuable input to help ensure that the TSC effectively meets the needs of the research community and continues to support the advancement of *Tetrahymena* as a platform for basic and translational research. Further community engagement is facilitated by TGD support for community annotation of the *Tetrahymena thermophila* genome. These interactions, plus the information and technical support provided TSC, strengthen the bond with the research community and promote interest in and use of the resource.

EDUCATIONAL OUTREACH

The TSC takes an active role in helping to promote the use of ciliates in the classroom, supporting *Tetrahymena*-based science education at the K-12 and undergraduate levels. Science activities for K-12 students have been developed through an NIH-funded Secondary Education Partnership Award program grant for Advancing Secondary Science Education with *Tetrahymena* (ASSET) (https://*Tetrahymena*asset.vet.cornell.edu). Innovative undergraduate activities have been implemented as part of the National Science Foundation-funded Ciliate Genomics Consortium (CGC) (http://faculty.jsd.claremont.edu/ewiley/about.php). Using *Tetrahymena* strains provided by the TSC, ASSET has developed a series of 20 laboratory modules for K-12 students that feature hands-on use of live *Tetrahymena* cells to teach core concepts in biology while stimulating student interest and enthusiasm for science.

In another TSC-linked activity, ASSET sponsors a collaborative research project that supports student scientists in determining the species identity of *Tetrahymena* strains collected from the wild. Students analyze three DNA regions, *cox*1, the

H3/H4 intergenic region, and the 5.8S rDNA (Doerder and Brunk 2012) for species identification. The CGC, a growing group of researcher/educators that engage undergraduate students in original *Tetrahymena*-based research, utilizes *Tetrahymena* and the TGD as a starting point for teaching students how to retrieve information from genomic databases, identify start and stop codons and introns and exons, use basic gene translation tools, launch BLAST searches and sequence alignments, understand the nature of functional domains and evolutionary sequence conservation, and undertake gene expression and protein localization studies. Students can web-publish their findings in a report database (http://suprdb.org) and annotate predicted genes in TGD. A promotional video highlighting TSC and ASSET educational activities entitled "Expedition: Science" (http://www.cornell.edu/video/expedition-science) was TSC's first entrée into video production and is a prototype for current efforts designed to market other TSC resources.

TSC MARKETING STRATEGIES

Informing the broader research community about the power of the *Tetrahymena* model system is central to the mission of the TSC. To this end, TSC has begun developing innovative marketing approaches to inform potential users of the possibilities inherent in the use of *Tetrahymena* as a model research organism. In 2016, the TSC produced a video, *Why Ciliates*?, highlighting the use of ciliates in basic and applied research (https://vimeo.com/191812936). The video was screened at *The Allied Genetics Conference* (TAGC) sponsored by the Genetics Society of America (GSA) in 2016 in a session entitled *Speaking Up for Model Organism Research*, where it was warmly received. Fast-paced and upbeat, the video was also featured by the GSA in their *Genes to Genomes* blog (http://genestogenomes.org/why-ciliates-making-a-video-introduction-to-a-model-organism). The digital format allows dissemination to wide audiences through web hosting services, such as *You Tube* and *Vimeo*, as well as through internet sites of professional societies, such as the GSA and International Society of Protozoology, providing an effective marketing tool for *Tetrahymena* and by extension for TSC activities. In addition to maintaining dedicated websites and e-mail addresses, TGC also utilizes strategic placement of information regarding TSC services on university websites and in university publications and interacts with university marketing and communication personnel to develop innovative marketing tools targeting the education community. Taken together, this diverse approach supports the continuing development of the TSC as a robust and expanding research an educational resource.

OTHER *TETRAHYMENA* COLLECTIONS

To our knowledge, the only other general service repositories for *Tetrahymena* are the American Type Culture Collection (ATCC) in the United States, and the Culture Collection of Algae and Protozoa (CCAP) in Scotland. The ATCC collection currently holds about 380 *Tetrahymena* strains, including 74 strains of *T. thermophila*, the majority of which are natural isolates collected by Drs. Ellen Simon and David Nanney. Another 17 *T. thermophila* strains are highly inbred clones from the collection of Dr. Joseph Frankel, and 18 are inbred strains developed by Dr. Sally Allen (Allen and Lee 1971). The majority of the remaining accessions are various other species collected from sites around the world by various investigators. The ATCC collection is an important and useful resource, providing many *Tetrahymena* species not available elsewhere. However, the ATCC collection is not designed to provide access to the many mutant and genetically engineered strains developed by the research community.

The CCAP collection includes 20 *Tetrahymena* strains, 7 of which are *T. thermophila*. The CCAP strains are not stored frozen, but are propagated by serial subculture, and are very likely to be infertile due to germline senescence. While providing a useful source of live protozoa for many research and teaching purposes, these strains are of limited value to researchers requiring integrity of the germline nucleus, and may be phenotypically heterogeneous due to mutations and their random assortment during long periods of prolonged vegetative reproduction. By supplying a large variety of well-characterized wild type, mutant, and genetically modified strains of *Tetrahymena* required by the *Tetrahymena* research community, the TSC greatly expands and complements the resources provided by ATCC and CCAP.

CONCLUSIONS

The TSC serves a primary function as a strain repository, providing a variety of strains that support the use of *Tetrahymena* as a versatile model organism for addressing diverse biological concepts. In addition, its TGD provides continually updated information on the *Tetrahymena* MIC and MAC genomes, BLAST and GBrowse interfaces, and the opportunity for community-wide gene annotation. TSC also offers a growing array of products and services, including plasmids, transformation services, and direct one-on-one technical support, which facilitate the use of *Tetrahymena* by any researcher. TSC additionally supports the educational use of *Tetrahymena* at both the K-12 and undergraduate levels through interactions with organizations such as the CGC and the ASSET K-12 educational outreach program. As the use of the resource grows, the TSC will continue its mission to support existing resources while identifying new and changing needs and evolving the best means to serve the research and educational communities.

REFERENCES

Akematsu, T., Y. Fukuda, J. Garg, J.S. Fillingham, R.E. Pearlman and J. Loidl. 2017. Post-meiotic DNA double-strand breaks occur in T*etrahymena* and require topoisomerase II and Spo11. *Elife* 6. doi:10.7554/eLife.26176.

Allen, S.L. 1963. Genomic exclusion in *Tetrahymena*: Genetic basis. *Journal of Eukaryotic Microbiology* 10:413–420.

Allen, S.L. 1967a. Genomic exclusion: A rapid means for inducing homozygous diploid lines in *Tetrahymena pyriformis*, syngen 1. *Science* 155:575–577.

Allen, S.L. 1967b. Cytogenetics of genomic exclusion in *Tetrahymena*. *Genetics* 55:797–822.

Allen, S.L. 2000. Genetic nomenclature rules for *Tetrahymena thermophila*. *Methods in Cell Biology* 62:561–563.

Allen, S.L. and I. Gibson. 1973. Genetics of *Tetrahymena*. In *Biology of Tetrahymena*, ed. A.M. Elliott, pp. 307–373. Stroudsburg, PA: Dowden, Hutchinson and Ross.

Allen, S.L. and P.H. Lee. 1971. The preparation of congenic strains of *Tetrahymena*. *Journal of Protozoology* 18:214–218.

Allen, S.L., M.I. Altschuler, P.J. Bruns et al. 1998. Proposed genetic nomenclature rules for *Tetrahymena thermophila*, *Paramecium primaurelia* and *Paramecium tetraurelia*. *Genetics* 149:459–462.

Allen, S.L., S.K. File and S.L. Koch. 1967. Genomic exclusion in *Tetrahymena*. *Genetics* 55:823–837.

Asai, D.L. and J.D. Forney. 2000. *Tetrahymena Thermophila*. San Diego, CA: Academic Press.

Bandrowski, A., M. Brush, J.S. Grethe et al. 2016. The resource identification initiative: A cultural shift in publishing. *Journal of Comparative Neurology* 524:8–22.

Borden, D., E.T. Miller, D.L. Nanney and G.S. Whitt. 1973c. The inheritance of enzyme variants for tyrosine aminotransferase, nadp-dependent malate dehydrogenase, nadp-dependent isocitrate dehydrogenase, and tetrazolium oxidase in *Tetrahymena pyriformis*, syngen 1. *Genetics* 74:595–603.

Borden, D., G.S. Whitt and D.L. Nanney. 1973a. Electrophoretic characterization of classical *Tetrahymena pyriformis* strains. *Journal of Protozoology* 20:693–700.

Borden, D., G.S. Whitt and D.L. Nanny. 1973b. Isozymic heterogeneity in *Tetrahymena* strains. *Science* 181:279–280.

Brickner, J.H., T.J. Lynch, D. Zeilinger and E. Orias. 1996. Identification, mapping and linkage analysis of randomly amplified DNA polymorphisms in *Tetrahymena thermophila*. *Genetics* 143:811–821.

Brunk, C.F. and P. Navas. 1988. Transformation of *Tetrahymena thermophila* by electroporation and parameters effecting cell survival. *Experimental Cell Research* 174:525–532.

Bruns, P.J. and D. Cassidy-Hanley. 2000. Methods for genetic analysis. *Methods in Cell Biology* 62:229–240.

Bruns, P.J. and T.B. Brussard. 1981. Nullisomic *Tetrahymena*: Eliminating germinal chromosomes. *Science* 213:549–551.

Bruns, P.J., H.R. Smith and D. Cassidy-Hanley. 2000. Long-term storage. *Methods in Cell Biology* 62:213–218.

Bruns, P.J., T.B. Brussard, E.V. Merriam. 1983. Nullisomic *Tetrahymena*. II. A set of nullisomics define the germinal chromosomes. *Genetics* 104:257–270.

Carlson, P.S. 1971. Mutant selection in *Tetrahymena pyriformis*. *Genetics* 69:261–265.

Cassidy-Hanley, D., H.R. Smith and P.J. Bruns. 1995. A simple, efficient technique for freezing *Tetrahymena thermophila*. *Journal of Eukaryotic Microbiology* 42:510–515.

Cassidy-Hanley, D., J. Bowen, J.H. Lee et al. 1997. Germline and somatic transformation of mating *Tetrahymena thermophila* by particle bombardment. *Genetics* 146:135–147.

Cassidy-Hanley, D., M.C. Yao and P.J. Bruns. 1994. A method for mapping germ line sequences in *Tetrahymena thermophila* using the polymerase chain reaction. *Genetics* 137:95–106.

Cassidy-Hanley, D., Y. Bisharyan, V. Fridman et al. 2005. Genome-wide characterization of *Tetrahymena thermophila* chromosome breakage sites. II. Physical and genetic mapping. *Genetics* 170:1623–1631.

Cassidy-Hanley, D.M. 2012. *Tetrahymena* in the laboratory: Strain resources, methods for culture, maintenance, and storage. *Methods in Cell Biology* 109:237–276.

Cech, T.R. 1990. Nobel lecture. Self-splicing and enzymatic activity of an intervening sequence RNA from *Tetrahymena*. *Bioscience Reports* 10:239–261.

Cervantes, M.D., E.P. Hamilton, J. Xiong et al. 2013. Selecting one of several mating types through gene segment joining and deletion in *Tetrahymena thermophila*. *PLoS Biology* 11(3):e1001518.

Chalker, D.L. and M.C. Yao. 2011. DNA elimination in ciliates: Transposon domestication and genome surveillance. *Annual Review of Genetics* 45:227–246.

Chalker, D.L., E. Meyer and K. Mochizuki. 2013. Epigenetics of ciliates. *Cold Spring Harbor Perspectives in Biology* 5:a017764. doi:10.1101/cshperspect.a017764.
Chan, H., Y. Wang and J. Feigon. 2017. Progress in human and *Tetrahymena* telomerase structure determination. *Annual Review of Biophysics* 46:199–225.
Chantangsi, C. and D.H. Lynn. 2008. Phylogenetic relationships within the genus *Tetrahymena* inferred from the cytochrome c oxidase subunit 1 and the small subunit ribosomal RNA genes. *Molecular Phylogenetics and Evolution* 49:979–987.
Chantangsi, C., D.H. Lynn, M.T. Brandl, J.C. Cole, N. Hetrick and P. Ikonomi. 2007. Barcoding ciliates: A comprehensive study of 75 isolates of the genus *Tetrahymena*. *International Journal of Systematic and Evolutionary Microbiology* 57:2412–2425.
Collins, K. 2012. Perspectives on the ciliated protozoan *Tetrahymena thermophila*. *Methods in Cell Biology* 109:1–7.
Corliss, J. 1973. History, taxonomy, ecology and evolution of species of *Tetrahyinena*. In *Biology of* Tetrahymena, ed. A.M. Elliot, pp. 1–55. Stroudsburg, PA: Dowden, Hutchinson and Ross.
Doerder, F.P. 2018. Barcodes reveal 48 new species of *Tetrahymena*, dexiostoma, and glaucoma: Phylogeny, ecology, and biogeography of new and established species. *Journal of Eukaryotic Microbiology* doi:10.1111/jeu.12642.
Doerder, F.P. and C. Brunk. 2012. Natural populations and inbred strains of *Tetrahymena*. *Methods in Cell Biology* 109:277–300.
Eisen, J.A., R.S. Coyne, M. Wu et al. 2006. Macronuclear genome sequence of the ciliate *Tetrahymena thermophila*, a model eukaryote. *PLoS Biology* 4:1620–1642.
Elliott, A. and D. Gruchy. 1952. The occurrence of mating types in *Tetrahymena*. In *Biological Bulletin*, Vol. 103, p. 301. Woods Hole, MA: Marine Biological Laboratory.
Flacks, M. 1979. Axenic storage of small volumes of *Tetrahymena* cultures under liquid nitrogen: A miniaturized procedure. *Cryobiology* 16:287–291.
Frankel, J., E.M. Nelsen and L.M. Jenkins. 1977. Mutations affecting cell division in *Tetrahymena pyriformis*, syngen 1. II. Phenotypes of single and double homozygotes. *Developmental Biology* 58:255–275.
Frankel, J., L.M. Jenkins, F.P. Doerder and E.M. Nelsen. 1976. Mutations affecting cell division in *Tetrahymena pyriformis*. I. Selection and genetic analysis. *Genetics* 83:489–506.
Furgason, W.H. 1940. The significant cytostomal pattern of the glaucoma-colpidium group and a proposed new genus and species, *Tetrahymena geleii*. *Archiv für Protistenkunde* 94:224–266.
Gaertig, J., L. Gu, B. Hai and M.A. Gorovsky. 1994. High frequency vector-mediated transformation and gene replacement in *Tetrahymena*. *Nucleic Acids Research* 22:5391–5398.
Galati, D.F., S. Bonney, Z. Kronenberg et al. 2014. DisAp-dependent striated fiber elongation is required to organize ciliary arrays. *Journal of Cell Biology* 207:705–715.
Gall, J.G. 1986, *Molecular Biology of the Ciliated Protozoa*. New York: Academic Press.
Hamilton, E.P., A. Kapusta, P.E. Huvos et al. 2016. Structure of the germline genome of *Tetrahymena thermophila* and relationship to the massively rearranged somatic genome. *Elife* 5. doi:10.7554/*eLife*.19090.
Hamilton, E.P., P.J. Suhr-Jessen and E. Orias. 1988. Pronuclear fusion failure: An alternate conjugational pathway in *Tetrahymena thermophila*, induced by vinblastine. *Genetics* 118:627–636.
Hill, D.L. 1972. *The Biochemistry and Physiology of Tetrahymena*. New York: Academic Press.
Howard-Till, R.A., A. Lukaszewicz, M. Novatchkova and J. Loidl. 2013. A single cohesin complex performs mitotic and meiotic functions in the protist *Tetrahymena*. *PLoS Genetics* 9(3):e1003418.

Jacobs, M.E., L.V. DeSouza, H. Samaranayake, R.E. Pearlman, K.W. Siu and L.A. Klobutcher. 2006. The *Tetrahymena thermophila* phagosome proteome. *Eukaryotic Cell* 5:1990–2000.

Kain, K.H. 2009. Telomeres and *Tetrahymena*: An interview with Elizabeth Blackburn. *Disease Models and Mechanisms* 2:534–537.

Karrer, K.M. 2000. *Tetrahymena* genetics: Two nuclei are better than one. *Methods in Cell Biology* 62:127–186.

Karrer, K.M. 2012. Nuclear dualism. *Methods in Cell Biology* 109:29–52.

Kher, C.P., F.P. Doerder, J. Cooper et al. 2011. Barcoding *Tetrahymena*: Discriminating species and identifying unknowns using the cytochrome c oxidase subunit I (cox-1) barcode. *Protist* 162:2–13.

Kidder, G.W. and V.C. Dewey. 1951. The biochemistry of ciliates in pure culture. In *Biochemistry and Physiology of Protozoa,* ed. A. Lwoff and, S.H. Hutner, pp. 323–400. New York: Academic Press.

Longcor, M.A., S.A. Wickert, M-F. Chau and E. Orias. 1996. Coassortment of genetic loci during macronuclear division in *Tetrahymena thermophila. European Journal of Protistology* 32:85–89.

Lwoff, A. 1923. Sur la nutrition des infusoires. *Journal of Protozoology* 176:928–930.

Lynch, T.J., J. Brickner, K.J. Nakano and E. Orias. 1995. Genetic map of randomly amplified DNA polymorphisms closely linked to the mating type locus of *Tetrahymena thermophila. Genetics* 141:1315–1325.

Lynn, D.H. and F.P. Doerder. 2012. The life and times of *Tetrahymena. Methods in Cell Biology* 109:9–27.

Martindale, D.W. and P.J. Bruns. 1983. Cloning of abundant mRNA species present during conjugation of *Tetrahymena thermophila*: Identification of mRNA species present exclusively during meiosis. *Molecular and Cell Biology* 3:1857–1865.

Maupas, E. 1889. Le rejeunissement caryogamique chez les cilies. *Archives de Zoologie Expérimentale et Générale* 7:149–517.

Mochizuki, K. and M.A. Gorovsky. 2004. Small RNAs in genome rearrangement in *Tetrahymena. Current Opinion in Genetics and Development* 14:181–187.

Mochizuki, K., N.A. Fine, T. Fujisawa and M.A. Gorovsky. 2002. Analysis of a piwi-related gene implicates small RNAs in genome rearrangement in *Tetrahymena. Cell* 110:689–699.

Nanney, D.L. 1953. Nucleo-cytoplasmic interaction during conjugation in *Tetrahymena. Biological Bulletin* 105:133–148.

Nanney, D.L. 1959. Genetic factors affecting mating type frequencies in variety 1 of *Tetrahymena pyriformis. Genetics* 44:1173–1184.

Nanney, D.L. and E.M. Simon. 2000. Laboratory and evolutionary history of *Tetrahymena thermophila. Methods in Cell Biology* 62:3–25.

Nanney, D.L. and J.W. McCoy. 1976. Characterization of the species of the *Tetrahymena pyriformis* complex. *Transactions of the American Microscopical Society* 95:664–682.

Nanney, D.L. and P.A. Caughey. 1953. Mating type determination in *Tetrahymena pyriformis. Proceedings of the National Academy of Sciences* (USA) 39:1057–1063.

Nanney, D.L., P.A. Caughey and A. Tefankjian. 1955. The genetic control of mating type potentialities in *Tetrahymena pyriformis. Genetics* 40:668–680.

Orias, E, E.P. Hamilton and J.D. Orias. 2000. *Tetrahymena* as a laboratory organism: Useful strains, cell culture, and cell line maintenance. *Methods in Cell Biology* 62:189–211.

Orias, E., E.P. Hamilton and M. Flacks. 1979. Osmotic shock prevents nuclear exchange and produces whole-genome homozygotes in conjugating *Tetrahymena. Science* 203:660–663.

Papazyan, R., E. Voronina, J.R. Chapman et al. 2014. Methylation of histone H3K23 blocks DNA damage in pericentric heterochromatin during meiosis. *Elife* 26;3:e02996.

Pucciarelli, S., P. Ballarini, D. Sparvoli et al. 2012. Distinct functional roles of beta-tubulin isotypes in microtubule arrays of *Tetrahymena thermophila*, a model single-celled organism. *PLoS One* 7(6):e39694.

Rabl, J., M. Leibundgut, S.F. Ataide, A. Haag and N. Ban. 2011. Crystal structure of the eukaryotic 40S ribosomal subunit in complex with initiation factor 1. *Science* 331:730–736.

Rasmussen, L. and E. Orias. 1975. *Tetrahymena*: Growth without phagocytosis. *Science* 190:464–465.

Sandoval, P.Y., P.H. Lee, X. Meng and G.M. Kapler. 2015. Checkpoint activation of an unconventional DNA replication program in *Tetrahymena. PLoS Genetics* 11(7):e1005405.

Sauvant, M.P., D. Pepin and E. Piccinni. 1999. *Tetrahymena pyriformis*: A tool for toxicological studies. A review. *Chemosphere* 38:1631–1639.

Scherbaum, O. and E. Zeuthen. 1954. Induction of synchronous cell division in mass cultures of *Tetrahymena* piriformis. *Experimental Cell Research* 6:221–227.

Sengupta, R.N., S.N. Van Schie, G. Giambasu et al. 2016. An active site rearrangement within the *Tetrahymena* group I ribozyme releases nonproductive interactions and allows formation of catalytic interactions. *RNA* 22:32–48.

Shabatura, S.K. and F.P. Doerder. 1981. Age-associated changes in the micronuclear cycle of *Tetrahymena thermophila* A III heterokaryons. A brief note. *Mechanisms of Ageing and Development* 15:235–238.

Shodhan, A., K. Kataoka, K. Mochizuki, M. Novatchkova and J. Loidl. 2017. A Zip3-like protein plays a role in crossover formation in the SC-less meiosis of the protist *Tetrahymena. Molecular Biology of the Cell* 28:825–833.

Simon, E.M. 1972. Freezing and storage in liquid nitrogen of axenically and monoxenically cultivated *Tetrahymena pyriformis. Cryobiology* 9:75–81.

Simon, E.M. 1982. Breeding performance of *Tetrahymena thermophila* following storage for 5 to 6 years in liquid nitrogen. *Cryobiology* 19:607–612.

Simon, E.M. and D.L. Nanney. 1979. Germinal aging in *Tetrahymena thermophila. Mechanisms of Ageing and Development* 11:253–268.

Simon, E.M. and M.V. Schneller. 1973. The preservation of ciliated protozoa at low temperature. *Cryobiology* 10:421–426.

Simon, E.M. and S.W. Whang. 1967. *Tetrahymena*: Effect of freezing and subsequent thawing on breeding performance. *Science* 155:694–696.

Simon, E.M., D.L. Nanney and F.P. Doerder. 2007. The *Tetrahymena pyriformis* complex of cryptic species. In *Protist Diversity and Geographical Distribution,* ed. W. Foissner and D. Hawksworth, pp. 131–146. Dordrecht, the Netherlands: Springer.

Sparvoli, D., E. Richardson, H. Osakada et al. 2018. Remodeling the specificity of an endosomal CORVET tether underlies formation of regulated secretory vesicles in the ciliate *Tetrahymena thermophila. Current Biology* 28:697–710.

Stover, N.A., C.J. Krieger, G. Binkley et al. 2006. *Tetrahymena* genome database (TGD): A new genomic resource for *Tetrahymena thermophila* research. *Nucleic Acids Research* 34:D500–D503.

Stover, N.A., R.S. Punia, M.S. Bowen, S.B. Dolins and T.G. Clark. 2012. *Tetrahymena* genome database wiki: A community-maintained model organism database. *Database* (Oxford) 20;2012:bas007.

Sugai, T. and K. Hiwatashi. 1974. Cytologic and autoradiographic studies of the micronucleus at meiotic prophase in *Tetrahymena pyriformis. Journal of Protozoology* 21:542–548.

Suryavanshi, S., B. Edde, L.A. Fox et al. 2010. Tubulin glutamylation regulates ciliary motility by altering inner dynein arm activity. *Current Biology* 20:435–440.

Thazhath, R., C. Liu and J. Gaertig. 2002. Polyglycylation domain of beta-tubulin maintains axonemal architecture and affects cytokinesis in *Tetrahymena. Nature Cell Biology* 4:256–259.

Tondravi, M.M. and M.C. Yao. 1986. Transformation of *Tetrahymena thermophila* by microinjection of ribosomal RNA genes. *Proceedings of the National Academy of Sciences* (USA) 83:4369–4373.

Turkewitz, A.P. 2004. Out with a bang! *Tetrahymena* as a model system to study secretory granule biogenesis. *Traffic* 5:63–68.

Wang, Y., X. Chen, Y. Sheng, Y. Liu and S. Gao. 2017. N6-adenine DNA methylation is associated with the linker DNA of H2A.Z-containing well-positioned nucleosomes in pol II-transcribed genes in *Tetrahymena. Nucleic Acids Research* 45:11594–11606.

Wickert, S. and E. Orias. 2000. *Tetrahymena* micronuclear genome mapping: A high-resolution meiotic map of chromosome 1l. *Genetics* 154:1141–1153.

Wilkes, D.E., H.E. Watson, D.R. Mitchell and D.J. Asai. 2008. Twenty-five dyneins in *Tetrahymena*: A re-examination of the multidynein hypothesis. *Cell Motility and the Cytoskeleton* 65:342–351.

Williams, J.G., A.R. Kubelik, K.J. Livak, J.A. Rafalski and S.V. Tingey. 1990. DNA polymorphisms amplified by arbitrary primers are useful as genetic markers. *Nucleic Acids Research* 18:6531–6535.

Wilson, D.N. and J.H. Doudna Cate. 2012. The structure and function of the eukaryotic ribosome. *Cold Spring Harbor Perspectives in Biology* 4. doi:10.1101/cshperspect.a011536.

Wood, C.R., R. Hard and T.M. Hennessey. 2007. Targeted gene disruption of dynein heavy chain 7 of *Tetrahymena thermophila* results in altered ciliary waveform and reduced swim speed. *Journal of Cell Science* 120:3075–3085.

Xiong, J., S. Gao, W. Dui et al. 2016. Dissecting relative contributions of cis- and trans-determinants to nucleosome distribution by comparing *Tetrahymena* macronuclear and micronuclear chromatin. *Nucleic Acids Research* 44:10091–10105.

Xue, Y., H. Li, C.Y. Ung, C.W. Yap and Y.Z. Chen. 2006. Classification of a diverse set of *Tetrahymena pyriformis* toxicity chemical compounds from molecular descriptors by statistical learning methods. *Chemical Research in Toxicology* 19:1030–1039.

Yao, M.C., J.L. Chao and Y.C. Cheng. 2014. Programmed genome rearrangements in *Tetrahymena. Microbiology Spectrum*. doi:10.1128/microbiolspec.

Zufall, R.A., K.L. Dimond and E.P. Doerder. 2013. Restricted distribution and limited gene flow in the model ciliate *Tetrahymena thermophila. Molecular Ecology* 22:1081–1091.

12 The National *Xenopus* Resource

Marcin Wlizla, Sean McNamara and Marko E. Horb

CONTENTS

Abstract: The clawed frogs of the *Xenopus* genus have a long history of use as a model in biomedical research, are a powerful system for study of complex question, and require relatively simple life support for maintenance in laboratory conditions. Historically, only wild type *Xenopus* frogs were used by most researchers and these were easily procured from commercial vendors. Development of new technologies has allowed for the creation of transgenic and mutant *Xenopus* lines, necessitating the creation of a national stock center to house and distribute these valuable lines. The National *Xenopus* Resource (NXR) was created in 2010 to help *Xenopus* researchers deal with this challenge by serving as a centralized repository maintaining a large number of diverse animal lines available for immediate distribution to the research community. It also serves as an educational facility holding advanced training workshops, and as a center where individual researchers can convene and collaborate on complex projects through the use of NXR animals and expertise. Here, we briefly discuss the history of *Xenopus* use in research and the considerable impact it has had in science. We then describe the aspects of NXR operations necessary for the fulfillment of its role as an animal repository and distribution center, educational facility, and a convening center.

A ROLE FOR *XENOPUS* IN BIOLOGICAL RESEARCH

Since the 1950s, frogs of the *Xenopus* genus, *Xenopus laevis* and *Xenopus tropicalis*, have become the most widely used amphibian model system for biological research. Historical, practical, and scientific reasons have made it a powerful system for the study of fundamental biological and disease mechanisms in neurobiology, physiology, molecular biology, cell biology, and developmental biology. The historical reasons for this have been previously reviewed in detail by Gurdon and Hopwood (Gurdon and Hopwood 2000), but highlighting some of the milestones in the domestication of this laboratory system will aid in understanding how these animals, indigenous to Southern and Central Africa, became common in laboratories around the world.

The earliest recorded description of *X. laevis*, the African clawed frog, is credited to François Marie Daudin. In 1802, he named it *Bufo laevis* in his *Histoire naturelle des rainettes, des grenouilles, et des crapauds* (Daudin 1802, Gurdon and Hopwood 2000). This was despite the fact that the provenance of his specimen was unknown, and the accompanying illustration ignored several features characteristic of the species, including the three claws present on each hind-foot, webbing between the hind-foot toes, and the distinctive dorsal pigmentation pattern. The genus name, *Xenopus*, was proposed by Wagler in a footnote to a published letter he received from Heinrich Boie in 1827 that described the finding of the "crazy frog with real nails" (Wagler 1827). For over a century, reports describing *X. laevis* development, systematics, breeding habits, habitat, and morphology would appear sporadically (Beddard 1894, Cuvier 1829, Gray 1864, Günther 1858, Leslie 1890).

The crucial breakthrough that aided in establishing colonies of *X. laevis* in American and European laboratories came from the work of comparative endocrinologist, Lancelot Thomas Hogben, working with Enid Charles and David Slome. While studying the effects of the pituitary on pigmentation, Hogben observed that hypophysectomised *X. laevis* females underwent ovarian involution, while injection of ox anterior pituitary extracts induced ovulation (Hogben 1930). In a following communication, Hogben compared his results to previous work done in mammals (Bellerby 1929, Fee and Parkes 1929, Hogben et al. 1931, Smith and Engle 1927, Zondek and Aschheim 1927) concluding that the anterior lobe of the pituitary controls the activity of the ovary in all vertebrates (Hogben et al. 1931). This observation, that ovulation in *X. laevis* can be induced by anterior pituitary extracts, together with recently developed animal-based pregnancy tests which involved injection of urine from pregnant women into the mouse or rabbit (Aschheim and Zondek 1928, Friedman and Lapham 1931) led others to speculate that female *X. laevis* could also be used to detect gonadotrophins present in urine of pregnant human females. By the late 1930s, induction of ovulation in *X. laevis* by injection of pregnant female urine became a reliable pregnancy test. It also gave results faster (within 8–12 h) than the mammalian tests which took several days to complete (Gurdon and Hopwood 2000). Demand for pregnancy testing in the 1940s and 1950s made *X. laevis* widely available in Europe and North America. The observation that female *X. laevis* could be induced to lay eggs all year round, together with the increasing availability of commercial preparations of human chorionic gonadotrophin (hCG) hormone, were instrumental in adapting this animal for biological research once demand for its use in pregnancy testing was replaced by novel immunological methods in the 1960s (Gurdon and Hopwood 2000). More recently, ovine luteinizing hormone was found to act as efficiently as hCG (Wlizla et al. 2017).

In addition to its ability to produce eggs and embryos year-round, numerous other practical and scientific reasons make *X. laevis* a powerful model for diverse fields of biological research. The adult animals are primarily aquatic, and relatively easy and inexpensive to maintain and breed (Wu and Gerhart 1991). A single female can easily produce between 2000 and 4000 eggs per spawning, providing abundant amounts of experimental material (Wlizla et al. 2017). The females maintain their fertility for over a decade, which provides for slower generational genetic drift than in some other common vertebrate systems. Mature *X. laevis* oocytes, or the egg progenitor cells, are commonly used in electrophysiology to study properties of exogenous membrane ion transporters and channels. Mature *X. laevis* oocytes are large (~1 mm in diameter), robust enough to permit microinjection and patch clamping, and are also fully functional cells that actively translate and express the exogenous proteins (Dascal 1987). Cytoplasmic extracts made from *X. laevis* eggs provide an intermediate between in vivo and in vitro cell-free models for the study of molecular processes involved in cell function and are the only cell-free system that permits full investigation of all DNA transactions related to cell cycle progression and DNA damage repair (Cross and Powers 2009, Hoogenboom et al. 2017).

The development of *X. laevis* is entirely external and stereotypically synchronous thus permitting even its earliest stages to be studied with ease. The early embryo is robust enough to handle quite drastic disruption, including cut-and-paste transplantation experiments instrumental in the study of inductive interactions between tissues (Gilbert and Barresi 2016, Slack 2012). The large size of early blastomeres, well established fate maps, and entirely holoblastic cleavage allow for microinjected tracers or molecules designed to disrupt gene activity to be precisely targeted to particular tissues, organs, or regions of the developing embryo (Bauer et al. 1994, Moody 1987a, 1987b). The effects of such disruptions can be rapidly observed as the embryos are transparent during organogenesis, and all the major body organs necessary for a healthy feeding tadpole are formed within 6 days following fertilization. The phylogenetic position of *Xenopus* makes it directly applicable to the study of human development and disease. Many molecular processes involved in early development are conserved between frogs and mammals (Sater and Moody 2017). Furthermore, *Xenopus* share features with mammals that are not found in zebrafish including the best characterized immune system outside mammals and chicken (Guselnikov et al. 2008, Zarrin et al. 2004), yet are evolutionarily distant enough to allow for identification of conserved non-coding regulatory regions that are obscured in genomic comparisons between human and other mammalian genomes.

The value of these experimental strengths of *Xenopus* is reflected in the numerous scientific breakthroughs achieved using it as a model system. These breakthroughs resulted in the awarding of three Nobel Prizes in the last two decades. In 2001, Sir Richard Timothy Hunt was awarded the Nobel Prize in Physiology or Medicine stemming from his initial discovery of *cyclin*, leading to major breakthroughs in the understanding of cell cycle checkpoints, and for demonstrating that cell cycle progression depends on proteolysis and de novo protein synthesis (Evans et al. 1983). In 2003, Peter Agre was awarded the Nobel Prize in Chemistry for his discovery of Aquaporins and for demonstrating (using *Xenopus* oocytes) that peptides serve as cell membrane water channels (Preston et al. 1992). Finally, in 2012, the prize in Physiology or Medicine was awarded to Sir John Bertrand Gurdon who, via transplantation of a nucleus from a frog intestinal cell into a frog oocyte, definitively demonstrated that nuclear genes are neither lost nor permanently inactivated during differentiation (Gurdon 1962). In fact, Gurdon's experiments generated the first successful clones of a vertebrate animal, approximately 30 years before the much-hyped mammalian clone, Dolly the sheep (Wilmut et al. 1997). Furthermore, his work was instrumental for the eventual identification of the Yamanaka factors, the four transcription factors necessary for induction of cell pluripotency. Shinya Yamanaka was a recipient of the 2012 prize in Physiology or Medicine.

SCIENTIFIC ACHIEVEMENTS USING *XENOPUS*

Although not recognized with the award of a Nobel Prize, additional profound scientific firsts and breakthroughs have been made using *Xenopus*, an abridged list of which includes:

- First use of mRNA injection as a rapid analysis of gene function via misexpression (Gurdon et al. 1971),
- First use of morpholino antisense oligonucleotides for the study of gene knockdown effects (Heasman et al. 2000),
- Demonstration that nuclear proteins are targeted to the nucleus by a part of their mature sequence (Gurdon 1970),
- Demonstration that mRNA is translated into protein (Gurdon et al. 1971),
- Purification of a single gene (Brown et al. 1971, McKnight 2012),
- Demonstration of the existence of maternally inherited mitochondrial DNA (Dawid 1966),
- Identification of the first eukaryotic transcription factor, and
- The first zinc finger (Engelke et al. 1980, Klug and Rhodes 1987), among others.

X. LAEVIS VS. *X. TROPICALIS*

Despite its many experimental strengths, the use of *X. laevis* as a genetic experimental system is handicapped by the fact that it is a tetraploid. Approximately 17–18 million years ago, a hybridization event between two distinct species resulted in the creation of a viable allotetraploid ancestor of *X. laevis*, which still carries over 56% of its genes as two homologous copies (Session et al. 2016). In the 1990s, *X. tropicalis*, a true diploid species, was proposed as an alternative to *X. laevis* for the modern genetic age in order to simplify the breeding logistics necessary for the generation of homozygous mutant animals. *Xenopus tropicalis* has proved to be a very beneficial model, particularly because it carries all the experimental advantages of *X. laevis*, including external development and production of thousands of eggs. Additionally, it is characterized by more rapid early development and a shorter time required to reach sexual maturity, thus reducing generation time. Also, out of the 29 currently recognized extant species in the *Xenopus* genus, *X. tropicalis* happens to be the only true diploid (Evans et al. 2015, Khokha 2012). Thus, *X. tropicalis* provides for a potentially more rapid generation of homozygous mutant lines than would be possible using *X. laevis*.

XENOPUS RESOURCE CENTERS FOR THE PRESERVATION OF ANIMALS OF HIGH SCIENTIFIC VALUE

The principal reason for the conservation of animals in the *Xenopus* genus stems from the need to preserve rare experimental stocks, that is, inbred, transgenic and mutant lines much like what has been done with other experimental models (i.e., *Caenorhabditis elegans*, *Drosophila melanogaster*, etc.). The need for the creation of a centralized stock and training center to address these issues was described in the 2000, 2003, 2006, and 2009 *Xenopus* Community White Papers (Khokha et al. 2009, Khokha 2012). Creation of a national stock center was anticipated to help provide a centralized resource to house, maintain, and distribute the various lines. This was recognized by the *Xenopus* research community. The need for the creation of a centralized stock and training center to address these problems was described in the 2000, 2003, 2006, and 2009 *Xenopus* Community White Papers (Khokha et al. 2009, Khokha 2012).

The U.S. National *Xenopus* Resource (NXR) was established in 2010 at the Marine Biological Laboratory with funding from the National Institutes of Health (NIH; Pearl et al. 2012). Since its inception, the NXR has grown to house approximately 3000 adult *X. tropicalis* and 5000 adult *X. laevis* frogs, comprising over 200 different strains and lines. The number of different frog lines continues to grow each year. In addition to its function as a centralized repository, the NXR serves as a training center in diverse technologies, including advanced imaging, bioinformatics, and genome editing.

In addition to the NXR, several other resource centers provide services that promote global research use of *Xenopus*. The European *Xenopus* Resource Center (EXRC) at the University of Portsmouth, England, stocks wild type, mutant, and transgenic animals and provides services similar to those provided by the NXR. The EXRC is also a repository for *Xenopus* cDNAs, fosmids, and antibodies (Pearl et al. 2012). Research Resource for Immunology (XIRRI) at the University of Rochester (New York) specializes in stocking genetically characterized inbred and transgenic lines made in-house and also antibodies, cell lines, and DNA libraries. The XIRRI takes advantage of the extensive characterization of the *Xenopus* immune system and its remarkable conservation with mammals in the study of infectious diseases and immune system function. Finally, the Hiroshima University Amphibian Research Center in Japan is part of the National BioResource Project and provides inbred and standard *X. tropicalis* frogs as well as training for researchers.

ANIMAL STOCKS AVAILABLE AT THE NXR

The NXR currently stocks nearly 225 different strains and lines of frogs, with that number continuously increasing as new lines are submitted by external researchers or generated in house. The majority of NXR frogs are transgenic, mutant, or inbred strains, since wild type animals can be purchased from commercial vendors. The inbred strains housed at the NXR include *X. laevis* J Strain and the *X. tropicalis* Nigerian and Ivory Coast strains. The J Strain and the Nigerian strains were used in sequencing the genomes of *X. laevis* and *X. tropicalis*, respectively (Hellsten et al. 2010, Session et al. 2016). The high inbreeding ratio of these two strains, the fact that these animals have been used for the sequencing of the genome, and the relatively slow generational genetic drift due to the fact that females remain fertile for well over a decade (Davis et al. 2012, McCoid and Fritts 1989) makes it very likely that the sequence of any site in the genome of animals held at the NXR exactly matches the published genomic sequence, thus making them an excellent choice for experiments involving targeted gene disruption.

The majority of transgenic frogs in the NXR have been obtained from the *Xenopus* research community. However, in the recent years the NXR has also been generating additional lines judged to be of potential high research impact. Transgenic lines were generated by random insertions of exogenous DNA to create animals expressing fluorescent reporter proteins in tissue-specific regions (Ishibashi et al. 2012a, 2012b, Kelley et al. 2012). Of the 105 transgenic lines housed at the NXR, each has been designed to study a specific biological question, including organ development, subcellular organelle function, fate mapping, and signaling pathway activity through the use of ubiquitous, cell type-specific, or inducible promoter driven expression of fluorescent proteins, as well as constitutively active and dominant negative mutants. Three examples of such transgenic lines are the green to red photo-convertible fluorescent protein line (*Xla. Tg(CAG:KikGR)*^{Flc}) generated by the Conlon lab, the kidney-specific reporter line (*Xla. Tg(Dre.cdh17:eGFP)*^{NXR}) generated by the NXR for the Miller lab, and the membrane GFP line (*Xla. Tg(CMV:memGFP, cryga:mCherry)*^{NXR}) generated by the NXR (Figure 12.1). A full list of available

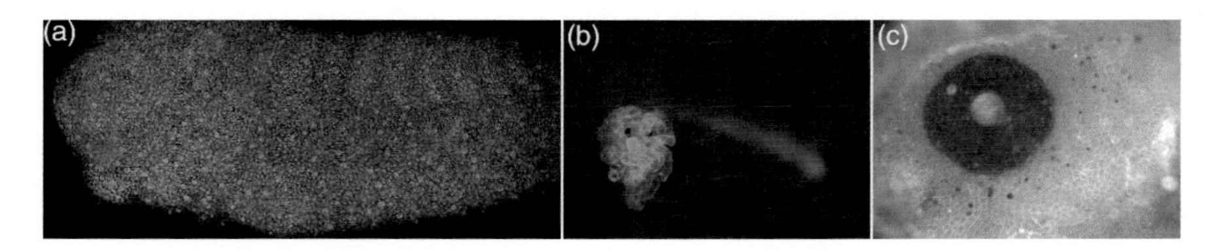

FIGURE 12.1 (See color insert.) Transgenic *Xenopus* frogs. (a) Stage 30 *Xla. Tg(CAG:KikGR)*^{Flc} transgenic embryo where the green fluorescent KikGR is expressed ubiquitously in the embryo. Green to red fluorescent photoconversion can be achieved in specific populations of cells using a 405 nm laser. (Courtesy of Mitch Butler in the 2015 National *Xenopus* Resource Advanced Imaging Workshop.) (b) Stage 46 *Xla. Tg(Dre.cdh17:eGFP)*^{NXR} transgenic tadpole illustrating (green fluorescent protein) fluorescence in the developing *Xenopus* kidney. (c) Stage 40 *Xla. Tg(CMV:memGFP, cryga:mCherry)*^{NXR} transgenic tadpole has membrane bound green fluorescent protein expressed in all cells.

transgenic lines can be found on the NXR webpage (http://www.mbl.edu/xenopus). Individual lines are maintained in separate tanks at a density (number of animals) that is dependent on the demand from the user community. All transgenic lines are available for purchase by the research community as live adults, embryos, or tadpoles. Stocks are also distributed as either isolated *X. laevis* testes that are shipped on ice (which maintain their viability for approximately a week) or as cryopreserved sperm from either species that is also shipped on dry ice and which can be stored indefinitely in liquid nitrogen (LN) or at −80°C.

ANIMAL MAINTENANCE PROCEDURES AT THE NXR

Animal maintenance involves a number of factors. First, attention to water quality and diet is necessary to ensure a viable and healthy frog colony. Second, as the number of animals housed increases, careful planning must be given to costs and space limitations. Finally, effective and efficient organization and accurate record keeping are essential when dealing with a large number of diverse animal lines.

In the NXR, all frogs are kept in recirculating aquatic systems which automatically measure and adjust temperature, pH, conductivity, and effluent. Additional parameters including alkalinity, as well as levels of ammonia, nitrate, and nitrite ions are measured manually by the animal support staff on a weekly basis. The systems exchange and replace 10% of water over the course of the day and are thus more economical and efficient than flow-through and static fill-and-dump system configurations. Adult frogs are kept in self-cleaning flood-and-flush tanks designed from a bell siphon to automatically clear any undissolved solids including leftover food and excretion products. This design decreases the time that the animal support staff would otherwise need to spend cleaning out each individual tank which is necessary for clear visual inspection of the housed animals when monitoring them for any signs of developing health problems. In contrast, the nursery system which houses tadpoles and newly metamorphosed froglets use tanks with an overflow bulkhead strainer configuration since flood-and-flush tanks risk flushing small animals and food out of the system. The use of overflow bulkhead tanks results in the food settling on the bottom. This is actually advantageous for a nursery as it lets the young animals feed throughout the day. The diet composition and feeding regiment is adjusted to the age and size of the animals, with the food size increasing and the feeding frequency decreasing with age. As such, tadpoles and young frogs are initially fed twice a day. Once they become fully grown adults, they are only fed two to three times a week (McNamara et al. 2018).

Since the NXR is a national distributor of frogs to the wider research community, all animals in the NXR are bred in the facility and are subject to strict procedures that to limit the likelihood of introducing any pathogens into the NXR colony. Adult stocks acquired from outside labs are kept in separate quarantine rooms and are never allowed to enter the main NXR facility. Embryos generated from these adult stocks undergo a surface sterilization procedure whereby early stage embryos are treated with 70% ethanol, followed by removal of the jelly coat using L-Cysteine, and a final incubation in 0.1 mg/mL thimerosal. They are then reared in the clean facility (Peng et al. 1991, Wlizla et al. 2018). Semi-annual health tests are performed on frogs in each of the facility's systems for common pathogens that affect frog colonies in captivity and in the wild.

The majority of the clean adult animals are kept in four large systems with centralized life support, each housing between 1500 and 2500 animals. Due to differences in water parameters, *X. laevis* and *X. tropicalis* are kept in separate systems (Figure 12.2). The systems are engineered to automatically notify the animal care staff, either via phone or e-mail, if any of the water quality parameters pass beyond an optimal range for the animals or if a system shut down occurs. Thus, the animal care staff are made aware of and can deal with most system operation problems as soon as they arise, even if the fault occurs outside of normal work hours.

There are logistical limits to the number of live animals that can be kept in a single facility. Larger numbers of animals require both additional space and animal care staff. However, there are a number of steps that can be taken to allow for the housing of a greater diversity of lines without the need to expand the facility size. First, the NXR relies on input from the research community and the internal tracking of animal sales to identify animal stocks that are of high interest and thus need to be maintained in greater numbers. The inventory number of those stocks that are in low demand can be reduced to free space for other lines. Second, lines that are not in high demand can be stored in LN or in a −80°C freezer as cryopreserved sperm (Pearl et al. 2017). Under these conditions, they can be stored essentially indefinitely with little additional maintenance necessary from the staff and with a drastically reduced footprint when compared to housing live animals. Third, availability of different

FIGURE 12.2 Modular housing systems for *Xenopus*. (a) *Xenopus laevis* frogs in the National *Xenopus* Resource are housed in tanks of different sizes. The picture shows one row of two different mutli-rack systems. (b) *Xenopus tropicalis* frogs in the National *Xenopus* Resource are also housed in multi-rack systems, but with smaller sized tanks. (c) Examples of independent stand-alone housing systems used in the National *Xenopus* Resource for both *X. laevis* and *X. tropicalis*.

animal housing modules with different tank sizes is essential for minimizing the footprint necessary for housing all the animals. Thus, low demand lines are stored in modules with smaller tanks while high demand lines are stored in modules with large tanks. Whereas a large number of animals emphasizes the concerns for space and staff size, a large diversity of different stocks emphasizes the need for effective organization and record keeping. The NXR uses Animal Bioware (Digital Paradigms Inc., Boston, MA) to keep track of all the different stocks currently available, and their location within the facility. This allows for quick and easy tracking to locate individual lines.

The *Xenopus* research community has recently agreed on a uniform set of rules for naming lines, identifying the genomic modifications present and the lab of origin. All lines housed at the NXR follow these nomenclature guidelines which aid in their identification and also prevents redundancy. Finally, the NXR participates in the Research Resource Identification (RRID) Initiative, the goal of which is to improve identifiability and scientific reproducibility (Bandrowski et al. 2016). As such, each individual stock is associated with a unique RRID number that is machine readable, free to generate and access, and consistent across publishers and journals. These RRID numbers are also associated with animal stock records available on Xenbase, the *Xenopus* model organism database, which contains extensive information about each line (James-Zorn et al. 2015, Karimi et al. 2018).

When propagating stored stocks, quality control procedures are necessary to confirm that all animal lines are correctly identified. For many transgenic lines, this is relatively simple, as they will typically carry fluorescent reporters that aid in their identification. To propagate these lines, in vitro fertilization or natural mating is used to generate embryos, and a fluorescent stereoscope is used to identify transgenic embryos that express the appropriate fluorescent protein at the corresponding stages. Only the embryos with the brightest (highest level of) expression are kept and grown to maturity in order to maintain the "best" transgenic line offspring (Wlizla et al. 2018). For mutant lines in which fluorescent reporters are not available, molecular based assays are necessary. Adult frogs containing specific mutations are genotyped to confirm the presence of specific mutations, and the heterozygous adults are kept in the NXR. When propagating inbred lines, it is important that the genome of offspring remain as true as possible to the published genome. As such, the NXR has a small number of breeding pairs set aside exclusively for use in the generation of inbred animals for distribution. The longevity of *Xenopus* and their long sexual viability allow for the same breeding pairs to be used over an extended period of time. This lowers the likelihood of generational genetic drift that could disrupt the similarity between the genomes of the housed animals and the animals that were initially used for sequencing the genome (McCoid and Fritts 1989).

STOCK DISTRIBUTION PROCEDURES AT THE NXR

Distribution of lines housed at the NXR to the research community is a simple process. If the line requested is available for distribution, the receiver and NXR staff work together to process a material transfer agreement, as well as deal with any health tests and additional paperwork that may be required by law or by the receiving institution. Coordination between the NXR and the receiver is crucial to prevent transport-induced stress on the animals, as well as any other potential problems. Adult animals are packed in moist peat moss for domestic orders and are shipped via common commercial couriers overnight and early in the week in case issues, which delay delivery, arise.

In certain cases, animals will be shipped with a specialized courier company that can assure that they will be kept in climate-controlled conditions. This is the case for all international shipments, as well as for shipments during times of the year when the ambient temperature at the destination or in Woods Hole is outside of the range tolerated by the frogs. This typically occurs throughout the winter and during the hottest days in the middle of summer. If requested, animals can also be shipped as embryos or tadpoles in a plastic container filled with 0.1 × Marc's Modified Ringers solution and secured within an insulated shipping box. As with adult animals, consideration is given to outside temperatures and courier selection. Finally, stocks are also distributed as either isolated *X. laevis* testes shipped on ice, which maintain their viability for approximately a week, or as cryopreserved sperm from either species shipped on dry ice, which can be stored indefinitely in LN or at −80°C. Testes and sperm shipments are done throughout the year.

IMPACT OF THE NXR ON CURRENT AND FUTURE RESEARCH GOALS

During its relatively short existence, the NXR has positively impacted the *Xenopus* research community in several ways. The study of complex biological processes increasingly requires access to a large number of diverse mutant and transgenic animal lines. However, researchers who are just starting their labs or who tend to run smaller operations may not have the space or the funds necessary to support all the animals that are important for moving their research forward. As a central repository of a large number of different *Xenopus* stocks, the NXR makes it possible for scientists to order animal lines on an as needed basis, effectively eliminating the need for them to invest into the infrastructure necessary to house the different lines that would be necessary for their research. Where appropriate, lines can be sent out as cryopreserved sperm or testes eliminating the need for housing adult males. Or, embryos can be generated in the NXR facility and shipped, thus making it possible for researchers who do not have the ability to house animals to use *Xenopus* in their work. Researchers who do not want to add extra frog lines to their colonies can come to the NXR for both short- and long-term visits and gain access to all the NXR's stocks and expertise as well as common laboratory equipment and supplies. This provides researchers with a cost-effective opportunity to run pilot experiments to identify lines housed at the NXR which may be of use to their studies, without having to first order and house them in their own facilities.

In addition to stocking lines, the NXR staff can generate custom mutant and transgenic animals. If a particular line of interest does not exist, a researcher can request to have it generated by the NXR. In fact, an increasing number of lines at the NXR have been made in house, and some have already been published (Corkins et al. 2018, Ratzan et al. 2017). Since its creation, with much input from the *Xenopus* research community, the NXR and the associated Horb laboratory have put significant effort into generating mutant lines of particular use for translational biology, or animals with mutations in genes associated with human disease. This is an ongoing process with additional F_0 founders being continuously generated. At the same time, some of the first lines generated at the NXR are now being bred as F_2 homozygous mutants for further phenotypic analysis by *Xenopus* researchers. It is very likely that in the near future mutant lines generated in the NXR and in the Horb laboratory will have a substantial impact on the current understanding of molecular mechanisms associated with many human diseases (Salanga and Horb 2015, Tandon et al. 2017).

The NXR has a broader role to play in promoting the use of frogs in research through the "conservation of ideas." This role is pursued in several different ways. First, the NXR is involved in enhancing uniformity and reproducibility of *Xenopus* research through participation in the RRID initiative, as well as by working closely with the other *Xenopus* stock centers and Xenbase, the *Xenopus* model organism database, to promote use of uniform stock nomenclature (Bandrowski et al. 2016, James-Zorn et al. 2015, Karimi et al. 2018, Pearl et al. 2012). Second, since its creation, the NXR has been hosting several workshops in which experts in the fields of genome editing, imaging, and bioinformatics in *Xenopus* teach cutting edge techniques to *Xenopus* researchers. These workshops are held once every 12–18 months and help assure uniformity in the research techniques used within the frog community. The NXR is also in the process of developing a husbandry workshop which will promote uniformity in how labs maintain their animal colonies. Third, visiting scientists can come to the MBL, embed within the NXR, and have access to all of the NXR's available animal lines and lab resources, as well as NXR's expertise in animal husbandry and mutagenesis. There is typically more than a single visiting scientist at the NXR at one time. This provides the basis for a collaborative environment within which researchers are free to exchange ideas and to develop projects that they would not typically do at their home institutions. Finally, the NXR also performs its own research (Ratzan et al. 2017, Wlizla et al. 2017). The focus of in-house projects is typically on developing resources, tools, and techniques that are broadly applicable for *Xenopus* biologists.

CONCLUSIONS

In its relatively brief existence the NXR has grown rapidly to serve as a centralized repository housing many diverse and some rare *Xenopus* lines available for distribution to *Xenopus* researchers. It has organized and held several workshops designed to teach advanced and innovative experimental techniques while also promoting standardization of protocols used throughout the *Xenopus* community. Finally, it provides a facility where researchers can visit and work with different animals before committing to bringing them into their facility, thus reducing the potential financial or logistical burden that they would have otherwise deal with. At the same time, these researchers can collaborate and exchange ideas with other visiting scientists. As a result of these and other activities, the NXR has had a considerable impact on promoting the use of *Xenopus* in biological research and will continue to strive toward ensuring that this research is scientifically rigorous and reproducible.

ACKNOWLEDGMENTS

We would like to thank the wider *Xenopus* community for their support and input in helping to make the NXR successful. Thanks to Jens H. Fritzenwanker for aid in translation of some of the references used here, from the original German. We also thank Mitch Butler for the KikGR transgenic image that was taken during the 2015 NXR Advanced Imaging Workshop. The NXR is supported by funds from the National Institutes of Health (OD010997) and the Marine Biological Laboratory.

REFERENCES

Aschheim, S. and B. Zondek. 1928. Die schwangerschaftsdiagnose aus dem harn durch nachweis des hypophysenvorderlappenhormons. *Klinische Wochenschrift* 7:1404–1411.

Bandrowski, A., M. Brush, J.S. Grethe et al. 2016. The resource identification initiative: A cultural shift in publishing. *Neuroinformatics* 14:169–182.

Bauer, D.V., S. Huang and S.A. Moody. 1994. The cleavage stage origin of Spemann's Organizer: Analysis of the movements of blastomere clones before and during gastrulation in *Xenopus*. *Development* 120:1179–1189.

Beddard, F.E. 1894. Notes upon the tadpole of *Xenopus laevis* (*Dactylethra capensis*). *Proceedings of the Zoological Society of London* 1894:101–107.

Bellerby, C.W. 1929. The physiological properties of anterior lobe pituitary extract in relation to the ovary. *The Journal of Physiology* 67:32–33.

Brown, D.D., P.C. Wensink and E. Jordan. 1971. Purification and some characteristics of 5S DNA from *Xenopus laevis*. *Proceedings of the National Academy of Sciences* (USA) 68:3175–3179.

Corkins, M., H. Hanania, V. Krneta-Stankic et al. 2018. Transgenic *Xenopus laevis* line for in vivo labeling of nephrons within the kidney. *Genes* 9:197. doi:10.3390/genes9040197.

Cross, M.K. and M.A. Powers. 2009. Learning about cancer from frogs: Analysis of mitotic spindles in *Xenopus* egg extracts. *Disease Model Mechanisms* 2:541–547.

Cuvier, G. 1829. *Le règne animal, distribué d'après son organisation, pour servir de base a l'histoire naturelle des animaux et d"introduction a l"anatomie comparée*. Paris, France: Chez Déterville.

Dascal, N. 1987. The use of *Xenopus* oocytes for the study of ion channels. *CRC Critical Reviews in Biochemistry* 22:317–387.

Daudin, F.M. 1802. *Histoire naturelle des rainettes, des grenouilles et des crapauds*. Paris, France: de Bertrandet.

Davis, J., M. Maillet, J.M. Miano and J.D. Molkentin. 2012. Lost in transgenesis: A user's guide for genetically manipulating the mouse in cardiac research. *Circulation Research* 111:761–777.

Dawid, I.B. 1966. Evidence for the mitochondrial origin of frog egg cytoplasmic DNA. *Proceedings of the National Academy of Sciences* (USA) 56:269–276.

Engelke, D.R., S.Y. Ng, B.S. Shastry and R.G. Roeder. 1980. Specific interaction of a purified transcription factor with an internal control region of 5S RNA genes. *Cell* 19:717–728.

Evans, B.J., T.F. Carter, E. Greenbaum et al. 2015. Genetics, morphology, advertisement calls, and historical records distinguish six new polyploid species of African clawed frog (*Xenopus*, Pipidae) from west and central Africa. *PLoS ONE* 10:e0142823.

Evans, T., E.T. Rosenthal, J. Youngblom, D. Distel and T. Hunt. 1983. Cyclin: A protein specified by maternal mRNA in sea urchin eggs that is destroyed at each cleavage division. *Cell* 33:389–396.

Fee, A.R. and A.S. Parkes. 1929. Studies on ovulation: I. The relation of the anterior pituitary body to ovulation in the rabbit. *The Journal of Physiology* 67:383–388.

Friedman, M.H. and M.E. Lapham. 1931. A simple, rapid procedure for the laboratory diagnosis of early pregnacies. *American Journal of Obstetrics and Gynecology* 21:405–410.

Gilbert, S.F. and M.J.F. Barresi. 2016. *Developmental Biology*. 11th ed. Sunderland, MA: Sinauer Associates.

Gray, J.E. 1864. Note on the clawed toads (Dactylethra) of Africa. *Proceedings of the Zoological Society of London* 458–464.

Günther, A. 1858. *Catalogue of the Batrachia Salientia in the Collection of the British Museum*. London, UK: Taylor & Francis Group.

Gurdon, J.B. 1962. The developmental capacity of nuclei taken from intestinal epithelium cells of feeding tadpoles. *Journal of Embryology and Experimental Morphology* 10:622–640.

Gurdon, J.B. 1970. Nuclear transplantation and the control of gene activity in animal development. *Proceedings of the Royal Society of London, B, Biological Sciences* 176:303–314.

Gurdon, J.B. and N. Hopwood. 2000. The introduction of *Xenopus laevis* into developmental biology: Of empire, pregnancy testing and ribosomal genes. *International Journal of Developmental Biology* 44:43–50.

Gurdon, J.B., C.D. Lane, H.R. Woodland and G. Marbaix. 1971. Use of frog eggs and oocytes for the study of messenger RNA and its translation in living cells. *Nature* 233:177–182.

Guselnikov, S.V., T. Ramanayake, A.Y. Erilova et al. 2008. The *Xenopus* FcR family demonstrates continually high diversification of paired receptors in vertebrate evolution. *BMC Evolutionary Biology* 8:148. doi:10.1186/1471-2148-8-148.

Heasman, J., M. Kofron and C. Wylie. 2000. Beta-catenin signaling activity dissected in the early *Xenopus* embryo: A novel antisense approach. *Developmental Biology* 222:124–134.

Hellsten, U., R.M. Harland, M.J. Gilchrist et al. 2010. The genome of the Western clawed frog *Xenopus tropicalis*. *Science* 328:633–636.

Hogben, L.T. 1930. Some remarks on the relation of the pituitary gland to ovulation and skin secretion in *Xenopus laevis*. *Transactions of the Royal Society of South Africa* 22(2):17–18.

Hogben, L.T., E. Charles and D. Slome. 1931. Studies on the pituitary. *Journal of Experimental Biology* 8:345–354.

Hoogenboom, W.S., D. Klein Douwel and P. Knipscheer. 2017. *Xenopus* egg extract: A powerful tool to study genome maintenance mechanisms. *Developmental Biology* 428:300–309.

Ishibashi, S., K.L. Kroll and E. Amaya. 2012a. Generating transgenic frog embryos by restriction enzyme mediated integration (REMI). *Methods in Molecular Biology* 917:185–203.

Ishibashi, S., N.R. Love and E. Amaya. 2012b. A simple method of transgenesis using I-SceI meganuclease in *Xenopus*. *Methods in Molecular Biology* 917:205–218.

James-Zorn, C., V.G. Ponferrada, K.A. Burns et al. 2015. Xenbase: Core features, data acquisition, and data processing. *Genesis* 53:486–497.

Karimi, K., J.D. Fortriede, V.S. Lotay et al. 2018. Xenbase: A genomic, epigenomic and transcriptomic model organism database. *Nucleic Acids Research* 46:D861–D868.

Kelley, C.M., D.A. Yergeau, H. Zhu, E. Kuliyev and P.E. Mead. 2012. *Xenopus* transgenics: Methods using transposons. *Methods in Molecular Biology* 917:231–243.

Khokha, M.K. 2012. *Xenopus* white papers and resources: Folding functional genomics and genetics into the frog. *Genesis* 50:133–142.

Khokha, M.K., J.B. Wallingford, E. Amaya et al. 2009. *Xenopus* Community White Paper. http://www.xenbase.org/community/static/xenopuswhitepaper/XWPFINAL.pdf.

Klug, A. and D. Rhodes. 1987. Zinc fingers: A novel protein fold for nucleic acid recognition. *Cold Spring Harbor Symposium on Quantitative Biology* 52:473–482.

Leslie, J.M. 1890. Notes on the habits and oviposition of *Xenopus laevis*. *Proceedings of the Zoological Society of London* 69–71.

McCoid, M.J. and T.H. Fritts. 1989. Growth and fatbody cycles in feral populations of the African Clawed Frog, *Xenopus laevis* (Pipidae), in California with comments on reproduction. *Southwestern Naturalist* 34:499–505.

McKnight, S.L. 2012. Pure genes, pure genius. *Cell* 150:1100–1102.

McNamara, S., M. Wlizla and M.E. Horb. 2018. Husbandry, general care, and transportation of *Xenopus laevis* and *Xenopus tropicalis*. *Methods in Molecular Biology* 1865:1–17.

Moody, S.A. 1987a. Fates of the blastomeres of the 16-cell stage *Xenopus* embryo. *Developmental Biology* 119:560–578.

Moody, S.A. 1987b. Fates of the blastomeres of the 32-cell-stage *Xenopus* embryo. *Developmental Biology* 122:300–319.

Pearl, E., S. Morrow, A. Noble, A. Lerebours, M. Horb and M. Guille. 2017. An optimized method for cryogenic storage of *Xenopus* sperm to maximise the effectiveness of research using genetically altered frogs. *Theriogenology* 92:149–155.

Pearl, E.J., R.M. Grainger, M. Guille and M.E. Horb. 2012. Development of Xenopus resource centers: The National *Xenopus* Resource and the European *Xenopus* Resource Center. *Genesis* 50:155–163.

Peng, H.B., L.P. Baker and Q. Chen. 1991. Tissue culture of *Xenopus* neurons and muscle cells as a model for studying synaptic induction. *Methods in Cell Biology* 36:511–526.

Preston, G.M., T.P. Carroll, W.B. Guggino and P. Agre. 1992. Appearance of water channels in *Xenopus* oocytes expressing red cell CHIP28 protein. *Science* 256:385–387.

Ratzan, W., R. Falco, C. Salanga, M. Salanga and M.E. Horb. 2017. Generation of a *Xenopus laevis* F_1 albino J strain by genome editing and oocyte host-transfer. *Developmental Biology* 426:188–193.

Salanga, M.C. and M.E. Horb. 2015. *Xenopus* as a model for GI/pancreas disease. *Current Pathobiology Reports* 3:137–145.

Sater, A.K. and S.A. Moody. 2017. Using *Xenopus* to understand human disease and developmental disorders. *Genesis* 55:e22997.

Session, A.M., Y. Uno, T. Kwon et al. 2016. Genome evolution in the allotetraploid frog *Xenopus laevis*. *Nature* 538:336–343.

Slack, J.M.W. 2012. *Essential Developmental Biology*. 3rd ed. Hoboken, NJ: Wiley-Blackwell.

Smith, P.E. and E.T. Engle. 1927. Experimental evidence regarding the rôle of the anterior pituitary in the development and regulation of the genital system. *American Journal of Anatomy* 40:159–217.

Tandon, P., F. Conlon, J.D. Furlow and M.E. Horb. 2017. Expanding the genetic toolkit in *Xenopus*: Approaches and opportunities for human disease modeling. *Developmental Biology* 426:325–335.

Wagler, J.G. 1827. Footnote to a letter of Heinrich Boie to Wagler. *Isis von Oken* 20:726.

Wilmut, I., A.E. Schnieke, J. McWhir, A.J. Kind and K.H. Campbell. 1997. Viable offspring derived from fetal and adult mammalian cells. *Nature* 385:810–813.

Wlizla, M., R. Falco, L. Peshkin, A.F. Parlow and M.E. Horb. 2017. Luteinizing hormone is an effective replacement for hCG to induce ovulation in *Xenopus*. *Developmental Biology* 426:442–448.

Wlizla, M., S. McNamara and M.E. Horb. 2018. Generation and care of *Xenopus laevis* and *Xenopus tropicalis* embryos. *Methods in Molecular Biology* 1865:19–32.

Wu, M. and J. Gerhart. 1991. Raising *Xenopus* in the laboratory. *Methods in Cell Biology* 36:3–18.

Zarrin, A.A., F.W. Alt, J. Chaudhuri et al. 2004. An evolutionarily conserved target motif for immunoglobulin class-switch recombination. *Nature Immunology* 5:1275–1281.

Zondek, B. and S. Aschheim. 1927. Das hormon des hypophysensvoederlappens. *Klinische Wochenschrift* 6:348–352.

13 *Xiphophorus* Fishes and the *Xiphophorus* Genetic Stock Center

Ronald B. Walter, Yuan Lu and Markita Savage

CONTENTS

Abstract: *Xiphophorus* are small, live-bearing, new world, freshwater fish found in freshwater habitats from northern Mexico into Guatemala and Honduras. The value of *Xiphophorus* in research stems from the extreme phenotypic variability found among the 26 known species and the ability to derive fertile progeny from interspecies crosses. *Xiphophorus* was the first model to show cancer was a heritable disease and is a valuable experimental model for studies of melanoma. The *Xiphophorus* Genetic Stock Center (XGSC) was established in 1939 and continues today maintaining and distributing *Xiphophorus* species and strains for distribution to the scientific community. The XGSC is currently housed at Texas State University within the Department of Chemistry and Biochemistry. The XGSC maintains 24 of the 26 known species of *Xiphophorus* and 54 pedigreed genetic lines, and provides other resources to assist scientists using these animals. The XGSC also hosts all versions of *Xiphophorus* genome assemblies.

THE *XIPHOPHORUS* FISHES—INTRODUCTION

Xiphophorus are a group of New World, viviparous (i.e., live-bearing), small freshwater fish, within the order Poeciliidae, class Cyprindontiformes. *Xiphophorus* fishes are commonly placed into two groups, the platyfish and swordtails, although these designations are somewhat subjective and used more for > convenience than taxonomic accuracy (Kallman and Kazianis 2006, Kazianis and Walter 2002, Walter 2011). In the wild, *Xiphophorus* fishes are found distributed from northern Mexico, extending south along the Atlantic versant of the Sierra Madre Oriental uplift through Mexico and into Guatemala, Belize, and Honduras (Kallman and Kazianis 2006). The geographic habitats in localized regions within this large range are highly variable. From relatively stagnant waters at sea level, to spring fed streams in the coastal plains, and fast flowing rivers in the mountainous regions of the rain forest, these extremely varied environments along the southerly path into Central America have given rise to many small and isolated *Xiphophorus* populations (e.g., *X. meyeri*, *X. couchianus*, and *X. gordoni*). However, there are also *Xiphophorus* species that are documented to have extensive and sympatric ranges nearly traversing Mexico (e.g., *X. variatus*, *X. hellerii*, and *X. maculatus*).

The genetic variability among the 26 *Xiphophorus* species reflects and an estimated ~6–8 million years of evolutionary divergence. The extreme morphological and genetic variability among *Xiphophorus* species is one attribute of these fishes that make them attractive as a model organism in biomedical research studies (Figure 13.1). Due to the geographic distribution in their native habitat, most *Xiphophorus* species are reproductively isolated; however, there is wide ranging sympatry for species such as *X. maculatus* and *X. hellerii*. In native sympatric regions, interspecies hybridization occurs only very rarely (Rosen 1979, Rosenthal 2003). This is likely due to the live-bearing nature of *Xiphophorus* reproduction that involves intricate species-specific courting

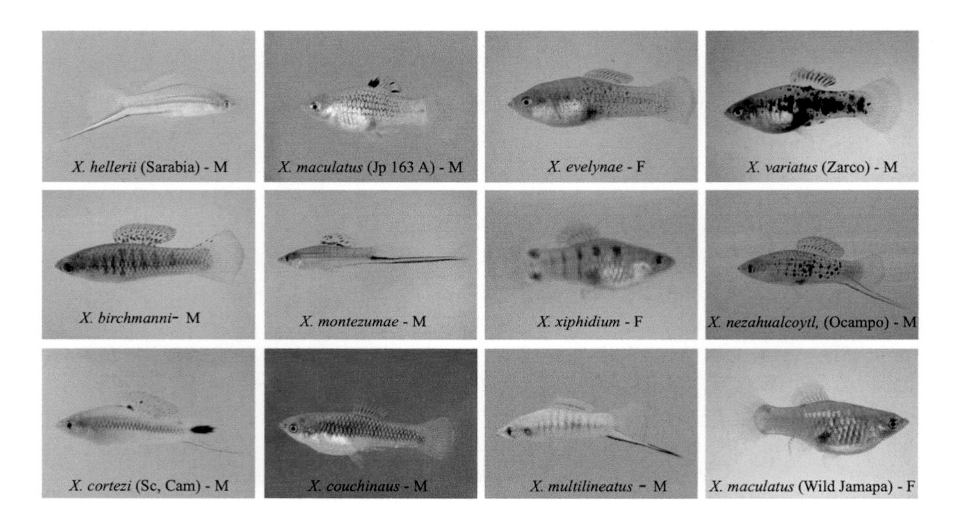

FIGURE 13.1 **(See color insert.)** Examples of the varied morphology among *Xiphophorus* species.

behaviors, and the development of the male reproductive structure (i.e., gonopodium) that favors sperm deposition in females of like species. These mechanisms keep sympatric *Xiphophorus* species reproductively isolated.

Xiphophorus varieties are commonly found in pet stores and are a perpetual favorite of aquaria hobbyists. Thus, many ornamental lines have been independently established by crossing different wild lines, and these have been disseminated through the aquaria trade. Such lines, although popular with aquarists, are so highly derived they are not appropriate for use as an animal model for controlled experimental studies. However, due to the early establishment of a genetic stock center for the production of pedigreed animals exclusively for research investigations, *Xiphophorus* fishes have enjoyed a long and storied history of use in scientific research and have led to significant contributions to current scientific thought across many diverse scientific disciplines. *Xiphophorus* fishes remain an important vertebrate model for scientific discovery as highlighted by more than 5000 peer-reviewed scientific works published using this model system (Figure 13.2). In the field of cancer genetics, *Xiphophorus* was the first model to show cancer was a heritable disease. Further, early *Xiphophorus* backcross hybrid studies are the first examples where tumorigenesis was shown to result from segregation of dominant genes (now termed oncogenes) from loci that kept these genes in check (now termed tumor suppressors). *Xiphophorus* remain a valuable experimental model for one of the most intractable cancers, melanoma.

In contemporary research, *Xiphophorus* fishes are actively used in many fields of study including evolution (Chalopin et al. 2015a, Cui et al. 2013, Culumber et al. 2015, Kang et al. 2013, Kawaguchi et al. 2015, Meyer et al. 2006, Schumer et al. 2018, Volff and Schartl 2002, Wang et al. 2015), behavior (Cummings and Ramsey 2015,

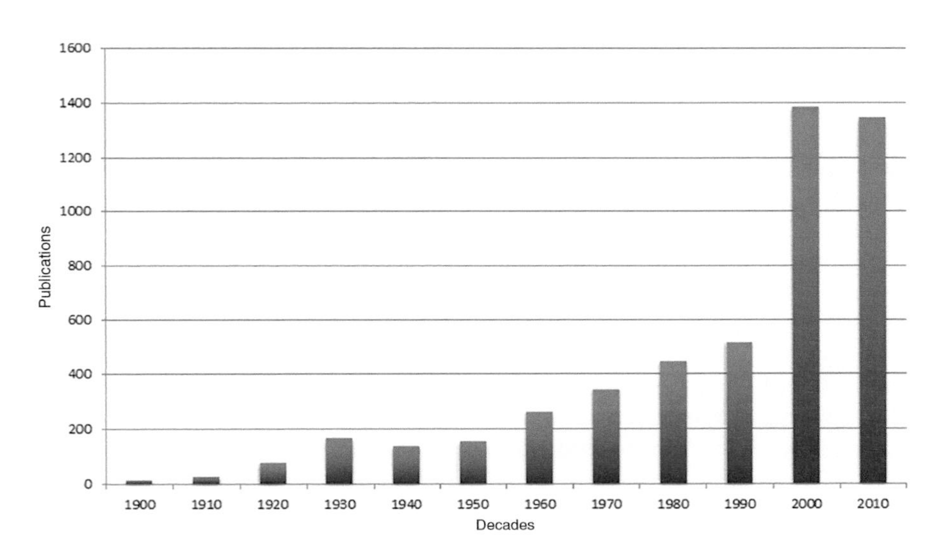

FIGURE 13.2 Number of *Xiphophorus*-related publications per decade. The number of *Xiphophorus* publications has steadily increased each decade since 1900. Data for 2010 is a tabulation of only 8.5 years of that decade.

D'Amorea et al. 2015), basic cell biology/physiology (Chang et al. 2015, Culumber and Tobler 2018, Lu et al. 2018a, Uribe et al. 2011, Walter et al. 2015, 2018, Yang et al. 2012a, Zhang et al. 2018), comparative biochemistry and genomics (Amores et al. 2014, Bastide et al. 2018, Boswell et al. 2018a, Schartl et al. 2013, Shen et al. 2016, Walter et al. 2014), sex determination (Alireza et al. 2016, Boswell et al. 2018b, Boulton et al. 2016, Chalopin et al. 2015b, Jie and Jian-Fang 2015, Kang et al. 2015, Khiabani et al. 2014, Luo et al. 2005), development (D'Amorea et al. 2015, Singh and Kumas 2016), endocrinology (Boulton et al. 2015), environmental biology (Culumber 2016, Culumber and Tobler 2016), regeneration (Costa et al. 2017), toxicology (Caixeta et al. 2016, Pereira et al. 2016, Silva et al. 2018, Zebral et al. 2018), cancer biology (Lu et al. 2018b, Regneri et al. 2015), parasitology (Hoshino et al. 2018, Trujillo-González et al. 2018), microbiology (Seyfahmadi et al. 2017), pathology (Bunnajirakul et al. 2015, Mirzaie 2015), and immunology (Hoseinifar et al. 2015, Liang et al. 2015, Mohr et al. 2015).

ATTRIBUTES OF *XIPHOPHORUS* FISHES—EMERGENCE OF A RESEARCH MODEL

Vertebrate animal models that enable research scientists to better understand complex (i.e. polygenic) traits and detailed molecular genetic mechanisms are of considerable importance. To be informative and useful, a research animal model should possess inherent attributes that make it amenable to controlled experimentation including ease of generation of large numbers of animals, the availability of both genetically characterized (i.e., pedigreed) inbred lines, conserved synteny in chromosomal structure allowing comparison with other models, robust genomic and molecular genetic resources, the ability to perform genetic crosses for mechanistic investigations, and enough genetic variability (i.e., polymorphisms) among different strains or lines such that genetic crosses will segregate identifiable trait inheritance patterns. The principle asset of using *Xiphophorus* in experimental science is the capability of producing fertile interspecies hybrid progeny. This provides a very powerful tool for investigation into the heritable bases of complex disease(s), and the genetic response to experimental or environmental stressors that are hallmarked by polygenic inheritance.

In a laboratory, most *Xiphophorus* species may be mated to produce fertile interspecies hybrid offspring by forced pairing in aquaria or through artificial insemination. Upon interspecies mating, the resulting interspecies hybrid progeny (i.e., F_1 hybrids) are nearly always fertile. Yet, at the molecular genetic level, they are extremely polymorphic (e.g., 1 bp change per 60 bp between *X. maculatus* and *X. hellerii*, and 1 bp per 80 bp between *X. maculatus* and *X. couchianus*) (Shen et al. 2016) and are capable of being successively backcrossed with one of the parental lines. Successive backcrossing of an interspecies hybrid to one parental line results in the loss of half the non-recurrent parental genome contribution in each backcross generation. Thus, one may assess the association between a complex phenotype (e.g., disease susceptibility) and inheritance of chromosomal markers from the non-recurrent parent genome, among backcross hybrid progeny. Such associations can then be utilized to determine genomic regions underlying the complex phenotype. An example of this type of genetic analysis, and the seminal observation driving development of *Xiphophorus* as a research model, is the inheritance of melanoma development.

XIPHOPHORUS IN CANCER RESEARCH AND TUMOR MODELS

In the 1920s, the American biologist Myron Gordon and German scientists Haussler and Kosswig independently discovered that interspecies hybrids between the platyfish *Xiphophorus maculatus* (Jp163 A) and the swordtail *Xiphophorus hellerii* developed cancers virtually identical to malignant melanomas in humans (Gordon 1931, Haussler 1928, Kosswig 1927; for review see Schartl and Walter 2016). Development of the *Xiphophorus* genetic system and its employment in varied scientific fields can be traced to these pioneering studies showing that select interspecies hybridizations produced backcross hybrid progeny that expressed spontaneous development of melanoma. Results from these early *Xiphophorus* tumor crosses, dating back to 1927 and following into the 1960s, established the presence of what are now termed "oncogenes" and were the initial indications that loss of gene function could be associated with tumorigenesis, thus indicating entities now known as tumor suppressor genes (Anders 1967, 1991). Parental species exhibit quantifiable differences in phenotypic expression. The Gordon-Kosswig (G-K) melanoma model serves as a paradigm for the role of dominant oncogenes and recessive tumor suppressor genes in tumor development (Schartl and Walter 2016, Walter and Kazianis 2001).

THE G-K MELANOMA MODEL

Since the 1930s, considerable research on tumorigenesis utilizing *Xiphophorus* has centered on the G-K melanoma model (Anders 1991, Klotz et al. 2018, Kneitz et al. 2016, Lu et al. 2017, Lu et al. 2018, Perez et al. 2012, Schartl 1990, 1995, Vielkind et al. 1989, Wittbrodt et al. 1989). This model involves the spontaneous development of malignant melanomas in backcross hybrids produced by the mating scheme *X. hellerii* × (*X. maculatus* Jp 163 A × *X. hellerii*). In this case, the *X. maculatus* X chromosome carries a macromelanophore pigment pattern marker designated *Sd* (spotted dorsal). Melanocytes derived from *Sd* become hyperplastic within a *X. hellerii* genetic background. In the backcross, half of the offspring inherit the *Sd* pigment pattern gene, and of those, half (25% of the progeny) exhibit hyperplastic pigment cell expression and half (25% of the progeny) develop malignant melanoma characterized by invasive nodular melanotic lesions starting at the dorsal fin and invading the musculature, eventually killing the animal (Meierjohann and Schartl 2006, Meierjohann et al. 2006,Vielkind et al. 1989, Zunker et al. 2006).

Comparative biochemical, histological, cytological studies and detailed genetic studies employing transcriptomic analyses of melanin-containing cells from melanoma bearing G-K backcross hybrid fish show a marked decline in the differentiated state of pigment cells in tumors and high similarity with human melanoma (Kneitz et al. 2016, Lu et al. 2018, Schartl and Peter 1988, Sobel et al. 1975). Within lightly pigmented BC_1 hybrids, cells differentiate only to an intermediate stage and are able to undergo limited cell division, whereas cells in malignant melanomas derived from heavily pigmented fish are obviously neoplastic and retain rapid, unlimited cell division (Adam et al. 1991, Vielkind 1976, Vielkind et al. 1989) and have been pathologically staged (Gimenez-Conti et al. 2001). This point is underscored by the observation that fish melanoma cells are able to proliferate in a manner virtually identical to human melanoma cells when transplanted into athymic mice (Manning et al. 1973, Schartl and Peter 1988).

The inheritance of Mendelian segregation of phenotypes in the G-K melanoma model predicates a two-hit genetic model: (a) inheritance of an *X. maculatus* derived oncogene tightly linked to the *Sd* pigment pattern gene, and (b) concomitant loss of *X. maculatus* tumor suppressor gene in backcross hybrid (BC_1) progeny that develop melanoma. Studies have demonstrated the *X. maculatus* X chromosome harbors both *Sd* pigment patterns tightly linked with two copies of Epidermal Growth Factor Receptor *(EGFR) (HER-1)*-related genes (Wittbrodt et al. 1989). These genes are termed *xmrk-1* and *-2* (Woolcock et al. 1994). Expression patterns have shown that one copy (*xmrk-2*) is overexpressed in hybrid melanoma tissues relative to the other copy (*xmrk-1*), which is expressed in most normal tissues (Adam et al. 1991, Wittbrodt et al. 1989, Woolcock et al. 1994). The melanoma-associated *xmrk-2* has been shown to harbor 2 critical mutations in the EGFR extracellular domain giving it ligand-independent, and constitutively active, growth promoting state (Gomez et al. 2001, Meierjohann et al. 2006). Proof of the oncogene came from studies showing that when the Xiphophorus *xmrk2* is expressed in pigment cells of transgenic medaka (*Oryzias latipes*), the medaka develop melanoma very early after hatching with 100% penetrance (Schartl et al. 2010, 2015). These and other results have unambiguously proven that *xmrk-2* is the primary oncogene in the Gordon-Kosswig melanoma model.

This *X. maculatus* tumor suppressor locus, termed R/*Diff*, is thought to keep the *xmrk-2/EGFR* in check, since parental *X. maculatus* fish only rarely develop melanoma, and even then only at an advanced age (2 + years). The R/*Diff* gene was localized to an area within *Xiphophorus* chromosome 5 by classical genetic mapping (Fornzler et al. 1991, Morizot and Siciliano 1983). A fish homologue in the cyclin-dependent kinase inhibitor gene family (*CDKN2,* aka- *p16, INK4*) was also mapped to the R/*Diff* position on chromosome 5. The inheritance of *cdkn2x* in melanoma-bearing hybrids established it as a primary determinant of melanin pigmentation phenotypes and the development of melanoma in a number of different *Xiphophorus* hybrid tumor models (Kazianis et al. 1998, 1999, 2000, Nairn et al. 1996a). Highlighting the relevance of this observation, studies within humans and rodents have also implicated a CDKN2 gene family member (*CDKN2A, p16*) as one of the most commonly mutant loci in familial melanomas (Ruas and Peters 1998). Thus, association of a related *Xiphophorus* gene with melanoma formation in this fish model was a provocative early discovery lending strength to the translational nature of *Xiphophorus* to model human disease.

These previous genetic linkage analyses relied on only a handful of genetic markers on chromosome 5. Although *cdkn2x* fits with the "two-hit" model and showed translational value in melanoma etiology research, continuing studies following these previous reports showed *cdkn2x* heterozygosity (for *X. maculatus* and *X. hellerii* alleles) in ~20% of tumor bearing BC_1, as well as *X. maculatus* allele biased expression in *cdkn2x* heterozygous tumor bearing individuals (Butler et al. 2007). Therefore, it is suggested that either inheritance of *cdkn2x* alone does not fully explain melanomagenesis in the G-K backcross model or, on the other hand, that *cdkn2x* is located close to, but may not be, the *R/Diff* gene.

Akin to a genome wide association study (GWAS) method, transcriptomes of *X. hellerii* × (*X. maculatus* × *X. hellerii*) backcross progeny that exhibited melanoma were sequenced using RNA-Seq. Expressed genes were genotyped by comparing the sequencing reads to the newly acquired genome sequence assemblies of the *X. maculatus* and *X. helleri* (Amores et al. 2014, Lu et al. 2017, Schartl et al. 2013, Shen et al. 2016). The *xmrk* locus on chromosome 21 (i.e., X chromosome) was found to be heterozygous for all melanoma-bearing animals, and a 5.8 Mbp region on chromosome 5 was found to be homozygous for *X. hellerii* alleles in all melanoma-bearing backcross progeny. This region contains 164 gene models and is forwarded as the *R/Diff* locus (Figure 13.3). The *cdkn2x* gene resides within this 5.81 Mbp region and is 0.42 Mbp to the end of the *R(Diff)* locus defined in this study. This observation supports the hypothesis that *cdkn2x* locates close to the core *R(Diff)* candidate gene(s) on a physical map. Current research interests focus on fine mapping this

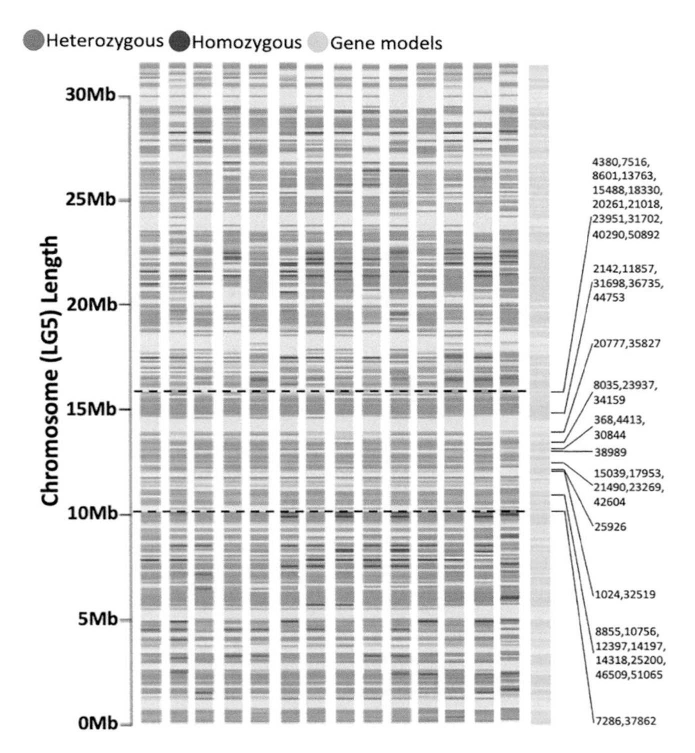

FIGURE 13.3 (See color insert.) The *R/Diff* locus on chromosome 5. The loci on chromosome 5 that are homozygous for *X. hellerii* alleles in BC_1 hybrid progeny bearing melanoma tumors defines the *R(Diff)* locus. Each line represents a gene, and shading of the lines represents the genotype (i.e., homozygous for *X. hellerii* allele, or heterozygous for *X. maculatus* and *X. hellerii* alleles). The region between the dashed lines is 5.8 Mbp long and contains 164 gene models, and also 40 RAD-tag markers established in a previous study. This region is forwarded as the *R/Diff* locus.

TABLE 13.1
Examples of Various *Xiphophorus* Tumor Models. The Usual Genetic Convention of Listing the Female Parent First Is Followed

Species Utilized in Cross	Tumor Type	Induction
X. hellerii (*X. maculatus* Jp 163 A *X. helleri*)	Melanoma	None (Spontaneous)
X. hellerii (*X. maculatus* Jp 163 B *X. hellerii*)	Melanoma	MNU & UV[a]
(*X. maculatus* Jp 163 A *X. andersi*) *X. andersi*	Melanoma	None (Spontaneous)
(*X. maculatus* Jp 163 B *X. andersi*) *X. andersi*	Several[b]	MNU
(*X. maculatus* Jp 163 A *X. couchianus*) *X. Couchianus*	Several[c]	MNU
(*X. maculatus* Jp 163 B *X. couchianus*) *X. Couchianus*	Melanoma	MNU & UV
X. hellerii (*X. variatus X. hellerii*)	Several[d]	MNU
X. corteziSc	Melanoma	Aging
X. variatus Pu2	Melanoma	Aging

[a] MNU—*N*-Methyl-*N*-Nitrosourea. UV—Ultra violet.
[b] Melanoma and renal adenocarcinoma.
[c] Fibrosarcoma, retinoblastoma, melanotic hyperpigmentation, and schwannoma.
[d] Fibrosarcoma, neuroblastoma.

5.8 Mbp *R/Diff* candidate locus using more backcross progeny exhibiting different macromelanophore phenotypes, and shortening the candidate region to 300 kb, or single gene definition.

Since the description of the G-K melanoma model, the genetic control of tumor susceptibility in *Xiphophorus* has been shown to exist in pure strains and particular interspecific hybrids for a variety of spontaneous and induced neoplasms, providing at least 8 melanoma models and a long list of less well-studied tumors such as neuroblastomas, neurofibromas, fibrosarcomas, and rhabdomyosarcomas (Kazianis et al. 2001a, 2001b; Nairn et al. 2001, Patton et al. 2010, Schwab et al. 1979). A partial list of *Xiphophorus* hybrid crosses leading to tumor susceptibility is presented in Table 13.1. The development of genetic models that require a inducing treatment (i.e., ultraviolet light exposure, Nairn et al. 1996b, Setlow et al. 1989) to express tumor development in a fraction of the BC_1 progeny, underscores the genetic complexity of carcinogenesis, and hallmarks a major strength of using the *Xiphophorus* fish genetic system in allowing identification of genetic mechanisms responsible for setting the stage, or genetically predisposing, individuals to tumorigenesis. This animal model was one of the first to prove that certain cancers were inherited diseases; after >85 years, these fish models remain useful in understating the complexities of tumor etiology.

Xiphophorus and the Discovery of Light Wavelength Induced Gene Expression

Evolution occurred for ~3 billion years exclusively under the full spectrum of sunlight. Thus, all solar wavelengths were represented, and each organism had the opportunity to adaptively pair genetic responses (i.e., gene expression patterns) with specific regions of the solar spectrum. In contrast to the solar spectrum, fluorescent light (FL) has only been in service for ≈60 years and these light sources emit a much narrower range than the visible spectrum, hallmarked by peaks and valleys in the wavelength intensities emitted. Research at the XGSC has employed *Xiphophorus* to characterize induced changes in gene expression patterns in the skin and other organs after exposure of the intact animal to various types of common FL (311 nm UVB, 4100 K or "cool white" FL, and 10,000K or "sunlight" FL) (Boswell et al. 2015, Chang et al. 2015, Lu et al. 2015, Walter et al. 2014, 2015, 2018, Yang et al. 2014) and to specific light wavebands (e.g., 50 nm wavebands between 300 and 600 nm) (Chang et al. 2015). Exposure of *X. maculatus* Jp163 A, zebrafish (*Danio rerio*), and medaka (*Oryzias latipes*) to common FL (4100 K) has been shown to promote modulated gene expression changes in the skin consistent with activation of the acute phase response signaling pathway concurrent with increased expression of cytokines promoting inflammation and innate immune responses (Boswell et al. 2018a).

Due to unexpected exposure results suggesting FL incites inflammation and immune response in the skin and other organs of three commonly utilized aquatic biomedical models, the XGSC has utilized a tunable xenon light source to examine gene expression profile effects in *Xiphophorus* organs after exposure to specific 50 nm wavebands of light from 300 to 600 nm (e.g., 300–350 nm, 350–400 nm, etc. to 600 nm). Since no fish would have ever seen a single light waveband exposure, until these experiments were performed, these studies tap into "hard-wired" light driven gene expression responses. For male *Xiphophorus* skin, 50 nm waveband effects on gene expression unexpectedly showed that two 50 nm light regions (350–400 nm and 500–550 nm) produced significantly higher differential gene expression than the other wavebands (Chang et al. 2015, Walter et al. 2018). Exposure at 500–550 nm appears to induce cellular stress as a major response. While exposure to either 350–400 or 500–550 nm wavebands leads to modulation of circadian gene activity and altered the activity of p53 gene targets, but via very different mechanisms (*atm* at 350–400 and *atr* at 500–550 nm; Chang et al. 2015).

In depth analyses of the gene expression patterns produced by the set of 50 nm exposures led to a surprising observation that specific 50 nm wavebands, only 100 nm apart, selectively modulated transcription of pathways involved in cell death. For example, *X. maculatus* males exposed to 350–400 nm light showed down modulated expression of 20 genes that collectively are expected to produce a decrease in necrosis. However, exposure of the same fish (i.e., highly inbred siblings) to the same dose of 450–500 nm waveband light led to substantial up-regulation of 19 necrosis function genes and an overall expected increase in necrosis. Similarly, 450–500 nm exposure up-regulated 32 genes leading to an overall decrease in apoptosis, but exposure to 550–600 nm light resulted in down-regulation of 49 genes that collectively are expected to produce an increase in apoptosis.

These novel findings are not specific for necrosis and apoptosis but have been extended to many other cellular functions where exposure to one waveband will activate a genetic pathway, while exposure to a different waveband serves to suppress the same pathway (Walter et al. 2018). These results support the concept that one may move the genetic state of skin toward a pre-selected gene expression state once all waveband specific responses have been characterized. Recent results indicate such waveband specific genetic responses do not just occur in skin, but also have been observed for internal organs (brain and liver). These experiments, once again, present the value of the *Xiphophorus* system as a genetic exploration tool to uncover new discoveries that translate to other vertebrates, perhaps including humans.

HISTORY OF THE *XIPHOPHORUS* GENETIC STOCK CENTER

Genetic stock centers are priceless resources in scientific research. Research experiments often require that special strains of genetically identical animals or plants be used to assure that results can be repeated in any laboratory, and that differences are not due to uncontrolled environmental factors or variations in natural populations. One of the oldest and best-defined groups of established genetic strains consists of the live-bearing *Xiphophorus* fishes currently housed at Texas State University.

Dr. Myron Gordon, as a student at Cornell University in the early 1930s, set out to identify the genetic factors responsible for melanoma development. Dr. Gordon realized that to precisely identify the genes responsible for development of cancer, genetically identical platyfish and swordtails would be needed. Therefore, in 1939, he formally established the *Xiphophorus* Genetic Stock Center (XGSC), first at Cornell University, and later at the New York Aquarium, which was located at the Battery in Castle Clinton, Manhattan, NY (reviewed in Kallman 2001). Over the ensuing years, Gordon made many trips to Mexico. In those days he traveled in a Model T ford, loaded with camping equipment and shotguns, carrying letters from various U.S. federal agencies as he explored Mexico looking for new *Xiphophorus* populations. When *Xiphophorus* were found, Gordon used metal milk canisters to ship fish back to New York via the railroad. Upon Gordon's death in 1959, and for the ensuing 30+ years, the center was overseen and greatly expanded by Dr. Klaus Kallman, who had trained under Gordon's supervision as a doctoral student. The rapidly growing XGSC collection moved again to the top floor of the American Museum of Natural History (Osborn Marine Laboratories) in New York. Kallman continued exploratory trips to Mexico and Central America and expanded the XGSC to represent 21 species and established over 50 pedigreed fish lines (Figure 13.4). In 1993, due to Kallman's retirement and the numerous researchers in Texas using these valuable animals, the XGSC was transferred to Texas State University (TSU) in San Marcos.

THE *XIPHOPHORUS* GENETIC STOCK CENTER TODAY

The XGSC is currently housed at Texas State University on the 4th floor (top) of Centennial Hall within the Department of Chemistry and Biochemistry, wherein the current XGSC Director holds academic appointment. The XGSC occupies about 130 m^2 (≈1400 sq ft) of aquaria space, with an additional 175 m^2 (1900 sq ft) of research space. The research space is often temporarily conscripted for added aquaria space as the XGSC is a dynamic "living collection." The XGSC maintains ≈1400 × 20-L (5 gal.), and ≈150, 40 to 200-L (10–50 gal), aquaria for stock perpetuation (Figure 13.5). The water needed for these aquaria is estimated at 32,176 liters (≈ 8500 gallons) with a dead weight of ≈32,130 Kg (≈70,890 lbs). The XGSC maintains 24 of the 26 known species of *Xiphophorus* and 54 pedigreed genetic

FIGURE 13.4 Past directors of the *Xiphophorus* Genetic Stock Center (XGSC) who established and greatly expanded the collection. (a) Prof. Myron Gordon established the XGSC in 1930. (b) The XGSC at the Castle Battery, NY. (c) Prof. Klaus Kallman in the XGSC in the New York Aquarium. (d) Prof. Kallman searching for new species and defining species' ranges in Southern Mexico.

FIGURE 13.5 Views of the current *Xiphophorus* Genetic Stock Center XGSC at Texas State University, San Marcos, TX.

Species Lines	Platyfish Strain Codes (i.d.)
X. maculatus	
Jp163A	**Jp163A**
XSrAr	**XSrAr**
Jp163B	**Jp163B**
Jp30R	**Jp30R**
Wild Jp (Jamapa)	**JpWild**
JpYlr	**JpYlr**
JpYBr	**JpYBr**
JpYlrBr	**JpYlrBr**
YSdSr	**YSdSr**
YSp	**YSp**
SpSr	**SpSr**
Nigra	**Nigra**
BpII (Belize Platy)	**BpII**
SR (sex reversal)	**SR**
Up-3 (Usamacinta)	**Up-3**
Pp (Papaloapan)	**Pp**
X. andersi	
andersi B	**andB**
andersi C	**andC**
X. gordoni	**gordoni**
X. meyeri	**meyeri**
X. milleri	**mil82**
X. evelynae	**eve**
X. couchianus	**Xc**
X. variatus	
Zarco	**Zarco**
Encino	**Encino**
Huichihuayan	**Huich**
X. xiphidium	
San Carlos	**Sc**
Ps	**Ps**
Rio Purification	**Rp**

Species Lines	Swordtails Strain Code (i.d.)
X. alvarezi	
Dolores	**DL**
X. birchmanni	**birchII**
X. cortezicortezi	
X. nigrensis	**nigrn**
X. multilineatus	**CoyIII** **multi**
X. pygmaeus	**pygIII**
X. signum	**signum**
X. clemanciae	
Finca II	**FincaII**
Grande	**Grand**
X. monticolus	
El Tejon	**Tej**
X. mayaemayae	
X. kallmani	**kallmani**
X. montezumae	
Ojo Caliente	**OjoCal**
Rascon	**Rascon**
X. nezahuacoyotl	
Ocampo	**Ocampo**
El Salto	**ElSalto**
X. helleri	
Cd	**Cd**
Belize	**Bel**
BxII	**BxII**
Hx	**HxII**
Jalapa	**Jalapa**
Lancetilla	**Lance**
Sarabia green	**Sara-gr**
Sarabia orange	**Sara-or**
Doce Millas	**Doce**
helleri Li	**HeLi**
helleri Alb	**HeAlb**
X. continens	**contiIV**

FIGURE 13.6 *Xiphophorus* species and lines maintained in the XGSC.

lines. Some of the original genetic strains of platyfish and swordtails developed by Dr. Gordon in the 1930s are still available at the XGSC; in some cases, the products of more than 110 generations of controlled brother-to-sister matings (Figure 13.6).

The XGSC produces pedigreed fish, and custom interspecies hybrid crosses or backcrosses, for a variety of research projects, on a case-by-case basis. The XGSC staff provide consultation on husbandry and genetic questions and have the capability to establish genotypes of research animals should questions of parentage arise. The XGSC assists in preparing contracts for fish supply and needed support letters for grant and contract proposals. In addition to fish and fish services, the XGSC has compiled genetic and genomic tools for the research community. Current information and *Xiphophorus* resources are provided via the https://www.xiphophorus.txstate.edu/ website and are briefly detailed below.

XIPHOPHORUS GENETIC STOCK CENTER RESOURCES AND OPERATION

Each *Xiphophorus* pedigreed line is carried forward by select matings between two parental fish that have been scored for all pertinent traits (i.e., tail patterns, pigment patterns, etc.). It takes about 2 years for XGSC technicians to learn all the phenotypes that need to be correctly scored to set up correct matings. Once the parental fish are chosen, they are placed together in a breeding tank and monitored until the female produces a brood. Once a female has produced a brood, she will generally drop new broods of 4–20 juveniles about every 30 days, for several (i.e. 3–5) successive months. *Xiphophorus* females can store sperm, which is deposited by the males in spermatophore packets, and may continue to produce broods in ≈30-day intervals after the male is removed from the tank. The broods are removed to a new tank, sex ratios and other traits scored and recorded, and then the brood growth is monitored until males and females are sexually dimorphic, but prior to sexual maturity. It is critical that sexed fish are split, males in one tank and females in another, prior to sexual maturity so the next generation is derived from specifically selected parental fish that have scorable phenotypes. Once the fish are mature, phenotypes are scored and the proper individuals selected to produce the next generations (Figure 13.7). Generally, a line has 3–4 matings and broods maintained at any given time, as back-ups, to offset random variations in any single breeding (i.e., biased sex ratios, etc.). Once fish are selected for the next generation matings, excess siblings from the same broods are maintained until the selected parents drop their own broods, then phenotypes of all siblings are recorded and they are euthanized. In this manner, records are kept on each fish in the XGSC from its birth to its selected fate (i.e., parent for next generation, back-up, entry into an experiment, shipped to another laboratory).

Once the new generation of a line has produced a brood, the grandparents giving rise to the next generation are ethanol preserved for reference in the event of later questions. Precautions are taken to insure genetic purity and the presence of select chromosomes in each fish stock. Birth dates, brood sizes, sex ratios, maturation rates, pigment patterns, sex chromosomes, and other phenotypic data are routinely collected for each stock maintained in the XGSC using computer routines and databases specifically developed for this purpose. Our intricate system of custom databases provide built in redundancy, as well as both computer and paper records, and enable quick access to pertinent pedigree data.

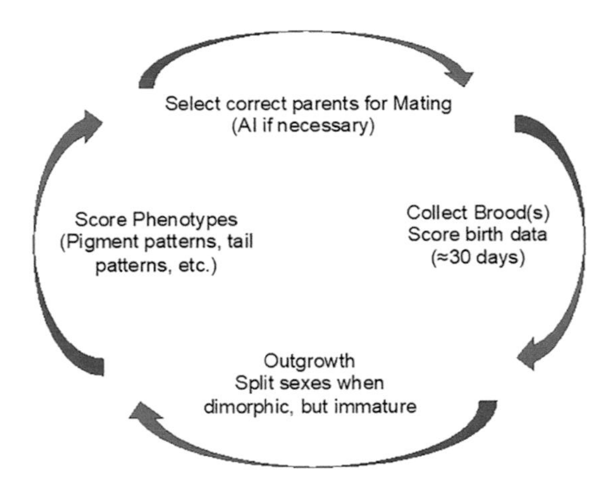

FIGURE 13.7 Standard circular process of *Xiphophorus* line maintenance.

The most important aspect of XGSC operation involves proper genetic management of each stock/line. Since all stocks are in live culture, and dynamic in principle, review of each stock and re-assessment of a management plan occurs at regular and timely intervals as a part of routine XGSC operation. During a stock review process the XGSC team discusses each concern related to a stock (e.g., biased sex ratios, disease issues, mating success, etc.). After discussion, the technician managing the stock writes a management plan for implementation.

Husbandry

All adult fish are fed twice daily, once in the morning with freshly hatched live brine shrimp (*Artemia nauplii*) and once in the afternoon with a Gordon's Liver paste that is made in the XGSC and composed of fresh beef liver, carrot extract combined with oat meal cereal, and other ingredients. New broods and small fish are fed three times per day, once with brine shrimp if they are large enough to eat them, and twice with flake food (Zeigler, Aquatox fish diet) that has been ground to a small size.

New tanks are set up simply with clean water (pH 8.1, total hardness at 290 ppm), with the tank bottom half covered with rinsed gravel, and a bit of Java moss is used inside each tank. There are no filters nor are the tanks aeriated. Each tank sits on a tiered rack with a glass lid having the top right corner cut to allow for feeding. Tanks are maintained at a constant water level by daily topping off water loss due to evaporation. Most tanks also undergo a half water change about every 1–2 months, after the tank sides have been scraped to remove accumulated algae and the gravel has been vacuumed to reduce detritus. The XGSC is maintained at a constant temperature of 25.5°C (78°F) and has both large windows and skylights in addition to multiple bands of fluorescent lights. The fluorescent fixtures each have two 10,000 K, T8m 232 watt, "sunlight" lamps (Coralife Inc.). Illumination ranges from 129.5 kJ/m^2 (bottom rack tier) to 179.6 J/m^2 (top tier), and the fluorescent light banks are on a 13 hr light, 11 hr dark cycle.

Animal Health

All fish are checked daily for disease, which is extremely rare since fish are kept in closed colonies (i.e., flow-through systems). The fish lines in the XGSC represent 24 species with site locales that span Mexico and into Central America. Thus, different species have variable susceptibilities to particular diseases. It is rare that more than one tank is affected with a diseased fish. If this occurs, the fish in the tank are euthanized, the tank dismantled and rinsed with chlorine bleach, rinsed, and allowed to dry. The gravel in the diseased tank is autoclaved, and any nets that were used to collect the fish are bleached. Any new fish entering the XGSC are kept in a quarantine room adjacent to the main laboratory and maintained in this room for at least 2 successive generations until it is determined that they are stable and healthy.

During normal operation, all nets, observation dishes, and other utensils that enter an aquarium are sterilized with a 1:4 dilution of bleach, followed by a dip in vinegar, and then with water. Nets are also soaked in methylene blue solution and rinsed with water before use in a new aquarium. To date, no epidemic viral or bacterial diseases have threatened the health of the colony in the 25 years the XGSC has been at Texas State University.

Stock Distribution

The XGSC routinely receives requests for information, live fish, derived tissues, and other materials from external investigators. The typical procedure, after initial contact from interested parties, is for the XGSC to inform and educate the researcher and assist in decisions regarding strain selection, sample size, and handling. The XGSC then ships the fish with an invoice and detailed pedigree information, including relevant fish numbers.

Aquaria space is always valuable. The XGSC maintains only enough fish of each stock to ensure its perpetuation. Thus, not all fish are likely to be available in large numbers at a specific moment in time without previous scheduling arrangements between investigators and XGSC staff. Most users are aware of this policy and contact the XGSC months before animals are required. The requested line is then "bredup" to produce sufficient animals to ensure that the user receives the exact fish, at the proper age, and in the needed numbers.

Record Keeping

A description of all species/strains and maintenance protocols, along with other pertinent information are available online at https://www.xiphophorus.txstate.edu/resources/ protocols.html. Also, available on the XGSC web page are short instructive videos showing protocols (e.g., artificial insemination, tumor inducing treatments, etc.). Pedigree records can be requested at any time, as well as ethanol fixed parental samples, if needed. Our pedigree database will provide information about a particular fish lineage; therefore, one can track the exact origin of the fish they have received, including scored data on the siblings of the fishes they may have. Such information can be of critical importance when addressing questions such as tissue transplantation experiments or when inbreeding coefficients need to be calculated, etc. Several *Xiphophorus* genetic stocks are highly inbred. For example, current pedigrees of *X. maculatus* Jp 163 A and Jp 163 B, and *X. couchianus* derive from 118th, 111th, and 87th inbred generations, respectively.

CRYOPRESERVATION OF *XIPHOPHORUS* SPERM FOR LONG-TERM STORAGE

In continuing collaboration, Dr. Terry Tiersch (Louisiana State University, Baton Rouge) has been devoted to establishing a repository of cryopreserved *Xiphophorus* germplasm (in the form of sperm) that represents all species and lines in the XGSC. When begun nearly 10 years ago, successful cryopreservation of sperm from any live-bearing fish species had not been accomplished. Thus, new methods needed to be developed as the sperm from these fish was found to behave quite differently than that from either marine or freshwater oviparous species (Dong et al. 2009, Huang et al. 2004a, 2004b, Uribe et al. 2011, Yang et al. 2006, 2007, 2009, 2012a–2012c).

Since the development of a successful base protocol for *X. maculatus*, protocols have been customized for the cryopreservation of all three major evolutionary groupings within the *Xiphophorus* genus. In addition, the National Institutes of Health (NIH) has entered into an agreement with the USDA National Animal Germplasm Program (USDA-AGP) in Ft. Collins, Colorado, to accept and maintain XGSC sperm samples for long-term storage. To date, the USDA-AGP has banked over 3316 cryopreserved straws representing 24 species and 53 of the 58 lines maintained at the XGSC. Most (45) of the XGSC lines have at least 15 straws in frozen long-term storage, the minimum level for backup to be used for reconstitution of a line (i.e., internal use by XGSC). These straws were frozen in batches of 15–20 with pooled sperm from 9 to 30 males each. Also, 23 lines in the XGSC have reached the target required (80 straws) and the frozen material has been archived as accessions in the USDA-AGP.

XIPHOPHORUS GENOME RESOURCES AND SUPPORTING DATA

Genome sequence assemblies of *Xiphophorus* fishes allows scientists to perform comparative genomic analyses to deconvolute the genetic interactions that give rise to spontaneous and inducible tumorigenesis, to delve into the evolution of live-bearing fishes, and also have served as important reference genomes for new studies aimed at determination of the mechanics behind complex life history traits such as maturation, aging, sexual selection, behavior, regeneration, and cell fate determination. From 2013 to 2018, the XGSC, though support of the National Institutes of Health (Office of Research Infrastructure Programs, Division of Comparative medicine), has been a driving influence in sequencing and assembly 4 *Xiphophorus* species genomes (i.e., *X. maculatus*, *X. hellerii*, *X. couchianus*, and *X. montezumae*). The genomes were initially sequenced using whole genome shotgun sequencing methodology (i.e., Illumina next generation sequencing). However, very recently three of these genomes (i.e., *X. maculatus*, *X. hellerii*, *X. couchianus*) were upgraded and further enhanced (Table 13.2) using state-of-the-art Pac-Bio (Pacific Bioscience,

TABLE 13.2
List of *Xiphophorus* Species and Genome Assembly Statistics

Species	Genome Size (Mb)	No. Chromosomes	Total # Contigs	Scaffold N50 (Mb)	Contig N50 (Mb)	No. Protein Coding Genes
X. maculatus	701	24	232	10.7	31	23,238
X. couchinaus	684	24	243	17.7	20	21,588[a]
X. hellerii	730	24	534	7.2	13	22,244[a]
X. montezumae	674[a]	24[a]	21,988	0.06[a]	1.3[a]	19,926[a]

Note: *Xiphophorus maculatus*, *X. couchianus*, and *X. hellerii* assembly statistics are based on PacBio SMRT sequencing. The *X. couchianus* and *X. hellerii* genomes are currently being annotated by NCBI. Thus, coding gene numbers are from older versions. The *X. montezumae* genome was sequenced using an illumina WGS sequencing strategy. (Data kindly provided by Dr. Wesley Warren, University of Missouri.)

[a] Statistics for Illumina-based sequencing only.

Menlo Park, CA) single molecule sequencing coupled with assembly employing BioNano (BioNano Genomics, Tulsa, OK) optical mapping methodologies (Amores et al. 2014, Schartl et al. 2013, Shen et al. 2016). Due to the highly inbred and pedigreed (e.g., *X. maculatus, X. couchianus*) or pedigreed (e.g., *X. hellerii*) maintenance of these lines by the XGSC, these new genome assemblies represent the most contiguous and complete genome assemblies available among vertebrates. The availability of these closely related genome assemblies makes *Xiphophorus* one of the very few cases where several species' genomes are available and the only system wherein high-resolution genome assemblies exist for species that may be crossed to produce fertile interspecies hybrids. This allows assessment of allele specific gene expression patterns within individual interspecies hybrid animals, enhancing the power of the *Xiphophorus* system to dissect individual genetic components leading to complex trait phenotypes.

The XGSC is currently hosting all versions of *Xiphophorus* genome assemblies (Table 13.2), including *De novo* contig assembly of *X. maculatus* genome, chromosome assemblies of *X. maculatus*, *X. couchianus*, *X. helleri*, and *X. montezumae*. Genome assemblies, as well as the corresponding genome annotation files in GFF or GTF formats are all available at the XGSC ftp site for download (https://viewer.xgsc.txstate.edu/data/). These data are also available through the Ensembl and NCBI genome databases.

In addition to genome sequences, the XGSC also hosts three types of data supporting research conducted using *Xiphophorus* models: chromosome structure, gene expression, and comparative genomics. For visualization of these data, two genome browser platforms have been adopted and configured on a dedicated server to run (a) an Integrated Genome Viewer (IGV), and (b) a University of California Santa Cruz (UCSC) genome browser. The IGV hosts the *X. maculatus* genome as a reference and it may be compared to all other available *Xiphophorus* genomes. An xml file was created to define genomic information, data, and features to be included during the IGV initiation, and a .jnlp file was created to generate downloadable session files based on the configurations in the.xml file IGV session files can be downloaded from the XGSC website to a local workstation and are able to launch a Java IGV session locally. The user can visualize data that is stored locally, in a remote site (user's own data server), or a data server that has been configured to share XGSC data with the public (See "Read me" files at http://www.xiphophorus.txstate.edu/genomeviewer.html). Databases have been built to allow the user to navigate through different chromosomes and localized genes of interest.

Several organ specific transcriptomic data sets constructed by RNA sequencing (RNA-Seq) are configured for the user to assess gene expression in different organs. In the past few years, research at the XGSC has resulted in the acquisition of hundreds of transcriptomic datasets from various *Xiphophorus* species. Some of these datasets are stored in Gene Expression Omnibus (Session number: GSE113622, GSE112473, GSE107449, GSE89561) at NIH-NCBI. All data are open to the public for download or can be made available in other forms upon request.

The *X. maculatus* genomic sequence has also been compared to several fish species, including medaka, Amazon molly, stickleback, cavefish, spotted gar, and Asian arowana, for comparative genomic studies. These sequence alignment data are available by querying the XGSC server. The UCSC browser is currently being updated in order to display the Pac-Bio enhanced *Xiphophorus* genomes. Databases for multiple species genome sequence comparisons have been built and tracks for gene models, transcriptome data, and multiple species sequence alignments have been configured. Updated versions of the Xiphophorus genome browser are available through the UCSC genome browser (https://genome.ucsc.edu/) track hub, and the XGSC genome resource webpage (https://www.xiphophorus.txstate.edu/genomeviewer. html). This browser runs on-site and has good data visualization features for comparative genomic applications.

SPECIES CONSERVATION OF *XIPHOPHORUS* IN THE WILD

The locations where *Xiphophorus* are found in Mexico and Central America represent some of the most species diverse areas in the world. However, as with the rest of the world, these regions are experiencing increased pressure on freshwater resources due to the demands of human population growth, urban expansion, and the necessity of industrialization. Of the freshwater fishes, the family Poeciliidae, which contain the genus *Xiphophorus*, is among the most exposed to habitat loss (Contreras-Balderas 2005, Contreras-MacBeath 2005). Of 86 species in the family Poeciliidae, 20 species are considered stressed or "at risk" with 8 endangered, 9 threatened, 2 under special protection, and one extinct. Of these, 5 species of *Xiphophorus* (Table 13.3) are listed as endangered, and at least one of these (*X. couchianus*) may be extinct in the wild (NORMA Oficial Mexicana NOM-059-SEMARNAT-2010). In addition, *X. andersi* and *X. evelynae* are endemic species to a very small area (a few kilometers) and should be considered as threatened. In general, the aggressive use of the environment in Mexico for agriculture and urban expansion suggest that most of the XGSC strains and species need protection in their site locales M. Schartl, personal communication, November 1, 2018).

Efforts on the part of governments to set up species protection regimens are often focused primarily on more popular or high-profile species such as sea turtles,

TABLE 13.3
List of *Xiphophorus* Species Registered as Threatened Based on Data from the NORMA Oficial Mexicana NOM-059-SEMARNAT-2010

Species	Distribution[a]	Class[b]	Threats[c]
X. couchianus	Springs near Monterrey	NP	Monterrey urban expansion
X. gorddoni	Cuartro Cienegas, Coahuila	NP	Pollution, exotic invader species
X. meyeri	Springs near Rio Salado, Coahuila	NP	Dewatering, pollution
X. milleri	Laguna de Catemaco, Veracruz	NP	Pollution, exotic invader species
X. clemenciae	Rio Coatzacoalcos, Veracruz and Oaxaca	SS	Agriculture and colonization

[a] Contrreras-MacBeath 2005.
[b] NP—Northern platyfish. SS—Southern swordtail.
[c] Contrreras-Balderas 2005, Kallman and Kazainis 2006.

jaguars, bears, and monkeys. Thus, the impacts of population growth and water management on the sustainability of freshwater fishes remains largely unaddressed. This worldwide problem will reach greater significance as regional water issues become more prominent on political and social agendas. The plight of *Xiphophorus* species represents one group already showing adverse effects. In addition, the changing environment may serve to further confound species protection mechanisms. A recent 15-model consensus projection of the regional impact of climate change on North America shows the Pacific drainages of northern and central Mexico may be "hot spots" for greater than normal effects in climate modulation (Kerr 2008). The larger responsiveness of these regions to climate change may exacerbate loss of freshwater habitats and hasten requirements for species preservation programs to be put in place.

ACKNOWLEDGMENTS

The authors would like to thank the staff of the NIH Office of Research Infrastructure Programs, Division of Comparative Medicine (grant awards R24-OD-011120 to RW and R15-CA-223964 to YL) that has provided support or the XGSC over the past 25 years. This support has been critical to the continued discoveries made in using the *Xiphophorus* genetic system. We cannot overstate our appreciation for the ORIP-DCM programmatic staff scientists and administrators who make the arguments, often to those with little understanding of the evolutionary continuum, of why a healthy national research enterprise requires maintaining a high diversity of animal models.

We also thank Texas State University for its ongoing support of the XGSC research mission and the scientific service it provides in hosting of the *Xiphophorus* Genetic Stock Center.

REFERENCES

Adam, D., W. Maueler and M. Schartl. 1991. Transcriptional activation of the melanoma inducing *Xmrk* oncogene in *Xiphophorus*. *Oncogene* 6:73–80.

Alireza, K., A. Hossein, M-S. Hamed and P. Cluj-Napoca. 2016. Effect of dietary administration of methyltestosterone and vitamin C on the sex reversal and survival of *Xiphophorus maculatus* (Cyprinodontiformes: Poeciliidae). *Poeciliid Research* 6:16–24.

Amores, A., J. Catchen, I. Nanda et al. 2014. A RAD-tag genetic map for the platyfish (*Xiphophorus maculatus*) reveals mechanisms of karyotype evolution among teleost fish. *Genetics* 197:625–641.

Anders, F. 1967. Tumour formation in platyfish-swordtail hybrids as a problem of gene regulation. *Experientia* 23:1–10.

Anders, F. 1991. Contributions of the Gordon-Kosswig melanoma system to the present concept of neoplasia. *Pigment Cell Research* 4:7–29.

Bastide, P., C. Solís-Lemus, R. Kriebel, K.W. Sparks and C. Ané. 2018. Phylogenetic comparative methods on phylogenetic networks with reticulations. *Systematic Biology* 67:800–820.

Boswell, M., W. Boswell, Y. Lu et al. 2018a. The transcriptional response of skin to fluorescent light exposure in viviparous (*Xiphophorus*) and oviparous (*Danio, Oryzias*) fishes. *Comparative Biochemistry and Physiology, Part C* 208:77–86.

Boswell, W., M. Boswell, J. Titus et al. 2015. Sex specific molecular genetic response to UVB exposure in *Xiphophorus maculatus* skin. *Comparative Biochemistry and Physiology, Part C* 178:76–85.

Boswell, W.T., M. Boswell, D.J. Walter et al. 2018b. Exposure to 4,100 K fluorescent light elicits sex specific transcriptional responses in *Xiphophorus maculatus* skin. *Comparative Biochemistry and Physiology, Part C* 208:94–104.

Boulton, K., E. Couto, A.J. Grimmer et al. 2015. How integrated are behavioral and endocrine stress response traits? A repeated measures approach to testing the stress-coping style model. *Ecology and Evolution* 5:618–633.

Boulton, K., G.G. Rosenthal, A.J. Grimmer, C.A. Walling and Wilson. 2016. Sex-specific plasticity and genotype × sex interactions for age and size of maturity in the sheepshead swordtail, *Xiphophorus birchmanni*. *Journal of Evolutionary Biology* 29:645–656.

Bunnajirakul, S., S. Pavasutthipaisit and D. Steinhagen. 2015. Pathological alterations due to motile *Aeromonas* infection in red swordtail fish (*Xiphophorus helleri*). *Tierarztkl Prax Ausg K Kleintiere Heimtiere.* 43:434–438.

Butler, A.P., D. Trono, R. Beard, R. Fraijo and R.S. Nairn. 2007. Melanoma susceptibility and cell cycle genes in *Xiphophorus* hybrids. *Molecular Carcinogen* 46:685–691.

Caixeta, E.S., C.F. Silva, V.S. Santos, O. de Campos Júnior E and B.B. Pereira. 2016. Ecotoxicological assessment of pyriproxyfen under environmentally realistic exposure conditions of integrated vector management for *Aedes aegypti* control in Brazil. *Journal of Toxicology and Environmental Health A*. 79:799–803.

Chalopin, D., J.-N. Volff, D. Galiana, J.L. Anderson and M. Schartl. 2015b. Transposable elements and early evolution of sex chromosomes in fish. *Chromosome Research* 23:545–560.

Chalopin, D., M. Naville, F. Plard, D. Galiana and J.-N. Volff. 2015a. Comparative analysis of transposable elements highlights mobilome diversity and evolution in vertebrates. *Genome Biology and Evolution* 7:567–580.

Chang, J., Y. Lu, W.T. Boswell, M. Boswell, K.L. Caballero and R.B. Walter. 2015. Molecular genetic response to varied wavelengths of light in *Xiphophorus maculatus* skin. *Comparative Biochemistry and Physiology, Part C* 178:104–115.

Contreras-Balderas, S. 2005. Conservation status of Mexican freshwater viviparous fishes. In *Viviparous Fishes*, ed. M.C. Uribe and H. Grier, pp. 415–423. Homestead, FL: New Life Publications.

Contreras-MacBeath, E.T.C. 2005. Analysis of the conservation of freshwater fish species in Mexico with emphasis on viviparous species. In *Viviparous Fishes,* ed M.C. Uribe and H. Grier, 401–414. Homestead, FL: Life Publications.

Costa, R.A., J.C. Cardoso and D.M. Power. 2017. Evolution of the angiopoietin-like gene family in teleosts and their role in skin regeneration. *BMC Evolutionary Biology* 17:14. doi: 10.1186/s12862-016-0859-x.

Cui, R., M. Schumer, K. Kruesi, R. Walter, P. Andolfatto and G.G. Rosenthal. 2013. Phylogenomics reveals extensive reticulate evolution in *Xiphophorus* fishes. *Evolution* 67:2166–2179.

Culumber, Z.W. 2016. Variation in the evolutionary integration of melanism with behavioral and physiological traits in *Xiphophorus variatus*. *Evolutionary Ecology* 30:9–20.

Culumber, Z.W. and M. Tobler. 2016. Spatiotemporal environmental heterogeneity and the maintenance of the tailspot polymorphism in the variable platyfish (*Xiphophorus variatus*). *Evolution* 70:408–419.

Culumber, Z.W. and M. Tobler. 2018. Correlated evolution of thermal niches and functional physiology in tropical freshwater fishes. *Journal of Evolutionary Biology* 31:722–734.

Culumber, Z.W., M. Schumer, S. Monks and M. Tobler. 2015. Environmental heterogeneity generates opposite gene-by-environment interactions for two fitness-related traits within a population. *Evolution* 69:541–550.

Cummings, M.E. and M.E. Ramsey. 2015. Mate choice as social cognition: Predicting female behavioral and neural plasticity as a function of alternative male reproductive tactics. *Current Opinions inBehavioral Sciences* 6:125–131.

D'Amorea, D.M., O. Rios-Cardenas and M.R. Morris. 2015. Maternal investment influences development of behavioural syndrome in swordtail fish, *Xiphophorus multilineatus*. *Animal Behaviour* 103:147–151.

Dong, Q., C. Huang, L. Hazlewood et al. 2009. Sperm cryopreservation of live-bearing fishes of the genus *Xiphophorus*. In *Methods in Reproductive Aquaculture: Marine and Freshwater Species,Section V: Protocols for Sperm Cryopreservation*, ed E. Cabrita, V. Robles and P. Herraez, pp. 339–344. Boca Raton, FL: CRC Press.

Fornzler, D., J. Wittbrodt and M. Schartl. 1991. Analysis of an esterase linked to a locus involved in the regulation of the melanoma oncogene and isolation of polymorphic marker sequences in *Xiphophorus*. *Biochemical Genetics* 29:509–524.

Gimenez-Conti, I., A.D. Woodhead, J.C. Harshbarger et al. 2001. A proposed classification scheme for *Xiphophorus* melanomas based on histopathology analyses. *Marine Biotechnology* 3:100–107.

Gomez, A., C. Wellbrock, H. Gutbrod, N. Dimitrijevic and M. Schartl. 2001. Ligand independent dimerization and activation of the oncogenic *Xmrk* receptor by two mutations in the extracellular domain. *Journal of Biological Chemistry*. 276:3333–3340.

Gordon, M. 1931. Hereditary basis of melanosis in hybrid fishes. *The American Journal of Cancer* 15:1495–1523.

Haussler, G. 1928. Uber melanobildung bei bastarden von *Xiphophorus helleri* und *Platypoecilus maculatus* var. *rubra*. *Klinische Wochenschrift* 7:1561–1562.

Hoseinifar, S.H., Z. Roosta, A. Hajimoradloo and F. Vakili. 2015. The effects of *Lactobacillus acidophilus* as feed supplement on skin mucosal immune parameters, intestinal microbiota, stress resistance and growth performance of black swordtail (*Xiphophorus helleri*). *Fish and Shellfish Immunology* 42:533–538.

Hoshino, É.M., M.D.F.G. Hoshino and M. Tavares-Dias. 2018. Parasites of ornamental fish commercialized in Macapá, Amapá State (Brazil). *Revista Brasileira de Parasitologia Veterinaria*. 27:75–80.

Huang, C., Q. Dong, L. Hazlewood, R.B. Walter and T.R. Tiersch. 2004b. Sperm cryopreservation of a live-bearing fish, the platyfish *Xiphophorus couchianus*.*Theriogenology* 62:971–989.

Huang, C., Q. Dong, R.B. Walter and T.R. Tiersch. 2004a. Initial studies on sperm cryopreservation of the live-bearing fish, green swordtail *Xiphophorus helleri*.*Theriogenology* 62:179–194.

Jie, M. and G. Jian-Fang. 2015. Genetic basis and biotechnological manipulation of sexual dimorphism and sex determination in fish. *Science China Life Sciences* 58:124–136.

Kallman, K.D. 2001. How the *Xiphophorus* problem arrived in San Marcos, Texas. *Marine Biotechnology* 3:S6–S16.

Kallman, K.D. and S. Kazianis. 2006. The genus *Xiphophorus* in Mexico and Central America. *Zebrafish* 3:271–285.

Kang, H.J., M. Schartl, R.B. Walter and A. Meyer. 2013. Comprehensive phylogenetic analyses of all species of swordtails and platies (genus *Xiphophorus*) uncover a hybrid origin of a swordtail fish, *Xiphophorus monticolus*. *BMC Evolutionary Biology* 13:25–44.

Kang, J.H., T. Manousaki, P. Franchini, S. Kneitz, M. Schartl and A. Meyer. 2015. Transcriptomics of two evolutionary novelties: How to make a sperm-transfer organ out of an anal fin and a sexually selected "sword" out of a caudal fin. *Ecology and Evolution* 54:848–864.

Kawaguchi, M., K. Tomita, K. Sano and T. Kaneko. 2015. Molecular events in adaptive evolution of the hatching strategy of ovoviviparous fishes. *Journal of Experimental Zoology. B. Molecular and Developmental Evolution* 324:41–50.

Kazianis, S. and R.B. Walter. 2002. Use of platyfishes and swordtails in biological research. *Laboratory Animal* 31:46–52.

Kazianis, S., D.C. Morizot, L.D. Coletta et al. 1999. Comparative structure and characterization of a *CDKN2* gene in a *Xiphophorus* fish melanoma model. *Oncogene* 18:5088–5099.

Kazianis, S., H. Gutbrod, R.S. Nairn et al. 1998. Localization of a *CDKN2* gene in linkage group V of *Xiphophorus* fishes defines it as a candidate for the DIFF tumor suppressor. *Genes, Chromosomes and Cancer* 22:210–220.

Kazianis, S., I. Gimenez-Conti, D. Trono et al. 2001a. Genetic analysis of neoplasia induced by N-nitroso-N-methylurea in *Xiphophorus* hybrid fish. *Marine Biotechnology* 3:S37–S43.

Kazianis, S., I. Gimenez-Conti, R.B. Setlow et al. 2001b. MNU induction of neoplasia in a platyfish model. *Laboratory Investigation* 81:1191–1198.

Kazianis, S., L.D. Coletta, D.C. Morizot, D.A. Johnston, E.A. Osterndorff and R.S. Nairn. 2000. Overexpression of a fish *CDKN2* gene in a hereditary melanoma model. *Carcinogenesis* 21:599–605.

Kerr, R.A. 2008. Climate change hot spots mapped across the United States. *Science* 321:909. doi:10.1126/science.321.5891.909.

Khiabani, A., H. Anvarifar, S. Safaeian, R. Tahergorabi. 2014. Masculinization of swordtail *Xiphophorus hellerii* (Cyprinodontiformes: Poeciliidae) treated with 17α-methyltestosterone and Vitamin E. *Global Research Journal of Fishery Science and Aquaculture* 1:21–25.

Klotz, B., S. Kneitz, M. Regensburger et al. 2018. Expression signatures of early-stage and advanced medaka melanomas. *Comparative Biochemistry and Physiology,Part C* 208:20–28.

Kneitz, S., R. Mishra, D. Chalopin et al. 2016. Germ cell and tumor associated piRNAs in the medaka and *Xiphophorus* melanoma models. *BMC Genomics* 17:357. doi:10.1186/s12864-016-2697-z.

Kosswig, C. 1927. Uber bastarde der teleostier *Platypoecilus* und *Xiphophorus.Zeitschrift fur Induktive Abstammungs und Vererbungslehre* 44:253. doi:10.1007/BF01740990.

Liang, X., L. Wang, R. Ou et al. 2015. Effects of norfloxacin on hepatic genes expression of P450 isoforms (CYP1A and CYP3A), GST and P-glycoprotein (P-gp) in swordtail fish (*Xiphophorus helleri*). *Ecotoxicology* 24:1566–1573.

Lu, Y., J. Reyes, S. Walter et al. 2018a. Characterization of basal gene expression trends over a diurnal cycle in *Xiphophorus maculatus* skin, brain and liver. *Comparative Biochemistry and Physiology, Part C* 208:2–11.

Lu, Y., M. Boswell, B. Boswell, K. Yang, M. Schartl and R.B. Walter. 2015. Molecular genetic response of *X. maculatus—X. couchianus* interspecies hybrid skin to UVB exposure. *Comparative Biochemistry and Physiology, Part C* 178:86–92.

Lu, Y., M. Boswell, W. Boswell et al. 2017. Molecular genetic analysis of the melanoma regulatory locus in *Xiphophorus* interspecies hybrids. *Molecular Carcinogenesis* 56:1935–1944.

Lu, Y., M. Boswell, W. Boswell et al. 2018b. Comparison of *Xiphophorus* and human melanoma transcriptomes reveals conserved pathway interactions. *Pigment Cell and Melanoma Research* 31:496–508.

Luo, J., M. Sanetra, M. Schartl and A. Meyer. 2005. Strong reproductive skew among males in the multiply mated swordtail *Xiphophorus multilineatus* (Teleostei). *Journal of Heredity* 96:346–355.

Manning, D.D., N.D. Reed and C.F. Schafer. 1973. Maintenance of skin xenografts of widely divergent phylogenetic origin of congenitally athymic (nude) mice. *Journal of Experimental Medicine* 138:488–494.

Meierjohann, S. and M. Schartl. 2006. From Mendelian to molecular genetics: The *Xiphophorus* melanoma model. *Trends in Genetics* 22:654–661.

Meierjohann, S., I. Wende, A. Kariss, C. Wellbrock and M. Schartl. 2006. The oncongenic epidermal growth factor receptor variant *Xiphophorus* melanoma receptor kinase induces motility in melanocytes by modulation of focal adhesions. *Cancer Research* 66:3145–3152.

Meyer, A., W. Salzburger and M. Schartl. 2006. Hybrid origin of a swordtail species (Teleostei: *Xiphophorus clemenciae*) driven by sexual selection. *Molecular Ecology* 15:721–730.

Mirzaie, M. 2015. Prevalence and histopathologic study of *Lernaea cyprinacea* in two species of ornamental fish (*Poecilia latipinna* and *Xiphophorus helleri*) in Kerman, South-East Iran. *Turkiye Parazitol Dergisi* 39:222–226.

Mohr, P.G., N.J. Moody, L.M. Williams et al. 2015. Molecular confirmation of infectious spleen and kidney necrosis virus (ISKNV) in farmed and imported ornamental fish in Australia. *Diseases of Aquatic Organisms* 116:103–110.

Morizot, D.C. and M.J. Siciliano. 1983. Linkage group V of platyfishes and swordtails of the genus *Xiphophorus* (Poeciliidae): Linkage of loci for malate dehydrogenase-2 and esterase- 1 and esterase-4 with a gene controlling the severity of hybrid melanomas. *Journal of the National Cancer Institute* 71:809–813.

Nairn, R.S., D.C. Morizot, S. Kazianis, A.D. Woodhead and R.B. Setlow. 1996b. Nonmammalian models for sunlight carcinogenesis: Genetic analysis of melanoma formation in *Xiphophorus* hybrid fish. *Photochemistry and Photobiology* 64:440–448.

Nairn, R.S., S. Kazianis, B.B. McEntire, L. DellaColetta, R.B. Walter and D.C. Morizot. 1996a. A CDKN2-like polymorphism in *Xiphophorus* LG V is associated with UV-B-induced melanoma formation in platyfish-swordtail hybrids. *Proceedings of the National Academy of Sciences* (USA) 93:13042–13047.

Nairn, R.S., S. Kazianis, L. Della Coletta et al. 2001. Genetic analysis of susceptibility to spontaneous and UV carcinogenesis in *Xiphophorus* hybrid fish. *Marine Biotechnology* 3(S1): 24–37.

Patton, E.E., D.L. Mitchell and R.S. Nairn. 2010. Genetic and environmental melanoma models in fish. *Pigment Cell Melanoma Research* 23:314–337.

Pereira B.B., E.S. Caixeta, P.C. Freitas, et al. 2016. Toxicological assessment of spinosad: Implications for integrated control of *Aedes aegypti* using larvicides and larvivorous fish. *J Toxicol Environ Health A* 79:477–481.

Perez, A.N., L. Oehlers, S.J. Heater, R.E. Booth, R.B. Walter and W.M. David. 2012. Proteomic analyses of the *Xiphophorus* Gordon-Kosswig melanoma model. *Comparative Biochemistry and Physiology* 155:81–88.

Regneri, J., J-N. Volff and M. Schartl. 2015. Transcriptional control analyses of the *Xiphophorus* melanoma oncogene. *Comparative Biochemistry and Physiology C. Toxicology and Pharmacology* 178:116–127.

Rosen, D.E. 1979. Fish from the uplands and intermontane basins of Guatemala: Revisionary studies and comparative geography. *Bulletin of the American Museum of Natural History* 162:271–375.

Rosenthal, G.G. 2003. Dissolution of sexual signal complexes in a hybrid zone between the swordtails *Xiphophorus birchmanni* and *Xiphophorus malinche* (Poeciliidae). *Copeia* 2003:299–308.

Ruas, M. and G. Peters. 1998. The p16INK4a/CDKN2A tumor suppressor and its relatives. *Biochimica et Biophysica Acta* 1378:F115–F177.

Schartl M. 1995. Platyfish and swordtails: A genetic system for the analysis of molecular mechanisms in tumor formation. *Trends in Genetics* 11:185–189.

Schartl, M. 1990. Homology of melanoma-inducing loci in the genus *Xiphophorus.Genetics* 126:1083–1091.

Schartl, M. and R.B. Walter. 2016. *Xiphophorus* and Medaka cancer models. In *Advances in Experimental Medicine and Biology: Cancer and Zebrafish Mechanisms, Techniques and Models*, ed D. Langenau, pp. 527–548. Cham, Switzerland: Springer Publishing.

Schartl, M. and R.U. Peter. 1988. Progressive growth of fish tumors after transplantation into thymus-aplastic (nu/nu) mice. *Cancer Research* 48:741–744.

Schartl, M., B. Wilde, J.A. Laisney, Y. Taniguchi, S. Takeda and S. Meierjohann. 2010. A mutated EGFR is sufficient to induce malignant melanoma with genetic background-dependent histopathologies. *Journal of Investigative Dermatology* 130:249–258.

Schartl, M., R.B. Walter, Y. Shen et al. 2013. The genome of the platyfish, *Xiphophorus maculatus*. *Nature Genetics* 45:567–572.

Schartl, M., Y. Shen, K. Maurus et al. 2015. Whole body melanoma transcriptome response in Medaka. *PLoS One* 10: e0143057.

Schumer, M., C. Xu, D.L. Powell et al. 2018. Natural selection interacts with recombination to shape the evolution of hybrid genomes. *Science* 360:656–660.

Schwab, M., G. Kollinger, J. Haas et al. 1979. Genetic basis of susceptibility for neuroblastoma following treatment with N-methyl-N-nitrosourea and X-rays in *Xiphophorus*. *Cancer Research* 39:519–526.

Setlow, R.B., A.D. Woodhead and E. Grist. 1989. Animal model for ultraviolet radiation-induced melanoma: Platyfish-swordtail hybrid. *Proceedings of the National Academy of Sciences* (USA) 86:8922–8926.

Seyfahmadi, M., S.R. Moaddab and A. Sabokbar. 2017. Identification of mycobacteria from unhealthy and apparently healthy aquarium fish using both conventional and PCR analyses of *hsp65* gene. *Thai Journal of Veterinary Medicine* 47:571–578.

Shen, Y., D. Chalopin, T. Garcia et al. 2016. *X. couchianus* and *X. hellerii* genome models provide genomic variation insight among *Xiphophorus* species. *BMC Genomics* 17:37–50.

Silva, D.C.V.R., C.V.M. Araújo, F.M. França et al. 2018. Bisphenol risk in fish exposed to a contamination gradient: Triggering of spatial avoidance. *Aquatic Toxicology* 197:1–6.

Singh, R.N. and A. Kumas. 2016. Beetroot as a carotenoid source on growth and color development in red swordtail (*Xiphophorus helleri*) fish. *Imperial Journal of Interdisciplinary Research* 2:637–642.

Sobel, H.J., E. Marquet, K.D. Kallman and G.J. Corley. 1975. Melanomas in platy/swordtail hybrids. In *The Pathology of Fishes*, ed W.E. Ribelin and G. Migaki, pp. 945–981. Madison, WI: University of Wisconsin Press.

Trujillo-González, A., J.A. Becker, D.B. Vaughan and K.S. Hutson. 2018. Monogenean parasites infect ornamental fish imported to Australia. *Parasitology Research* 117:995–1011.

Uribe, R-C., H. Yang, J. Daly, M.G. Savage, R.B. Walter and T.R. Tiersch. 2011. Production of F_1 offspring with vitrified sperm from a live-bearing fish, the green swordtail *Xiphophorus hellerii*. *Zebrafish* 8:167–179.

Vielkind, J.R., K.D. Kallman and D.C. Morizot. 1989. Genetics of melanomas in *Xiphophorus* fishes. *Journal of Aquatic Animal Health* 1:69–77.

Vielkind, U. 1976. Genetic control of cell differentiation in platyfish-swordtail melanomas. *Journal of Experimental Zoology* 196:197–204.

Volff, J.N. and M. Schartl. 2002. Sex determination and sex chromosomes evolution in the medaka, *Oryzias latipes*, and the platyfish, (*Xiphophorus maculatus*). *Cytogenetic Genome Research* 99:170–177.

Walter, D.J., M. Boswell, S. Volk de García et al. 2014. Characterization and differential expression of DNA photolyases in *Xiphophorus*. *Comparative Biochemistry and Physiology, Part C* 163:77–85.

Walter, R.B. 2011. *Xiphophorus* fish: Varieties and resources. In *Advances in World Aquaculture—Cryopreservation in Aquatic Species,* 2nd ed., ed T.R. Tiersch and C.C. Green, pp. 796–808. Baton Rouge, LA: World Aquaculture Society.

Walter, R.B. and S. Kazianis. 2001. *Xiphophorus* interspecies hybrids as genetic models of induced neoplasia. *Laboratory Animal Research* 42:299–321.

Walter, R.B., D.J. Walter, W.T. Boswell et al. 2015. Exposure to fluorescent light triggers down regulation of genes involved with mitotic progression in *Xiphophorus* skin. *Comparative Biochemistry and Physiology, Part C* 178:93–103.

Walter, R.B., M. Boswell, J. Chang et al. 2018. Waveband specific transcriptional control of select genetic pathways in vertebrate skin. *BMC Genomics* 19:355–373.

Wang, S., M. Cummings and M. Kirkpatrick. 2015. Coevolution of male courtship and sexual conflict characters in mosquitofish. *Behavioral Ecology* 26:1013–1020.

Wittbrodt, J., D. Adam, B. Malitschek et al. 1989. Novel putative receptor tyrosine kinase encoded by the melanoma- inducing *Tu* locus in *Xiphophorus*. *Nature* 341:415–421.

Woolcock, B.W., B.M. Schmidt, K.D. Kallman and J.R. Vielkind. 1994. Differences in transcription and promoters of *Xmrk-1* and *Xmrk-2* genes suggest a role for *Xmrk-2* in pigment pattern development in the platyfish, *Xiphophorus maculatus*. *Cell Growth and Differentiation* 5:575–583.

Yang, H., L. Hazelwood, R.B. Walter and T.R. Tiersch. 2006. Effect of immobilization on refrigerated storage and cryopreservation of sperm from green swordtail *Xiphophorus helleri*. *Cryobiology* 52:209–218.

Yang, H., L. Hazelwood, R.B. Walter and T.R. Tiersch. 2009. Sperm cryopreservation of a live-bearing fish, *Xiphophorus couchianus*: Male to male variation in post-thaw motility and production of F_1 hybrid offspring. *Comparative Biochemistry and Physiology* 149:233–239.

Yang, H., L. Hazlewood, S.J. Heater, P.A. Guerrero, R.B. Walter and T.R. Tiersch. 2007. Production of F_1 interspecies hybrid offspring with cryopreserved sperm from a live-bearing fish, the swordtail *Xiphophorus helleri*. *Biology of Reproduction* 76:401–406.

Yang, H., M.G. Savage, L. Hazlewood, R.B. Walter and T.R. Tiersch. 2012b. Offspring production with cryopreserved sperm from a live-bearing fish *Xiphophorus maculatus* and implications with females. *Comparative Biochemistry and Physiology* 155:55–63.

Yang, H., R. Cuevas, M. Savage, R.B. Walter and T.R. Tiersch. 2012c. Sperm cryopreservation in live-bearing *Xiphophorus* fishes: Offspring production from *Xiphophorus variatus* and strategies for establishment of sperm repositories. *Zebrafish* 9:126–134.

Yang, H., R.B. Walter and T.R. Tiersch. 2012a. Current status of sperm cryopreservation in the genus *Xiphophorus*. In *Viviparous Fishes* II, ed. M.C. Uribe and H. Grier, pp. 403–414. Homestead, FL: New Life Publications.

Yang, K., M. Boswell, D.J. Walter et al. 2014. UVB-induced gene expression in the skin of *Xiphophorus maculatus* Jp163 B. *Comparative Biochemistry and Physiology, Part C* 163:86–94.

Zebral, Y.D., I.S.A. Anni, S.B. Afonso, S.I.M. Abril, R.D. Klein and A. Bianchini. 2018. Effects of life-time exposure to waterborne copper on the somatotropic axis of the viviparous fish *Poecilia vivipara*. *Chemosphere* 203:410–417.

Zhang, D., P. Hu, T. Liu et al. 2018. GC bias lead to increased small amino acids and random coils of proteins in cold-water fishes. *BMC Genomics*. 19:315. doi:10.1186/s12864-018-4684-z.

Zunker, K., J.T. Epplen and M. Schartl. 2006. Genomic stability in malignant melanoma of *Xiphophorus*. *Melanoma Research* 16:105–113.

14 ATCC
The Biological Resource Center for the Future

Marco A. Riojas, Samantha L. Fenn, Manzour Hernando Hazbón, Frank P. Simione and Raymond H. Cypess

CONTENTS

Abstract: The American Type Culture Collection (ATCC) is a non-profit biological resource center (Cypess 2003) with a mission to acquire, authenticate, preserve, develop, standardize, and distribute biological materials and information for the advancement and application of scientific knowledge. The ATCC supports scientific advancements in academic, government, biotechnology, pharmaceutical, food, agricultural, and industrial sectors globally with industry-standard products, services, and solutions. Its services and custom solutions include cell and microbial culturing and authentication, development and production of controls and derivatives, proficiency testing, and biomaterial deposit services (Simione and Cypess 2012). The ATCC is a leader in providing innovative customer-focused solutions and standards through its research and development program.

As a BRC and with nearly 100 years of continuous operation, the ATCC plays a critical role in the conservation of biodiversity and biological information for model organisms, type strains, and many other biological materials. The ATCC's rigorously authenticated biomaterials and standards are produced using methods designed to minimize passage number and genetic mutation, thus conserving the phenotypic and genotypic characteristics of the material. In addition to its long history of providing quality biomaterials and reliable physical standards, the ATCC intends to offer highly authenticated reference-grade whole genome sequence in support of future scientific and technological innovations.

INTRODUCTION

The American Type Culture Collection (ATCC), a non-profit organization, is the world's most diverse biological resource center (BRC). ATCC supports public health initiatives for the US government, academia, pharmaceutical industry, and research foundations. ATCC has provided biological specimens from humans, animals, and microbes to scientists around the world since its founding in 1925. Many of these biological specimens are cited as standards by the US Food and Drug Administration, US Department of Agriculture, AOAC International (Association of Official Analytical Chemists), Clinical and Laboratory Standards Institute, US Pharmacopeia, World Health Organization (WHO), and other organizations and agencies involved in public health, diagnostics, food safety and clinical, and therapeutic product development. ATCC distributes about 300,000 biological specimens to more than 150 countries each year. ATCC's cold-chain management and regulatory compliance expertise enables the distribution of all hazard classes of sensitive biological materials anywhere in the world.

HISTORY

The collection that would eventually form the beginnings of ATCC was created by C.E.A. Winslow, a charter member of the Society of American Bacteriologists (SAB)—now the American Society for Microbiology. In 1911, Winslow established the Bacteriological Collection and Bureau for the Distribution of Bacterial Cultures at the American Museum of Natural History in New York. Researchers responded positively: by 1912, the collection had already grown to over 500 donated items (Taylor 2016).

The SAB agreed to assume responsibility for the collection, and it was transferred to the Army Medical Museum in Washington, DC, in 1922. Later, a

National Academy of Sciences committee, seeking to add stability and greater recognition to the collection, formally established ATCC in 1925. Shortly afterward, the collection was transferred to its first official home at the John McCormick Institute for Infectious Diseases in Chicago. ATCC published its first catalog in 1927. The catalog's second edition in 1929 included 650 new cultures. Upon further examination, a significant number of the original cultures were found to be atypical and were thus judged unsuitable for distribution as reliably identified organisms. These were removed from the catalog (Taylor 2016). This illustrates the emphasis that ATCC places on rigorous quality control, a characteristic for which ATCC is still known.

ATCC relocated to Washington, DC, initially at Georgetown University, in 1937. In the following years, the collection continued to increase in size, and the available space could not meet the demands of the growing collection. This prompted a move from Georgetown to a larger facility on M Street in 1947, then to an even larger facility also on M Street in 1954. Eventually even that location was insufficient. In 1964, ATCC constructed its first dedicated facility, a 35,000 sq. ft. building which opened in nearby Rockville, Maryland (Taylor 2016).

Led by current CEO and President Dr. Raymond Cypess, ATCC moved again in 1998 to its present-day corporate headquarters in Manassas, Virginia, a 126,000 sq. ft. building with 18,000 sq. ft. of repository space containing 200 freezers including vapor-phase liquid nitrogen freezers, mechanical freezers, and cold rooms, as well as 35,000 sq. ft. of laboratory space (Figure 14.1). More recently, a new facility was

FIGURE 14.1 American Type Culture Collection headquarters in Manassas, Virginia.

opened in Gaithersburg, Maryland, adding 24,180 sq. ft. of laboratory and office space and 6640 sq. ft. of repository space. Under Cypess' direction, ATCC has shifted its business model from being a passive culture collection to an active and forward-looking BRC that anticipates the needs of the scientific community in an effort to be more responsive (Cypess 1995, 1996).

ACQUISITION PRACTICES AND POLICIES

The acquisition of biological materials by a BRC necessitates paying careful attention to the potential use and value of the new resources. While collecting specimens simply to preserve biological diversity is an ideal goal, it is often not economically feasible and may in some instances be a disservice to the scientific community. The challenge is to assure that relevant biological materials are available when needed. Committing resources to the collection of materials solely for genetic conservation, without a focus on the current and anticipated future needs of the research community, is a questionable (though sometimes rewarding) practice.

ATCC's acquisition practices focus on identifying and procuring items of current and potential interest to its scientific constituency. Most newly developed biological resources evolve from the laboratories of practicing scientists who are either reluctant or unable to share them with their colleagues around the world. To assist in ensuring that these novel items are broadly available under equitable sharing terms and conditions, ATCC developed the Biomaterial Contributor Network (BCN). The BCN includes agreements in place with major universities and several government agencies that allow scientists to avoid time-consuming negotiations for each of their contributions. The BCN enables scientists and their supporting institutions to share their resources without concern about proprietary rights, which helps in accelerating the progress of life science research and development.

HOLDINGS

ATCC's collections include a wide range of biological materials for research, including cell lines, molecular genomics tools, microorganisms, and bioproducts. The organization holds an extensive collection of:

- Over 18,000 bacterial strains in over 750 genera, useful in industry applications, assay development, quality control, and environmental studies,
- Microbial panels for research related to infectious diseases, antibiotic resistance, food and environmental testing, and population studies,
- More than 3000 human and animal viruses isolated from various sources,
- More than 3400 continuous cell lines available by species, tissue/disease types, and signaling pathways,
- Tumor cell and molecular panels for cancer research, annotated with gene mutations and molecular profiles,
- Ethnic and gender diverse induced pluripotent stem cell (iPSC) collections derived from normal/disease tissues, as well as adult human mesenchymal stem cells and mouse embryonic stem cells,

- Donor diverse normal human primary cells and hTERT immortalized cells,
- Fungi and yeast representing over 7600 species,
- Taxonomically diverse protists including parasitic protozoa and algae,
- More than 1000 genomic and synthetic nucleic acids, as well as certified reference materials, and
- Over 500 microbial cultures recommended as quality control reference strains.

Collections

ATCC has a variety of collections spanning the biological spectrum. ATCC Microbiology Collection contains bacteria, viruses, fungi, yeasts, protozoa, and parasites, as well as molecular biology resources for studying these. ATCC Cell Biology Collection is the most comprehensive bioresource in the world, consisting of over 3600 cell lines from over 150 different species, including continuous, hTERT-immortalized, and iPS cell lines as well as culture media and associated reagents. One of ATCC's newest departments is the Cell Derivation Unit (CDU). The CDU produces cultures from primary cells, providing the capability to study tissue-specific characteristics on a cellular level *in vitro*, and to develop more biologically relevant models. Healthy cells, as well as cells derived from diseased tissues such as asthma, COPD, pulmonary fibrosis, diabetes, and many other pathologies have been produced.

In addition to its diverse corporate collections, ATCC has significant experience managing and supporting numerous specialized contract reagent resource centers in wide-ranging, high-impact areas including malaria, biodefense and emerging infections, influenza, cancer, and molecular epidemiology.

Patent Depository (PD)

ATCC functions as a depository for biological material in conjunction with patent applications. ATCC began accepting deposits for patent purposes in 1949. In 1977, ATCC played a key role when the World Intellectual Property Organization negotiated the Budapest Treaty on the International Recognition of the Deposit of Microorganisms for the Purposes of Patent Procedure to simplify the process of depositing biological materials in support of patent applications. In 1981, ATCC was named the first International Depository Authority under the Budapest Treaty.

ATCC PD accepts a wide variety of biological material, such as algae, bacteria, bacteriophages, cell lines, fungi, hybridomas, plant tissue cultures, protozoa, genomics materials (clones, vectors, libraries, etc.), agricultural seeds, viruses, and yeasts. All deposits are strictly confidential until the relevant patent is issued. As a standard aspect of patent deposits, ATCC provides depositors with notification of patent material receipt, viability testing, certificate of deposit, storage for a minimum of 30 years, release of deposited material for distribution according to deposit rules, regular notification to depositors when deposited material is provided to a third party, and regulatory compliance reviews for permits and distributions. ATCC PD currently contains more than 40,000 strains of biological materials in a facility specifically designed for secure long-term storage (Figure 14.2).

FIGURE 14.2 Cryopreservation facilities at the American Type Culture Collection.

QUALITY CONTROL (QC)

As part of its commitment to ensuring that distributed biomaterials remain as true as possible to the originally deposited materials, ATCC production follows a workflow similar to the master cell bank/working cell bank model. Upon initial production of deposited material, the item is grown at large scale and two batches of material are created. The first (larger batch) is to be distributed to customers. The second (smaller batch) is retained internally and used for replenishment of future batches of material to be delivered to customers. All batches produced are carefully quality controlled to ensure that the characteristics of the material are consistent with both the original deposit as well as previous batches. Using this methodology, ATCC is able to widely distribute (Figure 14.3) biomaterials to many customers while ensuring quality and minimizing factors such as genetic mutations, to the greatest extent possible.

Quality control comprises all the various processes through which biomaterials are evaluated against specified criteria. Based upon these results, the material can be either accepted or rejected for accession or distribution. ATCC applies QC to its biomaterials upon their initial accessioning as well as upon creation of subsequent batches for distribution.

CHARACTERIZATION AND AUTHENTICATION OF RESOURCES

As part of the rigorous authentication of its biomaterials, ATCC employs polyphasic (phenotypic and genotypic) testing to confirm the identity and to characterize its cultures, in addition to viability and screening for contaminants. In order to provide additional assurance of genetic stability, characteristics that are unique or essential to the material are evaluated frequently (Cypess 2003).

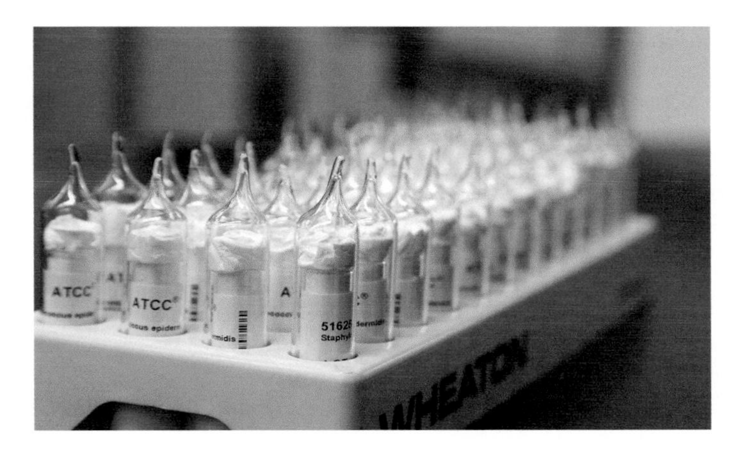

FIGURE 14.3 Vials containing preserved biological resources at the American Type Culture Collection.

Bacteria

Phenotypic characterization incudes, at a minimum, colony morphology, cellular morphology, Gram stain, motility, and an array of biochemical assays. Genotypic testing is performed using 16S rRNA sequencing as a standard process. In cases where the 16S gene may not be sufficiently differential, other genes such as *hsp65* or *rpoB* may be sequenced, or whole genome sequencing may be utilized.

Cell Lines

Phenotypic characterization includes, at a minimum, cellular morphology and growth properties (adherent, suspension, etc.). Genotypic testing includes characterization by short tandem repeat (STR) analysis, cytochrome c oxidase subunit 1 (CO1), and/or fluorescence *in situ* hybridization (FISH). Additionally, cell lines undergo sterility testing to ensure cultures are free from all microbial contaminants, particularly mycoplasmas.

ATCC understands that maintaining sample integrity is paramount to ensure that downstream studies are not adversely impacted and thus maintains strict control over all samples in its purview during any manipulations, storage, and shipping. Quality control activities are performed continually to verify that deliverables are of high quality and meet appropriate quality standards. Quality assurance (QA) activities focus on the products and processes used to manage and evaluate overall project performance on a regular basis.

CERTIFICATIONS AND ACCREDITATIONS

ATCC is an ISO 9001:2015 certified, ISO 13485:2016 certified, ISO 17025:2005 and ISO Guide 34:2009 accredited organization. ISO accreditations have allowed ATCC to evolve from a traditional culture collection to a BRC with an infrastructure that supports the future of life science and biotechnology. An experienced quality management team ensures that acquisition, authentication, preservation, development, and distribution of biological materials are performed under, and conform to, a documented quality management system.

In 2007, ATCC became the first BRC organization to become accredited by the American National Standards Institute (ANSI) as a Standards Developing Organization. This ANSI certification led to the development of the groundbreaking consensus standard for STR-based cell authentication by ATCC.

IMPACT

A 2004 Brookings Institution study examined the economic impact of BRCs, highlighting ATCC's impact in particular. The study showed that institutions such as ATCC function as "knowledge hubs," which assist their colleagues within the field in acquiring, authenticating, and accessing the knowledge and materials that are necessary to spur innovation and build upon previous knowledge. Other examples of knowledge hubs include open-access libraries, scientific journals, scientific societies, and standards setting bodies. An additional finding from the Brookings' study notes the direct impact of utilizing authenticated biomaterials as the basis of scientific research. Published papers studying ATCC materials were more than twice as likely to receive citations in the work of other scientists than were papers with no connection to ATCC materials (Stern 2004, Taylor 2016).

One of the unpredictable facets of a BRC is that the immediate downstream application(s) of a deposit are not always apparent. Many of the cultures deposited with ATCC were likely deposited with the intent of preserving biodiversity. While this is a noble goal in and of itself, an even broader benefit can be realized when novel unforeseen applications arise from research into preserved strains or when items suddenly become medically relevant. These important discoveries can occur even decades later. This section describes three important biomedical and biotechnological applications of organisms that were unanticipated when they were initially deposited with ATCC.

TAQ POLYMERASE

Perhaps the most unforeseen but serendipitous discovery from a deposit initially thought to be unremarkable was the isolation of a bacterial enzyme that has in large part formed the basis of the modern biotechnology revolution.

In 1967, a Gram-negative nonsporulating thermophilic bacterium was isolated from a hot spring in Yellowstone National Park in Wyoming. It was described in a 1969 publication that assigned it the name *Thermus aquaticus* (an etymological nod to its isolation from hot water), and it was deposited as ATCC® 25104™ in conjunction with the publication (Brock and Freeze 1969).

Nearly 20 years later, a creative scientist named Kary Mullis envisioned a method by which strands of DNA could be exponentially replicated *in vitro* by mimicking the natural process of DNA replication. However, because the initial steps of the process involve heating the DNA to very high temperatures, its convenient application required the use of an enzyme that would be able to tolerate conditions that would otherwise destroy most enzymes. For that, Mullis screened a number of microbes trying to find a thermostable DNA polymerase. One of the bacteria that was investigated was ATCC® 25104™ *Thermus aquaticus*. Because hot springs are the bacterium's native habit, it had evolved molecular machinery capable of withstanding extreme temperatures including a thermostable DNA polymerase that proved ideal for Mullis' technique. This enzyme became known as Taq polymerase (from the name *Thermus aquaticus*), and the technique became known as the polymerase chain reaction, or PCR.

Taq-enabled PCR was first implemented in 1985, and just 8 years later, Kary Mullis was awarded the Nobel Prize for Chemistry for his invention of PCR (Stern 2004). From the beginning, many different applications for the PCR technique became apparent, and it has now become an essential part of modern biotechnology, with applications in fields from diagnostics, epidemiology, forensics, and DNA sequencing. This remarkable advancement may not have been possible without the accessioning and preservation of seemingly unremarkable strains by BRCs such as ATCC.

Crithidia luciliae

In 1958, a novel protozoan was isolated from the green bottle fly (*Lucilia sericata*) (Wallace and Clark 1959). This trypanosome was given the name *Crithidia luciliae* and deposited as ATCC® 14765™ (now ATCC® 30258™). Similar to *Thermus aquaticus*, it likely initially generated little interest outside its immediate research community. However, nearly 20 years later, research on ATCC® 14765™ yielded a novel detection method for the autoimmune disorder lupus (Sontheimer and Gilliam 1978). Antibodies against double stranded DNA (dsDNA) have proven highly effective markers for systemic lupus erythematosus, the most common form of lupus. A characteristic of the protozoan family Trypanosomatidae is a large mitochondrion known as a kinetoplast which contains a network of dsDNA but no other nuclear antigens (Slater et al. 1976). *Crithidia luciliae* has been shown to be effective when used in indirect immunofluorescence (IIF) assays because this kinetoplastic DNA serves as a substrate for anti-dsDNA antibodies (allowing high detection sensitivity) while the non-nuclear localization prevents cross-reaction with other nuclear components (providing greater detection specificity). These criteria led scientists to use the *Crithidia luciliae* kinetoplast as the DNA substrate in IIF assays. They concluded that it is a very effective substrate for detecting the anti-dsDNA antibodies that are associated with lupus (Sontheimer and Gilliam 1978).

Zika Virus

More recently, a previously little-known virus burst onto the world stage and emerged as a significant threat to public health. The Zika virus is a flavivirus that was first isolated from the Zika Forest in Uganda in 1947. Of relatively niche interest at the time, it was deposited as ATCC® VR-84™ in 1953 (Dick et al. 1952) with no indication that it would emerge as an important medical pathogen a few decades later.

In 2015, an outbreak of Zika fever in North and South America (primarily Brazil) was linked to neurological symptoms and birth defects, particularly microcephaly. In 2016, the WHO declared the Zika epidemic a Public Health Emergency of

International Concern. As a result, the scientific community's interest in ATCC's strains of Zika virus grew overnight. It quickly jumped from relative obscurity to one of ATCC's most requested viruses.

RESEARCH AND DEVELOPMENT (R&D)

In addition to its role as a BRC committed to helping the scientific community achieve its research goals, ATCC also has an active R&D program of its own. The main focuses of ATCC's research is designed to further its core mission: finding better ways to identify, characterize, authenticate, and preserve the biological resources that it produces.

Taxonomic Reclassifications

A recent ATCC research effort aimed at better authenticating our biomaterial was to use whole genome sequencing (WGS) of the type strains of the species in the *Mycobacterium tuberculosis* complex (MTBC) to conduct a phylogenomic analysis with the objective of determining the relationships among these species. Using digital DNA-DNA hybridization, the study showed that the "species" in the MTBC are not sufficiently different from one another to be considered unique species. As a result of this study, all the members of MTBC were reclassified as *Mycobacterium tuberculosis* and are now recognized as variants of *M. tuberculosis* (Riojas et al. 2018).

ATCC is currently conducting similar phylogenomic investigations of numerous bacterial genera, including *Mycobacterium*, *Bacillus*, *Yersinia*, *Escherichia*, and *Bifidobacterium*. We expect that broadening the use of such WGS-based phylogenomic techniques will enhance the accuracy of biomaterial authentication and will help modernize taxonomic determination with techniques that examine the entirety of the genome rather than just a single gene (such as 16S rRNA) or via multi-locus sequence typing.

Cell Line Identification

Cell line misidentification is known to be a serious problem (American Type Culture Collection Standards Development Organization Workgroup ASN-0002) that can result in research being questioned and sometimes invalidated. For example, the cell line MDA-MB-435 has been a popular model for studying breast cancer. MDA-MB-435 was, until recently, thought to have originated from a ductal carcinoma (breast cancer) in a female patient. However, a recent collaboration that included the International Cell Line Authentication Committee and ATCC showed that MDA-MB-435 was actually derived from a melanoma in a male patient. Although the cell line has an XX karyotype (which ostensibly indicates its origin from a female patient), Korch et al. (2018) demonstrated that this is in fact due to the loss of the Y chromosome followed by a duplication of the X chromosome. Through efforts such as this, as well as advocating for the use of STR profiling to verify the identity of cell lines being used in research (American Type Culture Collection Standards Development Organization Workgroup ASN-0002), ATCC continues to push for proper authentication of biomaterials for the scientific community.

Patents

ATCC's commitment to innovative internal research is evidenced by the fact that it has been granted six patents by the US Patent and Trademark Office. Two patents (4,672,037 and 4,879,239) were issued for advances in methods of culturing freeze-dried microorganisms. Two patents (5,693,467 and 7,872,116) were issued for PCR assays to detect contamination of cell cultures with Mollicutes including Mycoplasma. Two additional patents (7,951,382 and 7,951,776) were issued for methods of treatment for type 2 diabetes and type 1 diabetes, respectively.

GENOMIC RESOURCES

As ATCC approaches its 100-year anniversary, biology finds itself firmly ensconced in the era of bioinformation. From whole genome sequencing to clustered regularly interspaced short palindromic repeats (commonly known as CRISPR) to synthetic biology and genetic circuits, the field of biology is being revolutionized by the "omics" (genomics, proteomics, transciptomics, etc.) and bioinformation. ATCC is committed to being at the forefront of this revolution and continuing to serve the scientific community as a BRC for the twenty-first century. ATCC is currently preparing a variety of genomic resources that will enable researchers to advance into cutting-edge areas of biotechnology and bioengineering.

ATCC intends to participate in numerous ambitious international collaborations with the aim of providing the scientific community with the whole genome sequences of all bacterial type strains. ATCC is planning to participate in both the Global Collection of Microorganisms sequencing project (GCM 2.0) (Wu et al. 2018) and the Global Encyclopedia of Bacteria and Archaea (GEBA) 10 K sequencing project (Kyrpides et al. 2014).

Finally, ATCC is currently developing a catalog of high-quality reference-grade whole genome sequences from our highly authenticated collection of organisms. These sequences will be assembled using the latest hybrid (short- and long-read) sequencing and assembly methods and will provide the highest confidence in the genome sequences.

CONCLUSIONS

ATCC's extensive history and expertise have allowed it to become the reputable and well-known BRC that exists today. From its beginnings in a small section of a museum to its current corporate headquarters located in Manassas, Virginia, ATCC has continued to grow and evolve to meet the demands of its customers by remaining scientifically relevant. The biodiverse collection of organisms that ATCC has acquired over its long history has provided researchers many opportunities to make high-impact contributions in academia, government, biopharmaceutical, and many other biotechnological applications worldwide. As new diseases and pandemics emerge, as diagnostic and clinical testing procedures are developed and optimized, and as biology moves into the synthetic and bioengineering realms, the collections of ATCC will remain at the forefront of enabling major scientific advances. As ATCC prepares to enter its second century, the leadership of the organization will continue to implement strategic plans to remain scientifically relevant and increase ATCC's impact in the future.

REFERENCES

American Type Culture Collection Standards Development Organization Workgroup ASN-0002. 2010. Cell line misidentification: The beginning of the end. *Nature Reviews Cancer* 10:441–448.

Brock, T.D. and H. Freeze. 1969. *Thermus aquaticus* gen. n. and sp. n., a nonsporulating extreme thermophile. *Journal of Bacteriology* 98:289–297.

Cypess, R.H. 1995. ATCC: Facing today's challenge. *American Society for Microbiology News* 61:274–275.

Cypess, R.H. 1996. The American type culture collection. In *Resource Sharing in Biomedical Research*, ed. K.I. Berns, E.C. Bond and F.J. Manning, pp. 23–32. Washington, DC: National Academies Press.

Cypess, R.H. 2003. *Biological Resource Centers: Their Impact on the Scientific Community and the Global Economy*. Manassas, VA: American Type Culture Collection.

Dick, G.W.A., S.F. Kitchen and A.J. Haddow. 1952. Zika virus (I). Isolations and serological specificity. *Transactions of the Royal Society of Tropical Medicine and Hygiene*. 46:509–520.

Korch, C., E.M. Hall, W.G. Dirks et al. 2018. Authentication of M14 melanoma cell line proves misidentification of MDA-MB-435 breast cancer cell line. *International Journal of Cancer* 142:561–572.

Kyrpides, N.C., P. Hugenholtz, J.A. Eisen et al. 2014. Genomic encyclopedia of bacteria and archaea: Sequencing a myriad of type strains. *PLoS Biology* 12(8):e1001920.

Riojas, M.A., K.J. McGough, C.J. Rider-Riojas, N. Rastogi and M.H. Hazbon. 2018. Phylogenomic analysis of the species of the *Mycobacterium tuberculosis* complex demonstrates that *Mycobacterium africanum*, *Mycobacterium bovis*, *Mycobacterium caprae*, *Mycobacterium microti* and *Mycobacterium pinnipedii* are later heterotypic synonyms of *Mycobacterium tuberculosis*. *International Journal Systematic and Evolutionary Microbiology* 68:324–332.

Simione, F.P. and R.H. Cypess. 2012. Managing a global biological resource of cells and cellular derivatives. In *Management of Chemical and Biological Samples for Screening Applications*, ed. M. Wigglesworth and T. Wood, pp. 143–164. Weinheim, Germany: Wiley-VCH Verlag GmbH & Co. KGaA.

Slater, N.G., J.S. Cameron and M.H. Lessof. 1976. The *Crithidia luciliae* kinetoplast immunofluorescence test in systemic lupus erythematosus. *Clinical and Experimental Immunolgy* 25:480–486.

Sontheimer, R.D. and J.N. Gilliam. 1978. An immunofluorescence assay for double-stranded DNA antibodies using the *Crithidia luciliae* kinetoplast as a double-stranded DNA substrate. *Journal of Laboratory and Clinical Medicine* 91:550–558.

Stern, S. 2004. *Biological Resource Centers: Knowledge Hubs for the Life Sciences*. Washington, DC: Brookings Institution Press.

Taylor, D. 2016. *Transformation of an Icon: ATCC and the New Business Model for Science*. Manassas, VA: American Type Culture Collection.

Wallace, F.G. and T.B. Clark. 1959. Flagellate parasites of the fly, *Phaenicia sericata* (Meigen). *The Journal of Protozoology* 6:58–61.

Wu, L., K. McCluskey, P. Desmeth et al. 2018. The global catalogue of microorganisms 10 K type strain sequencing project: Closing the genomic gaps for the validly published prokaryotic and fungi species. *Gigascience* 7(5). doi:10.1093/gigascience/giy026.

Index

Note: Page numbers in italic and bold refer to figures and tables, respectively.

B

C

D

E

F

G

H

I

J

K

L

M

N

O

P

Q

R

S

X

Z